COSMIC MAGNETIC FIELDS

Magnetic fields pervade the Universe and play an important role in many astrophysical processes. However, they require specialised observational tools, and are challenging to model and understand. This volume provides a unified view of magnetic fields across astrophysical and cosmological contexts, drawing together disparate topics that are rarely covered together. Written by the lecturers of the XXV Canary Islands Winter School of Astrophysics, it offers a self-contained introduction to cosmic magnetic fields on a range of scales. The connections between the behaviours of magnetic fields in these varying contexts are particularly emphasised, from the relatively small and close ranges of the Sun, planets and stars, to galaxies and clusters of galaxies, as well as on cosmological scales. Aimed at young researchers and graduate students, this review uniquely brings together a subject often tackled by disconnected communities, conveying the latest advances as well as highlighting the limits of our current understanding.

Jorge Sánchez Almeida is Research Professor at the Instituto de Astrofísica de Canarias, Spain. He has worked in several astrophysical fields including the cosmic microwave background, astronomical spectropolarimetry, solar magnetometry, galaxy formation and evolution, and big data science. He is author of more than a hundred refereed papers and has edited two books on solar magnetic fields. He has a wealth of experience teaching astronomical optics and solar magnetometry at graduate level.

María Jesús Martínez González is a Ramón y Cajal fellow at the Instituto de Astrofísica de Canarias. Her main scientific research deals with the magnetism of stars and the Sun. She has authored more than 60 research papers and was awarded the 2011 JOSO prize for the best paper by a young solar physicist. At present, she is the principal investigator of a research project on stellar magnetism and spectropolarimetry at the Instituto de Astrofísica de Canarias.

Canary Islands Winter School of Astrophysics

Volume XXV
Cosmic Magnetic Fields

Series Editor
Rafael Rebolo *Instituto de Astrofísica de Canarias*

Previous volumes in this series

I. Solar Physics
II. Physical and Observational Cosmology
III. Star Formation in Stellar Systems
IV. Infrared Astronomy
V. The Formation of Galaxies
VI. The Structure of the Sun
VII. Instrumentation for Large Telescopes: A Course for Astronomers
VIII. Stellar Astrophysics for the Local Group: A First Step to the Universe
IX. Astrophysics with Large Databases in the Internet Age
X. Globular Clusters
XI. Galaxies at High Rcdshift
XII. Astrophysical Spectropolarimctry
XIII. Cosmochemistry: The Melting Pot of Elements
XIV. Dark Matter and Dark Energy in the Universe
XV. Payload and Mission Definition in Space Sciences
XVI. Extrasolar Planets
XVII. 3D Spectroscopy in Astronomy
XVIII. The Emission-Line Universe
XIX. The Cosmic Microwave Background: From Quantum Fluctuations to the Present Universe
XX. Local Group Cosmology
XXI. Accretion Processes in Astrophysics
XXII. Asteroseismology
XXIII. Secular Evolution of Galaxies
XXIV. Astrophysical Applications of Gravitational Lensing

COSMIC MAGNETIC FIELDS

XXV Canary Islands Winter School of Astrophysics

Edited by

JORGE SÁNCHEZ ALMEIDA

Instituto de Astrofísica de Canarias, Tenerife

MARÍA JESÚS MARTÍNEZ GONZÁLEZ

Instituto de Astrofísica de Canarias, Tenerife

CAMBRIDGE
UNIVERSITY PRESS

CAMBRIDGE
UNIVERSITY PRESS

University Printing House, Cambridge CB2 8BS, United Kingdom

One Liberty Plaza, 20th Floor, New York, NY 10006, USA

477 Williamstown Road, Port Melbourne, VIC 3207, Australia

314–321, 3rd Floor, Plot 3, Splendor Forum, Jasola District Centre, New Delhi - 110025, India

79 Anson Road, #06-04/06, Singapore 079906

Cambridge University Press is part of the University of Cambridge.

It furthers the University's mission by disseminating knowledge in the pursuit of education, learning, and research at the highest international levels of excellence.

www.cambridge.org
Information on this title: www.cambridge.org/9781107097810
DOI: 10.1017/9781316160916

First published 2018

Printed in the United Kingdom by Clays, St Ives plc.

A catalogue record for this publication is available from the British Library.

Library of Congress Cataloging-in-Publication data
Names: Canary Islands Winter School of Astrophysics (25th : 2013 : La Laguna, Canary Islands) | Sánchez Almeida, J. (Jorge), editor. | Martínez González, María Jesús, editor.
Title: Cosmic magnetic fields / edited by J. Sánchez Almeida and M.J. Martínez González.
Other titles: Canary Islands Winter School of Astrophysics (Series) ; v. XXV.
Description: Cambridge, United Kingdom ; New York, NY : Cambridge University Press, 2018. | Series: Canary Islands Winter School of Astrophysics ; volume XXV | Lectures presented at the XXV Canary Islands Winter School of Astrophysics, held in La Laguna, Tenerife, Spain, November 11-22, 2013. | Includes bibliographical references and index.
Identifiers: LCCN 2017058591 | ISBN 9781107097810 (hardback ; alk. paper) | ISBN 1107097819 (hardback ; alk. paper)
Subjects: LCSH: Cosmic magnetic fields–Congresses.
Classification: LCC QB462.8 .C66 2018 | DDC 523.01/88 – dc23
LC record available at https://lccn.loc.gov/2017058591

ISBN 978-1-107-09781-0 Hardback

Additional resources for this publication at
www.cambridge.org/cosmicmagneticfields

Contents

List of Contributors — *page* viii

List of Participants — ix

Preface — xiii

Acknowledgments — xiv

1 Astrophysical Magnetic Fields: Essentials — 1
Jorge Sánchez Almeida and María Jesús Martínez González

2 Solar Magnetic Fields: History, Tragedy or Comedy? — 13
Philip G. Judge

3 Stellar Magnetic Fields — 47
Oleg Kochukhov

4 The Role of Magnetic Fields in AGN Activity and Feedback — 87
Rony Keppens, Oliver Porth and Hans (J. P.) Goedbloed

5 Magnetic Fields in Galaxies — 123
Rainer Beck

6 Primordial Magnetic Fields in the Early Universe and Cosmic Microwave Background Anisotropies — 161
Fabio Finelli and Daniela Paoletti

Contributors

Rainer Beck Max Planck Institut für Radioastronomie, Germany

Fabio Finelli INAF/IASF, Bologna, Italy

Hans (J. P.) Goedbloed FOM Institute DIFFER, Nieuwegein, The Netherlands

Philip G. Judge High Altitude Observatory, United States of America

Rony Keppens KU Leuven, Belgium

Oleg Kochukhov Uppsala University, Sweden

María Jesús Martínez González Instituto de Astrofísica de Canarias, Spain

Daniela Paoletti INAF/IASF, Bologna, Italy

Oliver Porth KU Leuven, Belgium & School of Mathematics, University of Leeds, UK

Jorge Sánchez Almeida Instituto de Astrofísica de Canarias, Spain

Participants

Participants (students and lecturers) of the XXV Canary Islands Winter School of Astrophysics, held in La Laguna, November 11th–22th, 2013.

Alonso, A.	Laboratorio Nacional de Fusión CIEMAT, Spain
Alsina, E.	Instituto de Astrofísica de Canarias, Spain
Alvarado-Gómez, J. D.	European Southern Observatory, Germany
Amouzou, E.	Montana State University, United States of America
Anderson, C.	University of Sydney, Australia
Barnes, D.	Mullard Space Science Laboratory UCL, United Kingdom
Beck, R.	Max Planck Institut für Radioastronomie, Germany (Lecturer)
Castellanos, A.	Instituto de Ciencias Nucleares UNAM, México
Cattaneo, F.	University of Chicago, USA (Lecturer)
del Pino Alemán, T.	Instituto de Astrofísica de Canarias, Spain
Dolag, K.	Max Planck Institut für Astrophysik, Germany (Lecturer)
Durrive, J.-B.	Institut d'Astrophysique Spatiale, France
Fabas, N.	Instituto de Astrofísica de Canarias, Spain
Fabbian, D.	Instituto de Astrofísica de Canarias, Spain
Filippova, E.	University of Geneva, Switzerland
Finelli, F.	Istituto di Astrofisica Spaziale e Fisica Cosmica, Italy (Lecturer)
García de Andrade, L.	State University of Rio de Janeiro, Brazil
Glushikhina, M.	Space Research Institute, Russia
González Pérez, L.	Instituto de Astrofísica de Canarias, Spain (LOC)
Judge, P. G.	High Altitude Observatory, United States of America (Lecturer)
Kaczmarek, J.	University of Sydney, Australia
Keppens, R.	KU Leuven, Belgium (Lecturer)
Kochukhov, O.	Uppsala University, Sweden (Lecturer)
Kuzmychov, O.	Kiepenheuer-Institut fuer Sonnenphysik, Germany
Marínez González, M.	Instituto de Astrofísica de Canarias, Spain (SOC-LOC)
Nóbrega Siverio, D.	Instituto de Astrofísica de Canarias, Spain
On, A.	Mullard Space Science Laboratory UCL, United Kingdom
Pastor Yabar, A.	Instituto de Astrofísica de Canarias, Spain
Paoletti, D.	INAF/IASF, Bologna, Italy (Lecturer)
Peláez Santos, A. E.	Instituto de Astrofísica de Canarias, Spain
Roettenbacher, R.	University of Michigan, United States of America
Rosén, L.	Uppsala University, Sweden
Ruiz Granados, B.	Universidad de Granada, Spain
Rusomarov, N.	Uppsala University, Sweden
Sánchez Almeida, J.	Instituto de Astrofísica de Canarias, Spain (SOC-LOC)
Stepanovs, D.	Max Planck Institute for Astronomy, Germany
Tramonte, D.	Instituto de Astrofísica de Canarias, Spain
Triviño Hage, A.	Instituto de Astrofísica de Canarias, Spain
Vignaga, R.	Instituto de Astrofísica de Canarias, Spain
Wisniewska, A.	Kiepenheuer-Institut für Sonnenphysik, Germany

1. A. Wisniewska
2. A. On
3. J. Kaczmarek
4. C. Anderson
5. J. Sánchez Almeida
6. F. Cattaneo
7. P. G. Judge
8. N. Rusomarov
9. A. Alonso
10. R. Keppens
11. O. Kochukhov
12. L. Rosén
13. E. Filippova
14. M. Glushikhina
15. J. D. Alvarado-Gómez
16. A. Castellanos
17. O. Kuzmychov
18. R. Roettenbacher
19. D. Barnes
20. E. Amouzou
21. E. Alsina
22. A. Pastor Yabar
23. J.-B. Durrive
24. N. Fabas
25. T. del Pino Alemán

1. N. Rusomarov
2. A. On
3. K. Dolag
4. C. Anderson
5. J. Sánchez Almeida
6. J. Kaczmarek
7. A. Wisniewska
8. R. Beck
9. L. Rosén
10. J.-B. Durrive
11. R. Roettenbacher
12. A. Castellanos
13. O. Kuzmychov
14. M. Glushikhina
15. L. García de Andrade
16. E. Filippova
17. E. Alsina
18. D. Barnes
19. R. Vignaga
20. D. Tramonte
21. J. D. Alvarado-Gómez
22. A. E. Peláez Santos
23. D. Stepanovs
24. B. Ruiz Granados

Preface

Magnetic fields play an important role in many astrophysical processes. They are difficult to detect and characterize because often their properties have to be inferred through interpreting the polarization of the light. Magnetic fields are also challenging to model and understand. Magnetized plasmas behave following highly non-linear differential equations having no general solution, so that every astrophysical problem represents a special case to be studied independently. Hence, magnetic fields are often an inconvenient subject that is overlooked or simply neglected (*the elephant in the room*, as they are dubbed on posters in the XXV Canary Islands Winter School of Astrophysics). Such a difficulty burdens the research on magnetic fields, which has evolved to become a very technical subject, with many small disconnected communities studying specific aspects and details. The school tried to amend the situation by providing a unifying view of the subject. The students had a chance to understand the behavior of magnetic fields in all astrophysical contexts, from cosmology to the Sun, and from starbursts to AGNs. The school was planned to present a balanced yet complete review of our knowledge, with excursions into the unknown to point out present and future lines of research.

The subject of *Cosmic Magnetic Fields* was split into seven different topics: *cosmic magnetic field essentials, solar magnetic fields, stellar magnetic fields, the role of magnetic fields on AGN feedback, magnetic fields in galaxies, magnetic fields in galaxy clusters and at larger scales,* and *primordial magnetic fields and magnetic fields in the early Universe.* The corresponding lectures were delivered by seven well known and experienced scientists that have played key roles in the major advances of the field during the last years: F. Cattaneo, P. Judge, O. Kochukhov, R. Keppens, R. Beck, K. Dolag, and F. Finelli. Their lectures were recorded and are freely available at the IAC website: http://iactalks.iac.es/talks/serie/19. Together with the reviews included in the present volume, they form a unique resource for both students and professional researchers. They provide a global view of this very compartmentalized, yet fundamental, field of research.

Each chapter of the book is self-consistent, so that they can be consulted and read independently. In particular, each one has its own notation and list of references. The introductory lectures by F. Cattaneo have been summarized by the Editors, and they give a quick overview of the fundamentals on magnetized plasmas, with abundant references to monographs for further reading. The book includes lecture notes on all the original topics, except for the lectures by K. Dolag on galaxy cluster magnetic fields. Some of the notes were written not only by the lecturers but also include collaborators.

Magnetic fields are usually associated with open questions. This fact is very well summarized in the introduction to the chapter by R. Beck. We expect that the school and these proceedings will help to clarify, and hopefully close, some of them.

The Editors

Acknowledgments

The organizers are indebted to the lecturers for the clarity and quality of the lectures and presentations, and to the participants, for their interest and for maintaining such a formidable atmosphere during the two weeks of the school.

The school was celebrated during the peak of the economical recession in Spain, which affected every aspect of life, including funding for science. We appreciate the efforts made by the IAC to keep the Winter School Series running. Our appreciation goes to the IAC members that participated in the organization and development of the school, starting with the Director, Rafael Rebolo, for his unequivocal support and for securing funding through the Severo Ochoa Program. We thank the secretary, Lourdes González Pérez, who made our school possible through her invaluable assistance with all the different aspects of the preparation and organization. The head of the graduate studies division, Francisco Grazón, was always most supportive. We thank Jorge Andrés Pérez Prieto for setting up the website (http://www.iac.es/winterschool/2013), and Gabriel Pérez for the design and preparation of the poster. Francisco Javier López and Diego Sierra made the video recording and archival of all lectures possible (http://iactalks.iac.es/talks/serie/19). In this sense, we express our gratitude to A. Pastor Yabar, who took care of the actual recording of the lectures. Miguel Briganti is responsible for the pictures included in the webpage, and many other people contributed to the success of the school, including the visits to *El Teide* and *Roque de los Muchachos* observatories, and the celebration of the public lecture on *Campos magnéticos en el cosmos* by R. Beck.

We feel indebted to Eduardo Battaner who helped us with the selection of lecturers, and so, the success of the school owes much to his advice. We would specially like to express our gratitude to Fabio Finelli for lecturing during these particularly sad days, and to Daniela Paoletti for taking up the task of completing his lectures with almost no time for preparation.

Finally, we need to thank CUP for the patience and understanding shown during the preparation of this book.

1. Astrophysical Magnetic Fields: Essentials

JORGE SÁNCHEZ ALMEIDA AND
MARÍA JESÚS MARTÍNEZ GONZÁLEZ

Abstract

Here we provide a general introduction to the physical aspects of astrophysical magnetic fields, presenting the widely used MHD approximation, and discussing concepts like advection, diffusion, and the operation of magnetic dynamos. During the Winter School (WS), the introductory lessons were delivered by Professor F. Cattaneo. Rather freely, these introductory notes roughly follow the thread of arguments he presented at the WS. Our *essentials* contain a number of references to classic books on astrophysical magnetic fields, so that the interested reader can expand the cursory description given here.

1.1 Astrophysical Plasmas

Only 5% of the energy density of the Universe is provided by baryonic matter (e.g., Fukugita & Peebles, 2004); however, it is the *luminous matter* and so it is our interface to study the Universe. Most of it is in the form of a plasma, where atoms and atom aggregates are partly ionized. Particles move under the influence of large-scale electromagnetic fields created collectively. The presence of magnetic fields in plasmas is pervasive, and one cannot be properly understood without the other. This holds from planets to the Universe as a whole, including stars and the interstellar medium, galaxies, clusters of galaxies, and the intergalactic medium. The conditions vary from object to object, and the resulting phenomena depend on the temporal and spatial scales of interest. Fortunately, a few equations capture (albeit in a simplified way) many of the physical processes that are characteristic of magnetized plasmas. They will be introduced here in preparation for the specific applications to be described in the forthcoming chapters. The main equations will be given without any proof or derivation; these can be found in many of the monographs on astrophysical plasmas and magnetohydrodynamics (MHD) existing in the literature, e.g., Parker (1979), Moffat (1978), Priest (2000), Cattaneo (1999), Kulsrud (2005), or Spruit (2013). The symbols used in this description are summarized later in the chapter, in Table 1.2.

• *How can you describe the plasma mathematically?* Plasmas are made out of neutrals, electrons, ions, and electromagnetic fields. By and large, a plasma can be characterized by the number density of each species n_i, the temperature of each species T_i, and the steady-state magnetic field $\vec{B}$. At equilibrium, the temperature must be the same for all species; however, the relaxation time to return to equilibrium may be long. Often one will have high energetic processes, localized in time and space (e.g., reconnection in a solar flare, or ejection of plasma blobs in magnetized jets), that move the system out of thermodynamic equilibrium. In returning to equilibrium, each species is often in equilibrium with its own temperature T_i, allowing us to assume a Maxwellian velocity distribution for each species separately. Nature provides situations where equilibrium is never reached, e.g., in turbulence.

1.1.1 *Characteristic Scales of Plasmas*

The physical properties and behavior of a plasma depend on a number of characteristic spatial and temporal scales. Their values in a number of representative astrophysical contexts are given in Table 1.1.

TABLE 1.1. Physical parameters characteristic of laboratory and astrophysical plasmas

Physical parameter	Magnetic fusion	Accretion disk (Sgr A)	Solar chromosphere	Interstellar medium	Intergalactic medium	Galaxy clusters
Electron density [cm^{-3}]	10^{14}	10^{6}	2×10^{10}	1	10^{-4}	5×10^{-2}
Gas temperature [K]	10^{8}	10^{11}	10^{4}	10^{4}	10^{5}	5×10^{7}
Magnetic field strength [G]	5×10^{4}	50	10	10^{-6}	10^{-7}	10^{-5}
Plasma parameter	10^{8}	5×10^{16}	10^{4}	10^{9}	5×10^{12}	10^{15}
Plasma frequency [Hz]	5×10^{11}	5×10^{7}	10^{10}	5×10^{4}	10^{3}	10^{4}
Mean-free path [cm]	10^{6}	10^{20}	50	10^{12}	10^{18}	5×10^{20}
Gyroradius electrons [cm]	5×10^{-3}	10^{2}	10^{-1}	2×10^{6}	10^{8}	10^{7}
Plasma beta	10^{-2}	10^{-1}	10^{-2}	50	5	10^{2}
Characteristic spatial scale [cm]	10^{2}	10^{13}	10^{8}	10^{20}	10^{26}	10^{24}

See, e.g., Schekochihin *et al.* (2009) or Alonso de Pablo & Sánchez Almeida (2013).

- *Debye length.* Most plasmas are quasi-neutral, i.e., they have the same number of free electrons as the number of positive charges in ions. However, this quasi-neutrality goes away when you average over the very small scales able to single out individual electrons or ions. The Debye length is the minimum scale at which the plasma looks neutral. A free charge in a neutral plasma creates an electric potential at a distance r from the charge $\phi(r)$, which is the electric potential in a vacuum $\phi_0(r)(\propto r^{-1})$ plus an exponential drop-off,

$$\phi(r) = \phi_0(r)\,\exp(-r/\lambda_D), \tag{1.1}$$

so that when $r > \lambda_D$ then $\phi \ll \phi_0$, and the effect of the point charge becomes negligible shielded by the plasma. The characteristic scale for shielding the charge is the Debye length,[1]

$$\lambda_D = \sqrt{\frac{kT_e}{4\pi\, n_e e^2}}. \tag{1.2}$$

In order to treat a plasma as quasi-neutral, one has to assume that the characteristic length of interest is much larger than λ_D.

- *Plasma frequency.* If, for any reason, ions and electrons are separated, their separation oscillates to try to recover the quasi-neutrality with a frequency called the plasma frequency, w_e

$$w_e = \sqrt{\frac{4\pi\, n_e\, e^2}{m_e}}. \tag{1.3}$$

The period associated with this frequency sets the timescale at which the plasma responds to quasi-neutrality, so that deviations from neutrality last less than this timescale. Note that the frequency given in Eq. (1.3) corresponds to the electron plasma frequency. Ions also respond to deviations from neutrality, but they oscillate at a much lower frequency ($w_p^2 \propto m_p^{-1}$), therefore, electrons set the overall characteristic timescale.

- *Plasma parameter.* The number of electrons in a volume with the radius of the Debye length is the plasma parameter Λ,

$$\Lambda = \frac{4\pi}{3}\,\lambda_D^3\, n_e = \frac{1}{3\sqrt{4\pi\, n_e}}\left(\frac{kT_e}{e^2}\right)^{3/2}. \tag{1.4}$$

$\Lambda \gg 1$ in astrophysical plasmas (see Table 1.1), i.e., there are many free electrons per Debye sphere. Each individual electron feels the Coulomb forces of many electrons simultaneously. Electrons do not interact strongly with each other, but feel the global potential created collectively by the plasma. Each electron random-walks with such small deflections that only a large number of them will produce significant changes in the original trajectory. The plasma is said to be weakly coupled. One can also show that the electrostatic potential energy is much smaller than the kinetic energy of the electrons (and the rest of the species). $\Lambda \gg 1$ implies that the plasma is "hot" and "tenuous" (high temperature and low density: see, e.g., Eq. (1.4)).

- *Gyro frequency and gyro radius.* An electron moving in a magnetic field follows a helical path around the axis set by the magnetic field vector. The gyro frequency of

[1] In this definition and throughout the chapter, equations are given in cgs-esu units following, e.g., Somov (2006).

this motion, Ω, is

$$\Omega = \frac{e\,B}{m_e\,c}, \tag{1.5}$$

with the gyro radius (or Larmor radius) given by

$$r_L = \frac{v_\perp m_e\,c}{e\,B}, \tag{1.6}$$

where $v_\perp$ is the component of the electron velocity perpendicular to the magnetic field. For the typical electron thermal energies and magnetic fields found in astronomy, the gyro radii are much smaller than the characteristic scale for the variation of the magnetic fields (see Table 1.1).

- *Mean free path.* The description of MHD plasmas requires having enough collisions between electrons, ions, and neutrals to grant the coupling of their properties. For this condition to be granted, the distance travelled between collisions must be much smaller than the characteristic scales for the variation of the macroscopic physical parameters such as magnetic field, temperature, or bulk velocity. The mean free path between collisions λ is

$$\lambda = \frac{1}{n_e\sigma}, \tag{1.7}$$

where σ is the Coulomb cross section, which is similar for electrons and ions (e.g., Kulsrud, 2005),

$$\sigma \simeq 10^{-12}\,\mathrm{cm}^2\,T_{\mathrm{eV}}^{-2}, \tag{1.8}$$

where T_{eV} stands for the temperature in eV ($1\,\mathrm{eV} \equiv 10^4\,\mathrm{K}$). The mean free paths in astrophysical plasmas are generally larger than the lengthscales characterizing the variation of macroscopic parameters; see Table 1.1.

- *Plasma β.* This is the ratio between the gas pressure P and the magnetic pressure,

$$\beta = \frac{8\pi P}{B^2}, \tag{1.9}$$

and it parameterizes whether magnetic forces ($\beta \ll 1$) or hydro forces ($\beta \gg 1$) dominate the behavior of the plasma.

Table 1.1 contains the typical physical parameters of various astrophysical plasmas. As a consequence of the large range of parameters, the physical behavior of the astrophysical plasma is very diverse. Before starting work on a particular subject, one has to work out the characteristic plasma scales, to be sure that the (approximate) equations to be used are appropriate for the problem at hand. For example, the atmospheres of collapsed objects such as white dwarfs and neutron stars do not have $\Lambda \gg 1$.

1.1.2 *Mathematical Description of Plasmas – MHD – Induction Equation*

The exact solution for the motion of each particle under the influence of local magnetic fields is extremely difficult to obtain. In practice, one uses approximate equations that involve averages of the exact solution. Depending on how such an averaging is carried out, one gets different approximations. Under the *Vlasov theory*, one averages over all particles of a given species with the same velocity at a given location. The properties of the plasma are then described in terms of distribution functions. In *multifluids theory* one averages over all particles of a given species for all velocities. One gets the mean densities and velocities for each species. Temperature and pressure emerge naturally. Finally, the *magnetohydrodynamic (MHD) approximation* is based on averages over all particles of all species and velocities. The properties of the plasma are then described in terms of mean density, velocity, and pressure at each spatial location.

Within MHD, one finds the equations for mass conservation, momentum conservation, and energy conservation typical of the fluid dynamics. Electro-magnetic fields are considered, starting from the Maxwell equations under the assumptions of (1) quasi-neutrality, (2) velocities that are small with respect to the speed of light, and (3) currents proportional to the electric field in the rest-frame of the current-carriers (see the references to textbooks given in the previous section). This approximation holds for plasmas where the different species are coupled by collisions, thus giving the character of a fluid to the plasma. With the above simplifications, one derives the *induction equation*, which is the fundamental equation in MHD,

$$\frac{\partial \vec{B}}{\partial t} = \nabla \wedge (\vec{v} \wedge \vec{B}) + \eta \nabla^2 \vec{B}, \tag{1.10}$$

where η is the magnetic diffusivity. This is the inverse of the electric (Ohmic) conductivity, and has units of a velocity times a length (it is a diffusion coefficient). The symbols $\vec{B}$ and $\vec{v}$ stand for the magnetic field and the bulk velocity of the plasma, respectively. This shows how the magnetic field evolves with time t given the velocity field and the diffusivity. The first term on the right-hand side of the induction equation creates magnetic fields from plasma motions, and contains the inductive and advective effects of the velocity on the magnetic field. The second term is a diffusive term that dissipates magnetic energy. The ratio between the two of them is the magnetic Reynolds number R_m,

$$\frac{|\nabla \wedge (\vec{v} \wedge \vec{B})|}{|\eta \nabla^2 \vec{B}|} \sim \frac{v\,B/\mathcal{L}}{\eta B/\mathcal{L}^2} \sim \frac{\mathcal{L} v}{\eta} = R_m, \tag{1.11}$$

where $v = |\vec{v}|$ and $\mathcal{L}$ is a characteristic lengthscale of the problem. R_m controls the relative importance of the advective and the diffusive processes. The properties of the magnetic fields satisfying the induction equation are analyzed in the next section.

1.1.3 *General Properties of the Induction Equation*

In many common astrophysical contexts the magnetic Reynolds number is very large $R_m \gg 1$, meaning that the last term in the right-hand side of Eq. (1.10) is negligible and the physics is advection dominated. Often this case is referred to as *ideal* MHD. In this case a number of interesting properties arise; in particular, the magnetic field lines are *frozen* in the plasma. Magnetic field lines are material lines, and they move by being dragged along with the fluid motion. The magnetic flux through any closed co-moving surface is constant.

- *Consequences of magnetic flux conservation.* Suppose that a star collapses to become a white dwarf conserving the magnetic flux, i.e., conserving $B\,\mathcal{L}^2$, where $\mathcal{L}$ is the size of the object. If a solar-size star ($\mathcal{L} \sim 10^{11}$ cm) collapses to become white-dwarf sized ($\mathcal{L} \sim 10^9$ cm), the magnetic field increases from, say, 10 G to 10^5 G. If the collapse goes all the way down to a neutron-star size object ($\mathcal{L} \sim 10^6$ cm), then the magnetic field booms up to 10^{11} G. These huge field strengths are indeed observed (e.g., Mereghetti, 2008).
- *Consequences of the flux freezing on the magnetic helicity.* These consequences are not as intuitive as for the magnetic flux conservation. The magnetic helicity $\mathcal{H}$ is defined as

$$\mathcal{H} = \int_u \vec{A} \cdot \vec{B}\, dV, \tag{1.12}$$

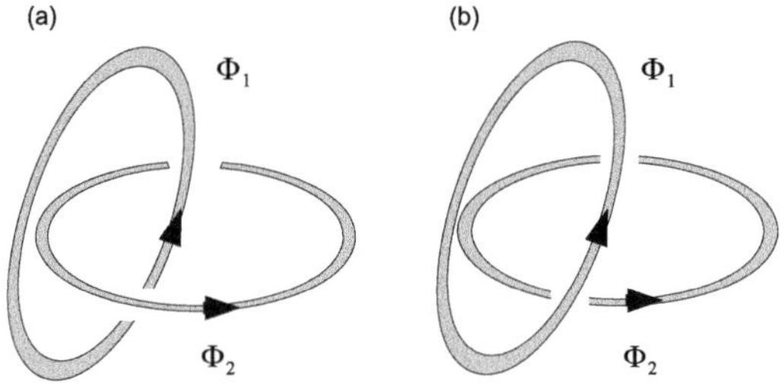

FIGURE 1.1. Schematic to illustrate the physical meaning of helicity and its conservation under frozen-in conditions. (a) The helicity of these two nested magnetic rings is $\mathcal{H} = 2\Phi_1\Phi_2 \neq 0$. The symbols Φ_1 and Φ_2 stand for the magnetic flux across the cross section of the rings 1 and 2, respectively. (b) In this case $\mathcal{H} = 0$ – the magnetic loops are similar to the previous ones but this time they are not intertwined.

with $\vec{A}$ the vector potential ($\vec{B} = \nabla \wedge \vec{A}$). The volume integral has to be bounded by a surface having $\vec{B}$ always parallel to the surface. (If $\vec{n}$ is the normal to the surface, then $\vec{B} \cdot \vec{n} = 0$.) Under the frozen-in condition, the helicity is conserved, so that

$$\frac{d\mathcal{H}}{dt} = 0. \tag{1.13}$$

How do we interpret this? The frozen-in condition does not allow the fluid evolution to modify the magnetic field topology. The situation is illustrated in Fig. 1.1. It can be proven that the helicity in Fig. 1.1a is the product of the fluxes in the two knotted magnetic fluxtubes (rings), i.e., $\mathcal{H} = 2\Phi_1\Phi_2$ (e.g., Moffat, 1978). In the case of Fig. 1.1b, $\mathcal{H} = 0$, implying that when the magnetic field is frozen in, it is imposible to change from the topology in Fig. 1.1a to topology in Fig. 1.1b. Diffusion is needed for the magnetic field lines to cross each other, and so for the helicity to change.

- *The effect of diffusion.* Without diffusion, the magnetic field would keep the topology from the moment when it was created at the origin of the Universe. Therefore, even if throughout space $R_m \gg 1$, there should be regions where $R_m \leq 1$, where the magnetic field is diffusive. These regions are critical to allow a change in magnetic field topology, and to allow creation and destruction of $\vec{B}$. (For example, to explain the magnetic cycles existing in stars like the Sun.) In the absence of velocity fields, the timescale for the magnetic field to diffuse τ is approximately given by

$$\tau \sim \frac{B}{|\partial\vec{B}/\partial t|} \sim \mathcal{L}^2/\eta, \tag{1.14}$$

where we have used the induction Eq. (1.10) without the advection term, with $|\nabla^2\vec{B}| \sim B/\mathcal{L}^2$. These timescales are huge in astronomical contexts, often larger than the age of the Universe, because the lengthscales $\mathcal{L}$ that are involved are astronomically large. Therefore this diffusion timescale is physically irrelevant. However, when velocity fields are considered, they are able to generate tiny lengthscales on the magnetic field – so small that diffusion occurs on timescales of the order of the dynamical timescale of the system, and become physically relevant. What is the lengthscale ζ at which diffusion balances advection? Setting $\partial\vec{B}/\partial t = 0$ in the induction equation,

$$\zeta \sim \mathcal{L}/R_m^{1/2}, \tag{1.15}$$

which implies that, for the typical $R_m \gg 1$, ζ is tiny. Therefore, the changes in the magnetic field topology have to occur at these tiny lengthscales for the diffusion to be effective. But the existence of these scales completely changes the timescale of the diffusion process, so that the timescale τ_ζ is no longer τ (Eq. (1.14)) but it turns out to be much shorter, as given by

$$\tau_\zeta \sim \zeta^2/\eta \sim \tau\, R_m^{-1}. \tag{1.16}$$

- *Turbulent diffusion.* When the velocity field has a turbulent component of amplitude v_t that varies over lengthscales l_t, the plasma behaves as if it has an effective diffusion called *turbulent diffusion*, given by

$$\eta_t \simeq l_t v_t. \tag{1.17}$$

  Often $\eta_t \gg \eta$, so that the turbulent diffusion dominates the diffusion processes in astrophysical plasmas.
- *The two competing roles of the plasma velocity.* The velocity field both speeds up the magnetic diffusion processes, and increases the magnetic field. The increase of magnetic field gradient leads to accelerated decay, and the stretching of field-lines leads to magnetic field growth. Both stretching and creation of gradients have exponential growth in time. Which one wins depends on the flow. If stretching wins, then we have a dynamo. When do we have it? There is no general rule and the answer has to be worked out on a case-by-case basis; however, any complicated velocity field at high R_m produces a dynamo (e.g., Childress & Gilbert, 1995). This is the context of many astrophysical plasmas.

1.2 Astrophysical Dynamos

If a magnetized plasma has no motions, then the magnetic field will decay with a timescale given by either Eq. (1.14) or Eq. (1.16). Fluid motions may increase these timescales to make them infinitely large. A velocity field $\vec{v}$ has *the dynamo property* if the energy of the magnetic field of the plasma does not decay with time, i.e., if

$$\int_{\text{Volume}} B^2\, dV \rightarrow \text{constant } \neq 0 \text{ for time } \rightarrow \infty. \tag{1.18}$$

Dynamos are supposed to be responsible for most astrophysical magnetic fields, from planets to the early Universe. Are dynamos universal objects? Is the mechanism that makes a dynamo work always the same? Unfortunately, dynamos are not universal and different mechanisms may operate to amplify magnetic fields. (Even though all complicated velocity fields tend to have the dynamo property.)

An important distinction to be made is the difference between large-scale and small-scale dynamos. If the typical lengthscale of the velocity is l_v, and the lengthscale of the generated magnetic field is l_B, the dynamo is small-scale if $l_B \ll l_v$. Otherwise, the dynamo is said to be large-scale. Take the Sun as an example. Understanding the dipolar component of the solar magnetic field involves a large-scale dynamo since there are no plasma motions with the scale of the full Sun (except for rotation). On the other hand, the small-scale magnetic field observed in the Sun (less than 100 km across) may be created by a small-scale dynamo if it is generated by solar granulation (1000 km size). Large-scale dynamos generate ordered magnetic fields, whereas small-scale dynamos produce tangled magnetic fields. For this reason, large-scale dynamos are said to generate magnetic flux, whereas small-scale dynamos generate magnetic energy.

- *Mean-field dynamo.* The mean-field theory, or mean-field electrodynamics, is described in many classical text books, e.g., Moffat (1978) or Krause & Rådler (1980). It is based on a two-scale approach. If this two-scale approximation fails, then the whole theory collapses. The two-scale approach is illustrated in Fig. 1.2. The magnetic field and

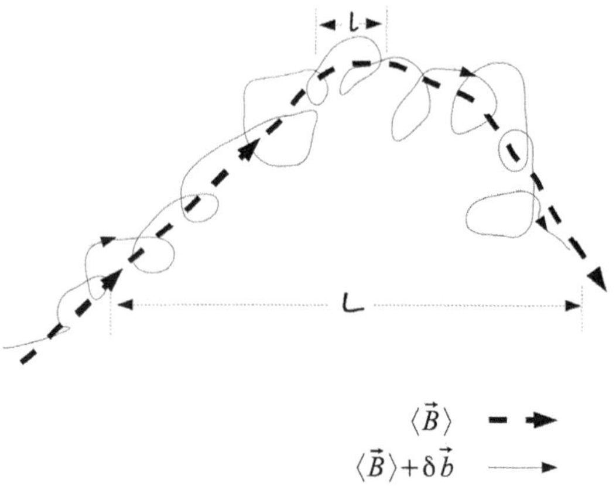

FIGURE 1.2. Schematic illustrating the two-scale approach to the mean-field dynamo theory. The magnetic field is divided into a mean field ($\langle\vec{B}\rangle$; the thick dashed line) and a fluctuating part ($\delta\vec{b}$), so that the total magnetic field (the thin solid line) is the sum of these two parts. The lengthscale for variation of the mean field (L) is much larger than the scale for the fluctuating part ($l \ll L$).

the velocity can be divided into a large-scale mean quantity and a fluctuating quantity,

$$\vec{v} = \langle\vec{v}\rangle + \delta\vec{v}, \tag{1.19}$$

$$\vec{B} = \langle\vec{B}\rangle + \delta\vec{b}, \tag{1.20}$$

with the angle brakets $\langle\ \rangle$ meaning averaging over a scale a in between the characteristic scale for the variation of the average quantities L, and the characteristic scale for the variation of the fluctuations l ($\ll a \ll L$). There is a gap in physical scales, and the average is carried out with a smearing scale between them. By definition, the fluctuations average out to zero,

$$\langle\delta\vec{v}\rangle = \langle\delta\vec{b}\rangle = 0. \tag{1.21}$$

Then the induction equation (Eq. (1.10)) yields,

$$\frac{\partial\langle\vec{B}\rangle}{\partial t} = \nabla\wedge\Big[\langle\vec{B}\rangle\wedge\langle\vec{v}\rangle + \langle\delta\vec{v}\wedge\delta\vec{b}\rangle\Big] + \eta\,\nabla^2\langle\vec{B}\rangle, \tag{1.22}$$

which is formally identical to the original Eq. (1.10) considering average quantities, plus an extra advective term created by the correlation of the fluctuating parts of the fields,

$$\epsilon = \langle\delta\vec{v}\wedge\delta\vec{b}\rangle. \tag{1.23}$$

In order to close the system of equations provided by the average induction equation (Eq. (1.22)), one needs to express ϵ in terms of the average magnetic field. The symbol ϵ is an electromotive force. It is possible to show that there should be a linear relationship between the electromotive force and the mean magnetic field. The most general expression is given by

$$\epsilon_i = \alpha_{ij}\langle B_j\rangle + \beta_{ijk}\frac{\partial}{\partial X_j}\langle B_k\rangle + \cdots \tag{1.24}$$

Since the average quantities are slowing varying functions of space coordinates X_j, the higher-order terms can be neglected. (This approximation is valid as long as the scale separation holds.) The tensors α_{ij} and β_{ijk} depend only on the velocity of the plasma and the conductivity, but not on $\langle B_j \rangle$. Assuming the velocity is turbulent but homogeneous (i.e., does not depend on position) and isotropic (i.e., does not depend on direction), then

$$\alpha_{ij} = \alpha\,\delta_{ij}, \qquad \beta_{ijk} = -\beta\,\epsilon_{ijk}, \tag{1.25}$$

where δ_{ij} is the Kronecker delta and ϵ_{ijk} is the completely antisymmetric rank-three tensor. If these expressions are plugged into the average induction equation (Eq. (1.22)), with $\langle \vec{u} \rangle = 0$ and with α and β constant for simplicity, one ends up with

$$\frac{\partial \langle \vec{B} \rangle}{\partial t} = \alpha\,\nabla \wedge \langle \vec{B} \rangle + (\eta + \beta)\,\nabla^2 \langle \vec{B} \rangle, \tag{1.26}$$

where $\nabla \wedge \langle \vec{B} \rangle$ is the mean current. So, there is a turbulent diffusion term (β) plus a source of magnetic field aligned to the main current (α term). If $\alpha \neq 0$, Eq. (1.26) tells us that dynamos are to be expected with the following order-of-magnitude argument: the diffusive terms drops as the square of the lengthscale L for the variation of the mean field ($|\nabla^2 \langle \vec{B} \rangle|$ is of the order of B/L^2), whereas the positive α-term drops as L ($|\nabla \wedge \langle \vec{B} \rangle|$ goes as B/L). Therefore, for $L \to \infty$, the source term wins, $\partial \langle \vec{B} \rangle / \partial t > 0$, and one has a dynamo.

How do we know that $\alpha \neq 0$? The parameter α must change sign under parity transformation as $\vec{B}$ does. This is just a necessary condition, but unfortunately no sufficient condition for $\alpha \neq 0$ is known to exist. Purely turbulent velocity fields do not fulfill the requirement since they lack reflexion symmetry. However, turbulence in rotating bodies does fulfill it, and thus rotation is an important ingredient in many astronomical dynamos. An intuitive account of the α effect in the Sun was put forward in a fundamental paper by Parker (1955). A velocity field is able to create currents parallel to the mean magnetic field, thus producing non-zero α. The mechanism is sketched in Fig. 1.3. The magnetic fields are stored at the bottom of the convection zone, and they can be dragged along by the convective plumes that continuously rise from the bottom toward the solar surface. Buoyancy plus Coriolis forces move up and rotate the plasma traveling through the solar convection zone, like the cyclonic events shown in Fig. 1.3. These cyclonic motions lift and twist the magnetic field lines as shown in Fig. 1.3a, creating a curled magnetic field with a current antiparallel to the mean field. Stronger cyclonic events produce currents that are parallel to the mean field (Fig. 1.3b). The collective effect of many cyclonic events produces the α effect.

For any velocity field to produce an α effect, there should be a non-trivial correlation between the velocity to go up and around, i.e., a non-trivial correlation between the velocity and its vorticity. In technical terms, its kinetic helicity K has to be different from zero, i.e.,

$$K = \langle \delta\vec{v} \cdot \nabla \wedge \delta\vec{v} \rangle \neq 0. \tag{1.27}$$

Given a velocity field, it is non-trivial to infer the value of α since this is equivalent to solving the induction equation for the magnetic field perturbations. It can be done in cases with, e.g., small magnetic Reynolds number, which are not relevant in the astrophysical context.

The mean-field dynamo may present a problem in astrophysical contexts, where R_m is large. The linear analysis of the growth rates of the various dynamo modes shows that the mode of maximum growth rate is the one that has the lengthscale of the fluctuating velocity field. This uncovers an internal inconsistency of the theory because the mean-field electrodynamics implies a two-scale approach that, ultimately, allows us to neglect all the high-order derivatives in the expansion (1.24). The importance of this problem is not clear yet. The mean-field dynamo works for diffusive plasmas (small R_m).

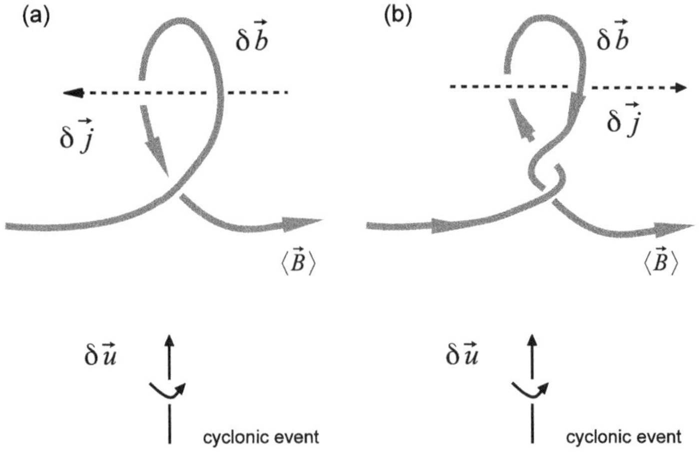

FIGURE 1.3. Parker's explanation for the α effect on the Sun. (a) Individual *cyclonic events* (velocity perturbation depicted at the bottom) move up and twist the magnetic field lines. It creates a fluctuation in the magnetic field $\delta\vec{b}$ with an associated current $\delta\vec{j}$, which is antiparallel to the mean field $\langle\vec{B}\rangle$, i.e., as required for the mean-field dynamo to operate. (b) If the cyclonic event is stronger, then the currents can become parallel to the mean field.

• *Small-scale dynamo.* The arguments mentioned above refer to large-scale magnetic fields, where their scale is larger than the characteristic scale of the velocity field producing the dynamo. What about small-scale fields? How do we generate magnetic energy, even in flows that cannot generate large scale magnetic fields? This is the small-scale dynamo, also extremely important in astrophysical contexts.

In an incompressible random plasma of large R_m, the magnetic field becomes locally transformed as it moves along with the flow as dictated by the Jacobian. If the flow is chaotic, the Jacobian corresponding to each parcel of the fluid is completely different so that, in the end, the magnetic field is modified according to the product of a large number of random matrices. One can prove that this process corresponds to an exponential expansion or stretching of the magnetic field in three orthogonal directions. The rate of expansion-stretching is given by the three Liapunov exponents

$$\lambda_1 > \lambda_2 > \lambda_3, \tag{1.28}$$

with the condition

$$\lambda_1 + \lambda_2 + \lambda_3 = 0 \tag{1.29}$$

needed to preserve the volume. If the magnetic field is stretched in one direction, it has to be compressed in other(s). One can think of λ_1 as the increase of length (per unit length) and of $|\lambda_3|$ as the increase of the gradients. The stretching increases the field strength whereas the gradients augment the diffusion, so that there will be a dynamo if the stretching wins, and this seems to happen very often when the velocity field is complex enough (e.g., Cattaneo, 1999), provided the magnetic Reynolds number is large.

The small-scale dynamo operates even when the viscosity is very small. The magnetic Prandtl number P_m, i.e., the ratio between the viscosity of the flow ν and the magnetic diffusivity η, is basically a measurement of the ratio between the lengthscale of $\vec{B}$, l_B, set by magnetic diffusion, and the lengthscale of $\vec{v}$, l_v, set by viscosity,

$$P_m = \frac{\nu}{\eta} = (l_v/l_B)^2, \tag{1.30}$$

where this expression considers that the timescales for changing $\vec{B}$ and $\vec{v}$ have to be similar (see Eq. (1.14); and keep in mind that the timescale for changing the velocities is formally similar to τ replacing η with ν). Consequently,

$$P_m \begin{cases} \ll 1 & \text{if random } \vec{v} \text{ at the characteristic scale of } \vec{B}, \\ \gg 1 & \text{if smooth } \vec{v} \text{ at the characteristic scale of } \vec{B}. \end{cases} \quad (1.31)$$

In the Universe, P_m is either very small (e.g., stellar interiors) or very large (e.g., interstellar medium). Only in simulations does P_m tend to be around one. Ten years ago it was found that, when P_m was significantly smaller than one, the turbulent dynamo in the numerical simulations switched off (e.g., Brandenburg, 2011, and references therein). For some time it was believed that small P_m dynamos were unfeasible, but this seems to be an artifact due to technical limitations. The computational box needed to capture the dynamo increases with the roughness of the velocity field. If you decrease P_m, everything else being constant, the roughness increases to a point that the computational domain is insufficient to maintain the dynamos, and the dynamo in the numerical simulations dies out. It can be proven using analytical solutions of the induction equation that dynamo action is possible at any P_m if R_m is large enough.

1.3 Symbols used in the chapter

TABLE 1.2. List of the main symbols used in the chapter

Symbol	Meaning
a	lengthscale for the averages in mean-field theory
$\vec{A}$	vector potential, i.e., $\nabla \wedge \vec{B}$
$\vec{B}$	magnetic field vector
B	magnetic field strength, i.e., $\lvert\vec{B}\rvert$
B_i	ith component of $\vec{B}$
β	plasma beta
$\delta\vec{b}$	fluctuating magnetic field in mean-field dynamo
$\delta\vec{u}$	fluctuating velocity field in mean-field dynamo
$\delta\vec{j}$	fluctuating current in mean-field dynamo
δ_{ij}	Kronecker delta
e	electron charge
ϵ	electromotive force
ϵ_{ijk}	completely antisymmetric rank-three tensor
η, η_t	magnetic diffusivity, turbulent magnetic diffusivity
$\mathcal{H}$	helicity
K	kinetic helicity
l_v	lengthscale for the velocity field
l_B	lengthscale for the magnetic field
l	lengthscale for the fluctuating part in mean-field theory
L	lengthscale for the average part in mean-field theory
λ_D	Debye length
ζ	lengthscale at which advection balances diffusion
Λ	plasma parameter
$\mathcal{L}$	characteristic lengthscale for the variation of the magnetic field
m_e	electron mass
m_p	proton mass
n_e	number density of electrons
n_i	number density of particles of species i
ν	viscosity

Ω	gyro frequency of electrons in a magnetic field
P	gas pressure
P_m	magnetic Prandtl number
r_L	gyro radius of electrons in a magnetic field
R_m	magnetic Reynolds number
R_e	Reynolds number
T_e	electron temperature
T_i	temperature of species i
τ	Ohmic diffusion timescale
τ_ζ	τ when B has structure of lengthscale ζ
$\vec{v}$	bulk velocity of the plasma
$v_\perp$	velocity perpendicular to $\vec{B}$
$\wedge$	vector product
V	volume
w_e	electron plasma frequency

Acknowledgments

We thank the CUP Astronomy Editor for understanding the various delays in writing this chapter. The introduction on Cosmic Magnetic Fields was given by F. Cattaneo. We have followed the syllabus and the organization of his presentation to write down these notes.

REFERENCES

Alonso de Pablo, A. & Sánchez Almeida, J., 2013, *Revista Española de Física*, **27**, 4

Brandenburg, A., 2011, *ApJ*, **741**, 92

Cattaneo, F., 1999, *Motions in the Solar Atmosphere* (Hanslmeier & Messerotti, Eds.), Kluwer, p. 119

Childress, S. & Gilbert, A. D., 1995, *Strech, Twist, Fold: the Fast Dynamo*, Springer-Verlag

Fukugita, M. & Peebles, P. J. E., 2004, *ApJ*, **616**, 643

Krause, F. & Rådler, K. H., 1980, *Mean-Field Magnetohydrodynamics and Dynamo Theory*, Akademie-Verlag

Kulsrud, R. M., 2005, *Plasma Physics for Astrophysics*, Princeton University Press

Mereghetti, S., 2008, *A&ARev*, **15**, 225

Moffat, H. K., 1978, *Magnetic Field Generation in Electrically Conducting Fluids*, Cambridge University Press

Parker, E. N., 1955, *ApJ*, **122**, 293

Parker, E. N., 1979, *Cosmical Magnetic Fields*, Clarendon Press

Priest, E. R., 2000, *Solar Magnetohydrodynamics*, Kluwer Academic Publisher

Schekochihin, A. A., Cowley, S. C., Dorland, W., *et al.*, 2009, *ApJS*, **182**, 310

Somov, B. V., 2006, *Plasma Astrophysics*, Part I, Springer

Spruit, H. C., 2013, *Essential Magnetohydrodynamics for Astrophysics*, ArXiv e-prints, http://adsabs.harvard.edu/abs/2013arXiv1301.5572S

2. Solar Magnetic Fields: History, Tragedy or Comedy?

PHILIP G. JUDGE

"Double, double, toil and trouble; Fire burn, and cauldron bubble!" – Three Witches, *Macbeth* (Shakespeare, 1623)[1]

Many children know that the Sun sometimes has spots. Sunspots were certainly known to ancient civilizations. Even so, it is sobering that we still do not have a unique answer to the simple question:

> *"Why, according to the basic laws of physics, is the Sun obliged to form a sunspot?"*
> E. N. Parker, *ca.* 1990
>
> (paraphrased)

This chapter summarizes how magnetic fields make their presence known in the spectra of solar plasmas, but the main purpose is to point to properties of solar magnetic fields that are surprising from the point of view of elementary physical ideas. These ideas include the physics behind the regeneration of magnetic fields that appears to require "dynamo" action and magnetic field dissipation, the role of magnetic field topology in highly conducting plasmas, and the energetic and dynamic effects of magnetic field rearrangement and dissipation that lead to nearly all phenomena of interest to modern human life. As a leitmotif, and in contrast to those who draw attention to obvious complexities in the Sun's magnetic field, I ask another question: Why does the Sun's magnetism show so much order in the presence of such chaos?

2.1 Introductory Remarks

When asked to review this subject for this Winter School, I confess to being a little surprised, for there are many solar physicists more qualified than I, owing to their particular areas of specialization, who can discuss both observations and theory relating to solar magnetism. In recent years I have tried to understand the physical problems presented by modern solar observations, and have offered some short courses and lectures that attempt to identify some elementary problems. By elementary, I do not mean "simple," but rather those building blocks of a complex non-linear system that a graduate student in physics might understand and retain. To me, it seems that modern solar physics boils down to one thing: the evolution of solar magnetic fields. So, this chapter is going to be concerned with explaining elementary aspects of the complex interaction of magnetic fields and plasma in the Sun, from the point of view of Newton, and Maxwell.

Right off the bat, I draw attention to a curious but central issue, captured in one of the best and exquisite images yet obtained of a sunspot (Figure 2.1). Sunspots exist in spite of the fact that 100% of the solar luminosity is carried by turbulent convection that extends across the outer 30% of the solar interior. The topmost convective layer is

[1] Credit for image of William Shakespeare used in this chapter: TonyBaggett/iStock/Getty Images Plus.

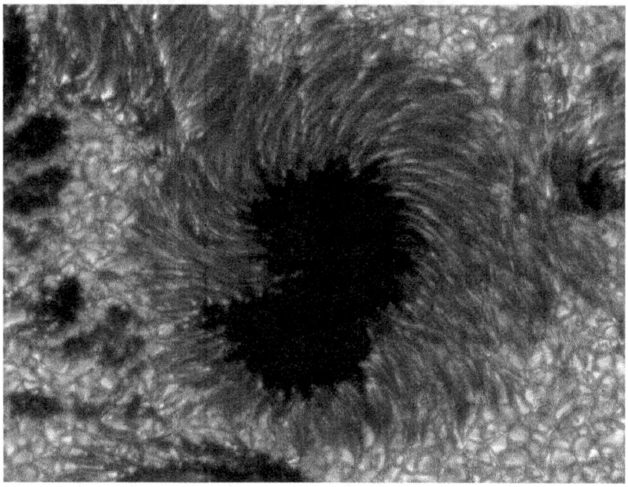

FIGURE 2.1. The mystery of sunspots is represented in this figure that appeared in *Sky and Telescope* magazine on July 23, 2003, an image from the Swedish Solar Telescope, a project led by G. Scharmer. Neither the nature of the umbra (darkest regions) nor penumbra (filamentary structures) is really understood, and the origin of why the magnetic field should emerge in such a concentrated form remains a mystery. The contrast between the large-scale order in the penumbral filament patterns and the turbulent convection outside the sunspot is dramatic.

seen clearly in the figure as cellular structures outside the sunspot itself. Why is the Sun obliged to produce such a thing, such order from chaos?

Sunspots exhibit order in many other ways. In the 1800s, Schwabe found that sunspots come and go cyclically with periods of 11 years. The magnetic nature of sunspots was discovered by Hale (1908), a decade or so after Zeeman discovered the effect of magnetic fields on spectral line intensities. Hale *et al.* (1919) later reported that sunspot magnetic fields change polarity from one cycle to the next, so that the Sun exhibits a 22-year cycling magnetic variation. Following the development of magnetohydrodynamics (MHD; e.g. Alfvén, 1950), sunspots are the most obvious manifestation of a large-scale MHD "dynamo." Various datasets show that the Sun's cycling magnetism is not atypical when compared to stars. Armed with powerful telescopes on the ground and in space and with recordings of solar magnetism encoded in paleo-climate records, we have many more observations of solar magnetic fields, from time scales exceeding 10^4 years to less than 1 second.

In the 1930s and 1940s, the Sun continued to surprise us, when spectroscopy showed that the temperature of plasma above the visible surface *increases* with distance from the Sun's center. This was in apparent violation of the second law of thermodynamics, and so not to be taken lightly. The obvious observational manifestation of this increased temperature is the solar corona. There is of course no such violation – instead we must invoke non-thermal processes for which thermodynamics does not apply, in contrast to the thermal modes of energy transport of radiative transport and convection present in the Sun's interior. In the latter half of the twentieth century, scientists have trodden a rather tortuous path towards the conclusion that the dissipation of energy associated with solar magnetism is intrinsically related to the formation of a corona.

Below we will look at the behavior of solar magnetism on large scales, meaning global scales on the order of the solar radius $R_\odot$, and those $\sim 0.01 - 0.1R_\odot$ that are characteristic of sunspots and active regions. Accordingly we will use the framework of MHD. This does not mean that we can ignore small scales – quite the contrary, one characteristic of

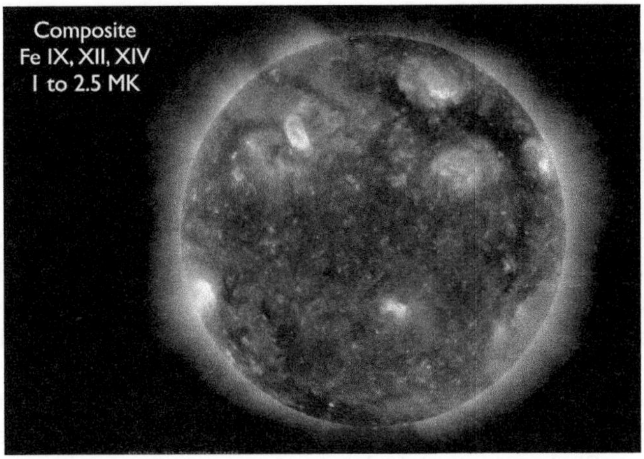

FIGURE 2.2. A black and white version of a three-color image taken by the AIA instrument on the SDO spacecraft on May 4, 2010. The corona is not only well organized magnetically (Figure 2.15) but also thermally, since this image shows darker regions (electron temperatures near 1 million K, 1 MK) and lighter regions (2 MK and 3 MK). Entire regions of the Sun somehow know how to be at given temperatures that are very different from other regions on the Sun. The dark features are coronal holes – regions of fast plasma outflow.

MHD (and hydrodynamics) is the development of small scales and a dynamical coupling across all physical scales. We speak of "turbulence."

Figure 2.2 shows more "order from chaos"– namely the relatively well-organized thermal structure of the solar corona on global scales. In part this results from the non-linear dependence of the energy loss due to electron heat conduction on electron temperature (Rosner *et al.*, 1978; Judge, 2002), but nevertheless, the degree of order in such images is remarkable given that the coronal heating mechanism is believed to occur on unobservably small scales (Section 2.7.3). Therefore, we have a puzzle: What is the origin of the large-scale order in the corona given that we know that dissipation occurs on very small scales?

For velocity field $\mathbf{u}$ in MHD, the evolution of the solar magnetic field is described by an "induction equation," the simplest form of which is

$$\frac{\partial \mathbf{B}}{\partial t} = \mathbf{curl}\,(\mathbf{u} \times \mathbf{B}) + \eta \nabla^2 \mathbf{B}, \tag{2.1}$$

where $\eta = 1/\mu_0\sigma$ is the magnetic diffusivity, with μ_0 the permeability of free space, and σ a scalar electrical conductivity. The velocity field $\mathbf{u}$ of the Sun is observed to contain large (rotation, circulation, solar wind) and small (convection, turbulence, waves) scales.

First consider the second term in Equation (2.1). Kinetic theory (Ohm's law) for collision dominated plasmas gives us values for the conductivity σ (e.g. Braginskii, 1965), from which we find that $\eta \sim 10^4$ cm^2 s^{-1} in the Sun's interior. Choosing a characteristic scale $\ell \sim R_\odot/3 = 2 \times 10^{10}$ cm we find that the diffusive term in Equation (2.1) has a time scale of $\ell^2/\eta \sim 10^9$ years. Yet the Sun exhibits magnetic variations many orders of magnitude faster. Next, note that the equation appears linear in $\mathbf{B}$, but this is misleading unless $\mathbf{u}$ is truly independent of $\mathbf{B}$. In real plasmas this is rarely the case since the plasma experiences a Lorentz force (effect of $\mathbf{j} \times \mathbf{B}$ on $\mathbf{u}$). MHD deals the coupling of the induction equation (or more complex versions of it) with equations of motion in which $\mathbf{u}$ and $\mathbf{B}$ and other fluid variables (density, pressure, temperature, etc.) evolve together.

The Sun is the archetypal object for study under the high magnetic Reynolds number regime, $R_M = u\ell/\eta \gg 1$, where u and ℓ are characteristic speeds and lengths of the motion of the fluid. In this regime, the first term of Equation (2.1) dominates. In the limit of zero diffusion η (the ideal MHD limit), tubes of magnetic flux are tied forever to the plasma that they entrain – "Alfvén's theorem." In this case the behavior of the plasma and fields must obey not only equations of motion but also an infinite number of topological constraints. It is a curious fact, almost a poetic tragedy, that ideal MHD leads to its own demise (Section 2.7.3). In "natural" systems like the Sun, as opposed to idealized models with high degrees of order, Parker (1972; 1994) has argued that topological constraints plus the equations of motion over-determine the solutions. The system evolves by trying to form mathematical "tangential discontinuities." But even in the presence of very small but finite plasma resistivities, in trying to become singular, the Maxwell stresses make steeper gradients until ideal MHD causes its own demise – small scales in ℓ develop so that the second term in Equation (2.1) eventually leads to non-ideal behavior. Shakespeare might have written:

"Fair is foul, and foul is fair." – Witches, Act I, scene I, *Macbeth* (Shakespeare, 1623)

Here then are some central issues discussed below: In terms of the generation of magnetic fields, *why is the Sun obliged to vary cyclically, flipping the large-scale fields every 11 years, and why must it appear most clearly as a sunspot?* In terms of the dissipation of magnetic field, *why does every solar-like star possess a corona, or chromosphere, which is a partially ionized region between the visible surface and corona? Why must flaring occur?* I will use some simple physical arguments to identify some cutting edge problems. I am guided by some recent arguments by Parker (2009) and Spruit (2011). Pedagogical articles by Casini and Landi Degl'Innocenti (2008) and Rempel (2009) are also recommended.

2.2 Solar Magnetism: So What?

Sunspots are small enough, and the Sun bright enough, that the occasional reports of them by ancient civilizations are of limited use. The first quantitative measurements of sunspots really required the invention of the telescope and the projection of solar images. Spots began to be counted in around 1610. It was not until 1908 that Hale proved that the spots were concentrations of strong magnetic fields, but his discovery meant that spots could then be used to probe solar magnetism back to the early 1600s, the time of Galileo.

Sunspots generally appear benign to the naked eye – so why should humanity care about these most obvious manifestations of a varying magnetic field on the Sun? The first rapid changes (minutes) seen in sunspots were reported by Carrington (1859) in a complex group of sunspots, confirmed independently by Hodgson. In the book by Young (1892), we find that this phenomenon

> ...was immediately followed by a magnetic storm of unusual intensity, the auroral displays being most magnificent on both sides of the Atlantic, and even in Australia.

Now, aurorae are accompanied by magnetic disturbances on the ground, affecting compasses. These disturbances occur because the changing magnetism on the Sun, which leads to flares including the one observed by Carrington, produces high-energy radiative

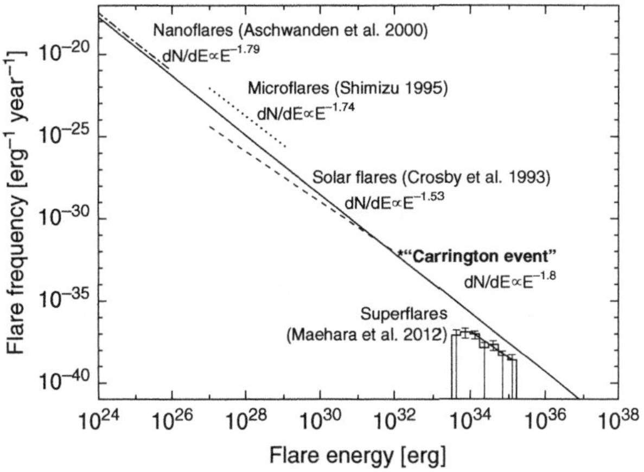

FIGURE 2.3. Distributions of flares for the Sun in comparison with stars, modified from Shibata *et al.* (2013) (see also references therein) to include the "Carrington event." The solid histogram shows the frequency distribution of superflares on G V-type stars with rotational period >10 d and effective temperatures of 5600–6000 K.

and particle disturbances at the Earth. In turn, these disturbances induce electrical currents in the Earth's magnetosphere and ionosphere. In 1859, telegraph wires hundreds of miles long were affected by currents along them induced through the changing magnetic fields in the ionosphere, such that signals could be transmitted with no applied electromagnetic field!

Towards the end of the twentieth century it became clear that flares are also associated with the large scale ejection of material in the magnetized corona, "coronal mass ejections" (CMEs). In many flaring events, solar energetic particles (vastly supra-thermal) are emitted by the evolving fields in the corona, and via shocks in interplanetary space induced by CMEs. All of these phenomena, driven by evolving solar magnetic fields, put at risk modern society, as we beome more and more dependent on spacecraft flying in the ionosphere, magnetosphere and in interplanetary space, and as our need for electrical power ever increases. Power grids and satellites, are just the most obvious "Achilles Heels" of our technological society. The huge geomagnetic effects of the "Carrington event" of 1859, (with energy estimated at $\approx 10^{32}$ erg by Shibata *et al.*, 2013, probably a lower limit), suggest that it is the largest flare recorded in history. The question arises as to how our infrastructure might be affected by such a strong flare, given that plasma ejection associated with smaller flares (X class between March 11 and 13, 1989) disrupted power transmission in Quebec. Given the spectacular aurorae that accompany such magnetic storms, a twentieth-century descendent of Macbeth might conceivably have said

"So foul and fair a day I have not seen." – Macbeth's opening line (Shakespeare, 1623)

Recent work combining stellar and solar data (Shibata *et al.*, 2013; see Figure 2.3) allows us to estimate the rate of occurrence of flares of Carrington amplitude and higher. Although solar data are based upon EUV and X-ray data, and stellar data are based upon visible wavelength photometry from the *Kepler* satellite, Shibata *et al.* (2013) find that a flare of 100 times the Carrington energy is expected roughly once in 800 years.

We must therefore attempt to understand why the Sun, in particular its magnetism, behaves as it does.

2.3 Measuring Solar Magnetic Fields

2.3.1 *Remote Sensing*

This section draws on the pedagogical article by Casini and Landi Degl'Innocenti (2008), another very nice article is that of Lites (2000) with several examples of solar measurements. Determining physical properties in solar plasmas is an exercise in "remote sensing." In astronomy, the origin of remote sensing is traced to Kirchhoff and Bunsen (1860), who identified spectral lines in the laboratory that coincided with Fraunhofer's dark lines in the solar spectrum recorded some four decades earlier. "Plasma spectroscopy" in fact began the modern era of astrophysics, and it remains a primary tool – without it we would know little about the Universe. Spectra of atoms, molecules and ions embedded in plasmas emit and absorb photons in a manner that encodes conditions in the plasma and any electric and magnetic fields threading the plasma. Simple examples are bulk flows of plasma reflected by Doppler shifts of the spectrum (expansion of the Universe), temperatures of stellar atmospheres are reflected in the relative strengths of lines belonging to molecules, neutral atoms or ions. Curiously, after Fraunhofer's work but before that of Kirchhoff and Bunsen, philosopher Auguste Comte (1835) wrote

> On the subject of stars... we shall never be able by any means to study their chemical composition... or even their density... I regard any notion concerning the true mean temperature of the various stars as forever denied to us.

I cannot resist the following quote:

"... But swords I smile at, weapons laugh to scorn, Brandish'd by man that's of a woman born." – Macbeth (Shakespeare, 1623)

Compte's "weapons," a tragic, perhaps comedic prediction of the future, serve as a warning to those of "us of a woman born" who try to predict the future of science (Macbeth, however, is killed quickly after these words by Macduff, not "born" of woman but delivered by cesarian section).

The effects of magnetic fields in spectra of astronomical plasmas are generally more subtle, because bright objects are usually hot and so spectral lines are broadened by thermal motions. Only when magnetic fields are strong can one use spectroscopy alone. More generally, one must also use "plasma polarization spectroscopy" because the magnetic fields are weaker. Magnetic fields break symmetries in current-carrying systems such as atoms, so it is not surprising that magnetically induced atomic polarization leads to polarized spectra.

Spectral line polarization originates in plasmas in essentially two ways (from Casini and Landi Degl'Innocenti, 2008). Firstly, magnetic substates may be unevenly populated, so the transition components ($JM \to J'M'$) no longer combine with the particular weights (natural populations) needed for the total polarization to vanish, such as in local thermodynamic equilibrium (LTE).[2] Unequal populations occur when *atomic excitation/de-excitation processes are anisotropic*. Secondly, even with atomic substates

[2] Here, J is the total angular momentum quantum number, M the projection of this quantum number on to a particular axis, such as along a magnetic vector $\mathbf{B}$.

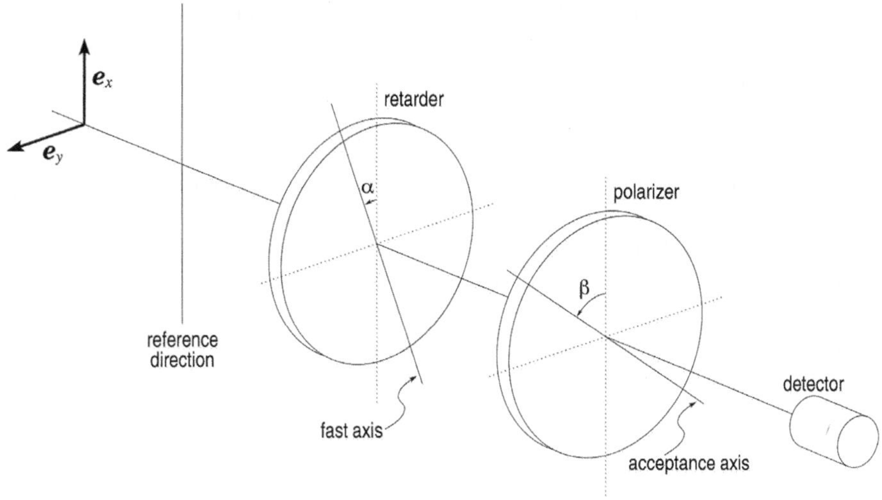

FIGURE 2.4. Typical configuration of a polarimeter. Different polarization states of the radiation passing through the instrument are measured by suitable combinations of the retarder and polarizer position angles, α and β. From Casini and Landi Degl'Innocenti (2008).

populated naturally, the substates may be *separated in energy*, so that a spectral analysis of atomic transitions reveals varying polarization properties with wavelength. This second case is most familiar to solar physicists, leading to the Zeeman and Stark effects.

The two sources of polarized spectral lines are not mutually exclusive. Hanle (1924) experimented with anisotropically excited atoms in the presence of magnetic fields, finding that magnetic fields can modify the zero-field polarization with field strengths far below those necessary to produce visible energy separations via the Zeeman effect. Such modifications are referred to as the Hanle effect.

2.3.2 *Measuring Polarization of Light*

Polarization of light is mathematically specified by the two complex Cartesian components, E_x and E_y of the radiation electric field with wavevector along the z axis. In a physical description a coherency matrix (or polarization tensor) of the radiation field (averaged over the acquisition time and elemental surface of the light detector) is specified,

$$C = \begin{pmatrix} \langle E_x^* E_x \rangle & \langle E_x^* E_y \rangle \\ \langle E_y^* E_x \rangle & \langle E_y^* E_y \rangle \end{pmatrix} \equiv \frac{1}{2} \begin{pmatrix} I+Q & U-iV \\ U+iV & I-Q \end{pmatrix}, \tag{2.2}$$

so that four independent parameters are needed in order to describe polarized radiation, either in terms of the electric fields or equivalently the Stokes parameters, I, Q, U, V (Casini and Landi Degl'Innocenti, 2008). Observationally, it is customary to use the four (real) Stokes parameters: I is the intensity, Q and U are the two independent parameters needed to describe linear polarization on the $x-y$ plane, and V is the circular-polarization parameter. Casini and Landi Degl'Innocenti (2008) adopt an operational definition for measuring polarization of light in terms of an ideal instrument, consisting of a linear polarizer and retarder (Figure 2.4). A retarder introduces a phase difference between orthogonal directions of polarization, examples being calcite crystals with different refractive indices for light with electric vectors along and perpendicular to certain directions (birefringence). For a $\lambda/4$ retarder, this "polarimeter" registers counts on a detector as functions of angles α and β between the retarder's fast axis, the polarizer's acceptance axis, and a reference direction as:

$$S(\alpha, \beta) = k\left[I + (Q\cos 2\alpha + U\sin 2\alpha)\cos 2(\beta - \alpha) + V\sin 2(\beta - \alpha)\right], \tag{2.3}$$

where k is a factor that includes detector gain, flat field etc. corrections. A sequence of measurements of $S(\alpha, \beta)$ with α and β varying with time can be used to determine I, Q, U, V with suitable choices of α and β. This involves a 4×4 matrix inversion coupling all four Stokes parameters. The needed sequence of observations is called "polarization modulation."

Several factors are critical in making accurate polarimetric measurements.

- Frequently, Q, U, V are significantly smaller than I, sometimes by orders of magnitude.
- Since complete measurement of all four Stokes parameters requires at least four measurements, sequential in time in most polarimeters, *any (spurious) time variations can lead to measurement errors.* Such errors include *crosstalk*, in which, say, variations in Q *entering the polarimeter from the telescope feed*, due to non-solar sources, induce spurious signals in, say U.
- A large source of error and crosstalk includes *atmospheric seeing* (e.g. Lites, 1987). Space polarimeters, such as the SP instrument (Lites *et al.*, 2001) on the Hinode spacecraft, avoid seeing problems (but do have some spacecraft jitter).
- In observations affected by seeing, $I \rightarrow Q, U, V$ crosstalk often dominates. The solution is then to split the beam, with suitable choices of $\beta - \alpha$, to measure effectively $I + aQ + bU + cV$ and $I - aQ - bU - cV$ simultaneously. This yields an 8×4 matrix inversion (Seagraves and Elmore, 1994) or its algebraic equivalent. The Hinode SP instrument in space has a dual beam capability, but results obtained with one beam are of comparable quality, as one might expect (Lites, private communication, *ca.* 2009).
- The polarimeter measures only the state of polarized light entering the instrument. If the telescope is "polarizing," in the sense that it changes the incoming Stokes array I, Q, U, V via off-axis asymmetries for example, via a matrix $\mathbf{T}$, then this matrix must be determined (a "polarization calibration" is required) and then inverted.

Sometimes we need to measure very small polarizations, say $10^{-4}I$, to determine magnetic fields in the Sun. Such measurements are extremely difficult to perform, but there are ways around them, such as by modulation/demodulation at frequencies far above any frequency induced by seeing or other factors. The ZIMPOL instrument does this with a combination of piezo-electric modulation by crystals and by charge caching detectors (e.g. Gandorfer and Povel, 1997).

2.3.3 *The Sun is Not Always Bright Enough for Polarimetry*

It should be recognized that spectropolarimetry at high angular resolution actually is a *photon starved* exercise. This surprising conclusion has been re-emphasized by Landi Degl'Innocenti (2013). At the diffraction limit of any telescope, with $\theta_{DL} = 1.22\lambda/D$ the flux density f_{DL} of photons is independent of telescope aperture D because

$$f_{DL} = I\pi\theta_{DL}^2 = I\pi(1.22\lambda/D)^2 \tag{2.4}$$

per unit area, where I is the solar disk intensity and we have assumed the flux is contained in one Airy disk. The total photon flux per detector pixel is $\pi D^2/2$ times this and independent of D. Using reasonable values for system efficiency, spectral resolution, wavelength, Landi Degl'Innocenti (2013) finds that one can accumulate only 10^6 photons per second per pixel, meaning that 100 s integrations are needed to achieve a statistical sensitivity of 10^{-4}. While not a problem for stars, the diffraction limit of large solar telescopes corresponds to tens of km at the solar surface, in 100 s the Sun's atmosphere can change dramatically on such scales. Trade-offs must be made by a judicious selection of angular resolution, wavelength, instrumental throughput, even with 4-m class telescopes.

Lastly, even with great care, it should be pointed out that systematic sources of error (fringes, crosstalk, calibration errors) generally dominate polarization measurements at high sensitivities. Spectropolarimetry presents interesting challenges.

"... that is a step on which I must fall down, or else o'erleap, for in my way it lies." – Macbeth (Shakespeare, 1623)

2.3.4 *Zeeman Effect*

The Zeeman splitting of atomic sub-levels is $h\nu_L \sim \mu_B B$, where ν_L is the Larmor frquency, μ_B is the Bohr magneton. The radiation emitted or absorbed by the associated "σ" ($\Delta M = \pm 1$) and "π" ($\Delta M = 0$) radiative components then depends on the magnitude and direction of the magnetic field vector. Requiring polarizations in excess of a few percent in most instruments, then, as we will see below, this means $\nu_L/\Delta\nu \gtrsim 2 \times 10^{-2}$, where $\Delta\nu$ is the characteristic line width. In the Sun's photosphere, Doppler broadening is usually dominant, with a value of a few km s^{-1}, $\sim 10^{-5}c$. Translated into frequency units, we find that the Zeeman effect is useful for magnetic fields of order

$$\nu_L \sim \frac{\mu_B B}{h} \gtrsim 2 \times 10^{-2} \times 10^{-5} \frac{c}{\lambda_0} \tag{2.5}$$

$$B \gtrsim 90 \, \frac{5000\,\text{Å}}{\lambda} \text{ G.} \tag{2.6}$$

Figure 2.5 shows the "classical" analog for the Zeeman effect which applies to $J = 0 \rightarrow J = 1$ transitions, from Lites (2000).

In the upper panel of the figure (longitudinal Zeeman effect), the circularly orbiting oscillators are seen "face on" so that they emit circularly polarized light in the observer's direction. Conversely, in the lower panel, the oscillators are seen "edge on" so that they emit linearly polarized light. Owing to symmetry, the "red" and "blue" shifted components are of the same amplitude, and they are shifted in frequency by the Larmor frequency of the atomic level ν_L in both cases.

In the longitudinal-field case, a difference in measurement of left versus right circularly polarized light (this is the definition of Stokes V) will reveal the V polarized profile as shown. The I profile shown is "fully split," i.e. Zeeman splitting exceeds line broadening from the plasma (thermal, bulk motions, collisions), and is symmetric. In the limit that the Zeeman splitting is small, the Stokes I profile is merely broadened a little – the two absorption features σ_R, σ_L in I in Figure 2.5 being blended into one feature. However, the Stokes V profile then survives with an amplitude that is proportional to the longitudinal field strength, as shown below.

Figure 2.6 shows profiles measured by the Advanced Stokes Polarimeter. The profiles shown are most readily understood using the excellent pedagogical article of Jeffries

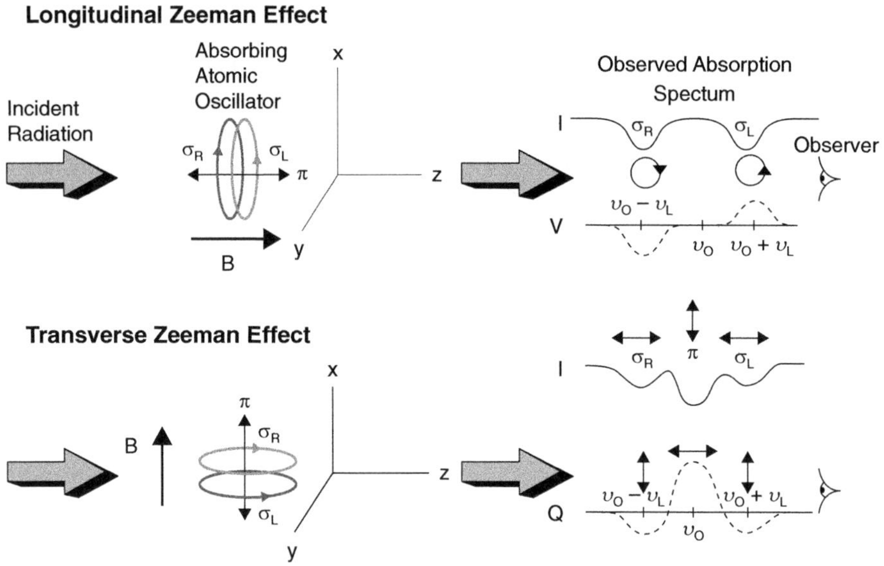

FIGURE 2.5. The classical analog of the Zeeman effect on an atomic system represented by classical oscillators. An atom absorbs a little of the incoming continuum light (from the left) with the magnetic field aligned along the z and x axes respectively, showing the longitudinal (top) and transverse (bottom) Zeeman effect. From Lites (2000).

et al. (1989), which focuses on the transport of polarized radiation in an atmosphere. These authors first derive emission and absorption coefficients for light passing through electrons bound to atoms and ions using equations of motion and dielectric theory, in the presence of arbitrarily oriented magnetic fields. The absorption and emission processes are assumed uncorrelated, such as occurs in thermal equilibrium and LTE, and the processes of damping and Larmor precession are assumed far smaller than the "resonance frequency" (i.e. frequency of the line radiation, $\sim 10^{15}$ Hz). They then derive the equations of radiation transport and solutions for simple cases. Jeffries *et al.* discuss in particular the "weak field limit" (Larmor frequency ν_L much smaller than the combined Doppler and natural widths $\Delta\nu$ of the lines). Their Equations (45) and (47), consistency relations related to their transfer equations, readily explain the Q, U, V profiles seen in Figure 2.6: the emergent Stokes V profile is a term that is first order in $\nu_L/\Delta\nu$, proportional to the first derivative of the intensity profile and so asymmetric around line center. The QU profiles are second-order terms and are symmetric about line center.

This different behavior originates from the dependence of the emitted dipolar radiation on the quantum numbers $\alpha JM \rightarrow \alpha' J' M'$ of the atomic transitions (refer to Figure 2.5). When $\Delta M = \pm 1$, labeled "σ" transitions, the atom emits or absorbs left- and right-circularly polarized light, the radiative transitions being shifted in frequency by $\pm\nu_L$. When $\nu_L/\Delta\nu \ll 1$, we can use a first-order expansion of the (left–right) polarization states in frequency or wavelength, yielding $V \propto \nu_L \phi'$, where $\phi' = d\phi/d\nu$. For the $\Delta M = 0$ "π" transitions, the atom emits or absorbs linearly polarized light unshifted in frequency. Unlike Stokes V, the linear polarization (Stokes Q, U) measurements "see" both the σ and π components. When the substate populations are equal (LTE), since they occur in the combination $\pi - (\sigma_{+1} + \sigma_{-1})/2$, the leading order term is second order in $\nu_L/\Delta\nu$, therefore yielding a frequency dependence $\propto \phi''$.

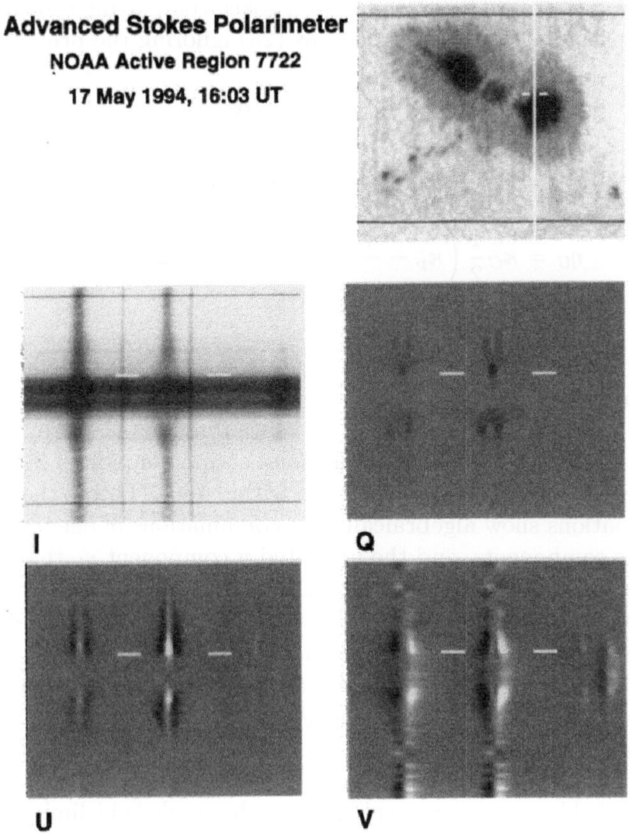

FIGURE 2.6. Spectropolarimetric line profiles of a sunspot with polarization generated by the Zeeman effect. The upper panel shows the position of the slit on a continuum image, other panels show Stokes profiles as a function of wavelength (abscissa) and position along the slit. The two broad dark vertical lines are a pair of 630 nm lines of Fe I, they show Zeeman splitting in I and also V over the sunspot umbra and penumbra. Away from the sunspot the Stokes QUV profiles are similar, being formed in the "weak field limit" (see text). From Lites (2000).

In the work of Jeffries *et al.* (1989), the Zeeman effect translates to the absorption coefficients for Q, U, V (unprimed terms in their Equation (28)). Taking account also of the geometry, Jeffries and colleagues present solutions to the simplest radiative transfer equation applicable to photospheric lines, the assumption that the source function varies with (continuum) optical depth τ as $S(\tau) = B_0 + B_1\tau$. Ignoring, for tutorial purposes, complications due to "magneto-optical effects,"[3] we can write their Equation (39) for the emergent Stokes parameters at any particular frequency as

$$I = B_0 + \frac{\mu B_1}{\Delta}(1 + \eta_I)^3 \tag{2.7}$$

$$Q, U, V = -\frac{\mu B_1}{\Delta}(1 + \eta_I)^2 \eta_{Q,U,V}, \tag{2.8}$$

[3] MO effects arise from phase changes introduced by non-unit real parts of the refractive index in the dielectric theory. In the notation of Jeffries *et al.* the MO effects are contained in the quantities having have a "prime" superscript in their standard transfer Equation (35) and subsequently in their variable ϱ.

where the $\eta_i = \kappa_i/\kappa_C$ are ratios of the line's Stokes component i absorption coefficient to that of the continuum κ_C (their Equations (28) ignoring the primed MO terms):

$$\eta_I = \kappa_C \frac{1}{2}\left[\left(\frac{\kappa_r + \kappa_l}{2}\right)(1+\cos^2\gamma) + \kappa_p \sin^2\gamma\right] \quad (2.9)$$

$$\eta_Q = \kappa_C \frac{1}{2}\left(\kappa_p - \frac{\kappa_r + \kappa_l}{2}\right)\sin^2\gamma\cos 2\chi \quad (2.10)$$

$$\eta_U = \kappa_C \frac{1}{2}\left(\kappa_p - \frac{\kappa_r + \kappa_l}{2}\right)\sin^2\gamma\sin 2\chi \quad (2.11)$$

$$\eta_V = \kappa_C \left(\frac{\kappa_r - \kappa_l}{2}\right)\cos\gamma\,. \quad (2.12)$$

Here, γ and χ are angles that magnetic fields make with the line of sight and on the plane of the sky defined according to the reference direction (Figure 2.4). Note that the azimuthal angle χ occurs only through a sine and cosine function of 2χ, thus the *azimuth* is determined by observations of Q, U only to within 180°. This is the "azimuthal ambiguity." These equations show algebraically the combinations of energetically shifted left- (κ_l), right- (κ_r) σ components, and the unshifted π component κ_p that lead immediately to the profiles seen in Figure 2.5. The Q, U terms are of the form $2 \times \pi$ component minus the sum of the σ components: this is a difference equation for the second derivative; the V term is simply the first difference between the σ components.

In the "weak field limit," magnetic information is contained only in the *amplitudes* of the Stokes profiles, and angular factors, the Q, U, V profile *shapes* being set by derivatives of I. It is not possible to discriminate between two configurations containing the same net magnetic flux per unit area unless additional measurements from, e.g., a line formed outside the weak field regime, are available. In the weak field limit the Zeeman effect merely measures the *magnetic flux density* (Maxwells per square cm) not the *magnetic field strength* (Gauss).

Lastly, it is worth emphasizing that because Zeeman-induced linear polarization is usually small, the bulk of the literature, indeed almost the entire literature that deals with *global* fields, is based upon circularly polarized data from the longitudinal Zeeman effect. Such measurements, called "longitudinal magnetograms," measure only the *net line of sight field component* of **B**. Regular observations began in the 1950s after the invention of the scanning magnetograph (Babcock and Babcock, 1952; Babcock, 1953). With the advent of the SOLIS ground-based instrument and HMI instrument SDO, "vector magnetograms" (using linear and circular polarization measurements) are becoming more commonplace.

2.3.5 *Hanle Effect*

Above, we found that the Zeeman effect requires field strengths in excess of a few tens of G, with the polarized light depending on the ratio of Larmor frequency to Doppler width frequency. In the Hanle effect the relevant parameter is the product

$$2\pi\nu_L A^{-1} \sim 1 \quad (2.13)$$

with A the spontaneous decay rate of an atomic level. An atom in a given state with a net magnetic moment μ gyrates around magnetic field lines with frequency $\nu_L = \mu B/h$. If $2\pi\nu_L A^{-1} \sim 1$, line photons are emitted while the atom's azimuthal angle around the magnetic field changes by one radian. In the vector model of the atom an initially polarized state will lose some "memory" of its initial direction, the emission will be *rotated*

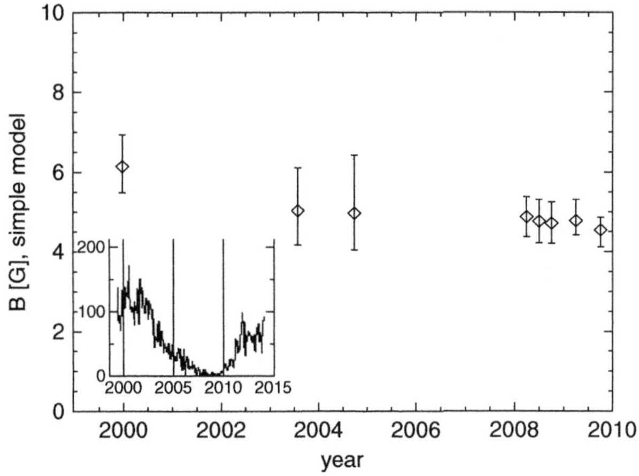

FIGURE 2.7. Field strengths measured from the Hanle effect over a decade, near the limb of the quiet Sun from Kleint *et al.* (2010). In contrast, the inset shows sunspot numbers from 1999 to the present, varying by a factor of 10–100 over the same period! Reproduced with permission © ESO.

and the magnitude of polarization *reduced*, since radiative decay is a stochastic process obeying a distribution of lifetimes $\propto \exp(-tA)$.

The Hanle effect is sensitive to weaker magnetic fields than the Zeeman effect. For electric dipole transitions in neutral species, $A \sim 10^8$ s^{-1} for a spectral line at $\lambda_0 = 5000$ Å. Then, with $\nu_0 = 6 \times 10^{14}$ Hz, we have $A/\nu_0 \sim 1.6 \times 10^{-7}$. If we allow for rotation of ~ 0.1 radians to be detectable, then the Hanle effect is sensitive to magnetic fields when

$$\nu_L \sim \frac{0.1 \to 1}{2\pi A}, \text{ or} \tag{2.14}$$

$$B \sim 1 \to 10 \text{ G} \quad \text{(permitted lines of neutrals)}. \tag{2.15}$$

The Hanle effect merely *alters existing polarization*; its use in solar magnetic field measurements is related to spectral lines that are polarized by other processes, most commonly by excitation by radiation fields that are anisotropic. This is explicitly then a non-LTE phenomenon since LTE implies detailed balance, isotropic excitation in particular. Unlike the Zeeman effect, Hanle rotation and depolarization are additive, no matter the sign of magnetic field. Therefore the effect is suited to detection of randomly oriented fields. In this context, Kleint *et al.* (2010) used the Hanle effect to investigate small-scale, disordered magnetic fields in the quiet Sun using lines sensitive to magnetic fields formed near the solar limb. Their results, shown in Figure 2.7, show no significant variations around 5 G between 2000 and 2010. In contrast, Zeeman measurements of solar magnetic fields associated with sunspots amount to average field strengths between 4 Mx cm^{-2} and 20 Mx cm^{-2} (Schrijver and Harvey, 1989) over the sunspot cycle.

The reader should refer to Casini and Landi Degl'Innocenti (2008) for an accessible pedagogical text on the (far more complex) quantal treatment of the Hanle effect.

2.3.6 *Natural Systems as Recorders of Solar Magnetism*

The large-scale (several AU) solar magnetic fields influence the Earth by modulating the incoming flux of cosmic rays; see McCracken *et al.* (2013) for a recent review. Radionuclides ^{10}Be and ^{14}C are produced in the atmosphere and sequestered in polar ice and tree rings. Neutrons are also created by cosmic ray interactions with Earth's atmosphere.

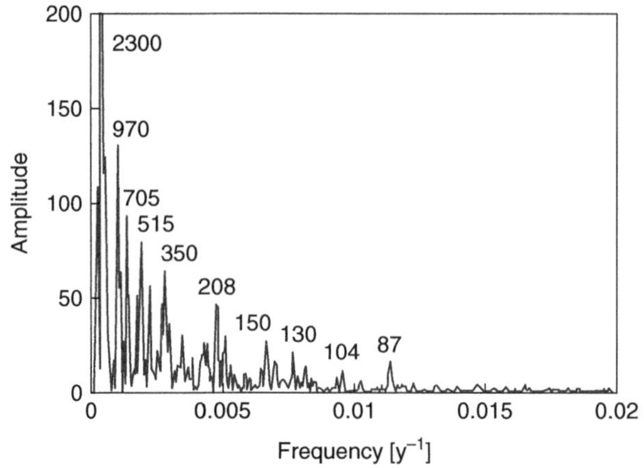

FIGURE 2.8. Fourier amplitude spectrum of the ^{10}Be ice core data versus frequency, for periods between 50 and a few thousand years, taken from McCracken *et al.* (2013). Periods corresponding to the major peaks are annotated in years.

Neutron fluxes show a clear 22 year modulation of a fundamental 11 year cycle as measured over the past 60+ years (e.g. Beer *et al.*, 2012). Taken together, this has allowed scientists to infer a "solar modulation" function for the past 10 000 years. The function, an indirect measure of heliospheric magnetic fields, is based upon theoretical models of cosmic ray transport (Parker, 1965).

As shown by McCracken *et al.* (2013), the independent ^{10}Be and ^{14}C records are mutually consistent over the past 10 000 years (the "holocene"). Figure 2.8 shows Fourier amplitudes of the ^{10}Be time series reported by McCracken and colleagues. The large-scale (heliospheric) solar magnetic field appears to show a quasi-periodic behavior on time scales between 50 and 10 000 years.

That the Earth is a detector of solar influences should be no surprise, the radionuclide traces of galactic cosmic ray modulation are one interesting example. The Earth responds to solar radiation and particles: aurorae caused by electromagnetic perturbations from solar charged particles (Birkeland and Muir, 1908) influence the magnetosphere, and are correlated with sunspot activity. Indeed Eddy (1976) cited the lack of aurorae in support of the reality of the Maunder Minimum in sunspots. The mere existence of an ionosphere and upper atmospheric ozone prompted Saha (1937) to conclude that the Sun must emit excess UV radiation before the conclusive evidence of a hot corona (Grotrian, 1939; Edlén, 1943).

2.4 The Observational Record

B. C. Low, a theoretician who studies physical and mathematical properties of the MHD equations, often reminds us that "solar physics is an observationally-driven science". I believe he means that, even in the case of the simplest theoretical model for coupled magnetic fields and plasmas – the MHD picture – the non-linear coupling between the governing equations admits such a broad variety of solutions, that we must be guided by observations. Indeed, the justification for our new facilities, such as the Daniel K. Inouye Solar Telescope, the first major facility for ground-based solar physics for the United States in almost 50 years, is to be able to study the non-linear interactions

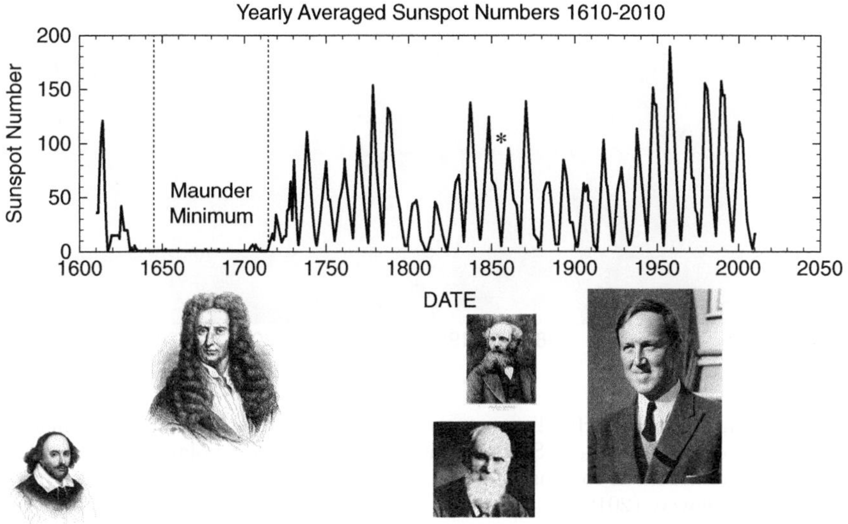

FIGURE 2.9. The historical record of sunspot numbers is shown along with some characters discussed in the text. The asterisk at 1859 marks the "Carrington event". From left, there is Shakespeare (credit: TonyBaggett/iStock/Getty Images Plus), Newton (credit: ThePalmer/Digital Vision Vectors/Getty Images), Carrington (lower), Maxwell, and Alfvén. Widths of the images corresponds to life spans.

between plasma and (electro-) magnetic fields threading them, on scales inaccessible to laboratories. So here I review landmark observations in a historical context.

2.4.1 *Early Hints of Variable Solar Magnetism*

Figure 36 of the book by Young (1892) shows spectra from 1870 of Fraunhofer's D lines. The figure is clearly recognizable today as a typical case of spectral lines split by magnetic fields – the Zeeman effect – but because the nature of sunspots was not known until much later these data were interpreted in terms of earlier work showing reversals in chromospheric lines; i.e. the origin of the reversal was interpreted as thermodynamic, not magnetic. During the 1878 total eclipse, during a deep minimum in sunspot number, it was noted that the corona appeared to trace magnetic lines of force.

From 1610 sunspots have been recorded essentially daily. While the data are from heterogeneous sources, several properties of sunspot numbers are robust. Figure 2.9 shows sunspot numbers as compiled by David Hathaway.[4] After 17 years of observations, Schwabe (1844) discovered the cyclic increase and decrease of sunspot numbers. In 1852, Sabine reported that the sunspot cycle period was "absolutely identical" to that of geomagnetic activity, Young (1892) shows data from 1772 to 1880.

2.4.2 *Large-Scale Properties of Solar Magnetic Fields*

Here I offer selected observations of relevance of the origin and dissipation of solar magnetic fields, some of these are already evident in Figure 2.10, showing surface magnetic fields as measured using longitudinal magnetographs. I follow lines of argument from Parker (2009) and Spruit (2011):

[4] This is a version of http://solarscience.msfc.nasa.gov/images/ssn_yearly.jpg.

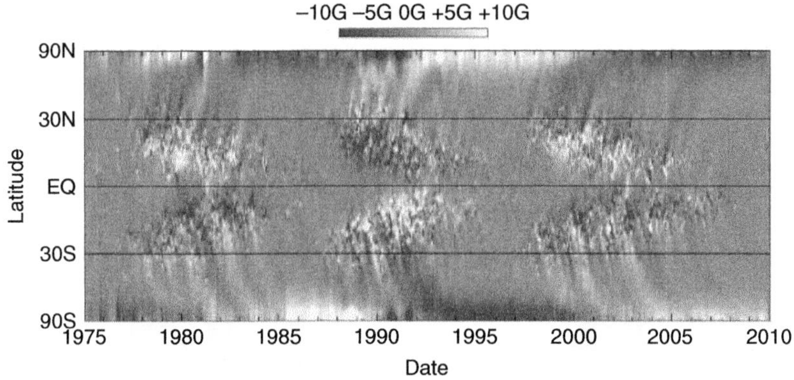

FIGURE 2.10. A magnetic butterfly diagram constructed from the radial magnetic field obtained from instruments on Kitt Peak and SOHO. This illustrates Hale's polarity laws, Joy's tilt law, polar field reversals in relation to sunspots, and the apparent transport of magnetic field toward the poles. From Hathaway (2010).

- spots – compact regions of high field strength – are the most obvious signature (Hale, 1908)
- spots emerge only below latitudes of 30°, their distribution of latitude with time follows the "butterfly wing" pattern as spots emerge closer to the equator with time (Maunder, 1904, reporting on work with wife Annie)
- leading polarities of spots emerge closer to the equator than following polarity, this "tilt" is larger the further spots emerge from the equator ("Joy's law"; Hale *et al.*, 1919)
- a 22 year magnetic cycle dominates (Hale *et al.*, 1919) but includes some stochasticity
- the large-scale (dipole, quadrupole) components vary on a 22 year cycle, somewhat out of phase with the spots (Babcock, 1961)
- brightness variations alone vary with a dominant 11 year period and a ~26 day rotational period
- spot emergence in N and S hemispheres is not necessarily in phase
- at least one period occurred where sunspots almost disappeared over 70 years (the Maunder Minimum; Eddy, 1976)
- large-scale (heliospheric) fields show stochastic and periodic magnetic variability over the last few thousand years (Figure 2.8) as recorded in ^{10}Be and ^{10}C radionuclide data and correlated with neutron monitor data (McCracken *et al.*, 2013)
- spots often re-emerge at specific longitudes
- the Sun's 11 year brightness variations resemble a "cycling" class of similar, slowly rotating G stars (Baliunas *et al.*, 1995; Judge and Thompson, 2012), although roughly as many stars do not show cycling activity as do on decadal time scales (Figure 2.11).

McCracken *et al.* (2013, and references in the paper) draw further conclusions from sunspot and radionuclide studies over the past 600 years:

- The Hale 22 year cycle of solar magnetism, and the heliospheric counterpart, continued throughout the Spoerer and Maunder Minima, and
- The amplitude of the 24/12-yr modulation of the paleo-cosmic ray record for the Spoerer Grand Minimum implies a . . . 0.5–2.5 variation in the heliospheric magnetic field near Earth.

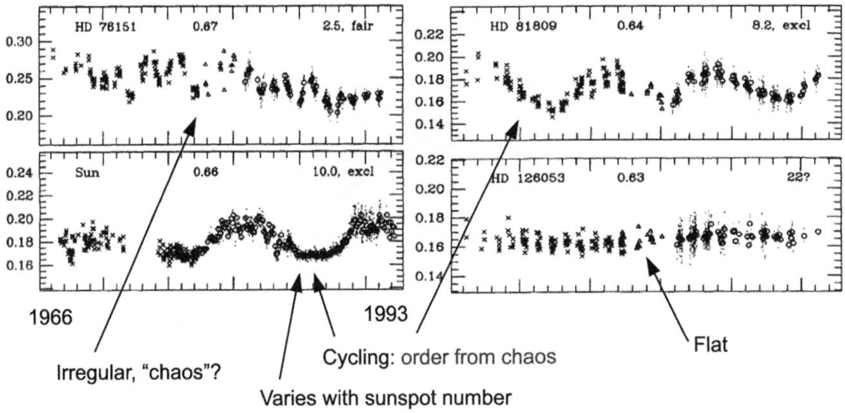

FIGURE 2.11. A selection of sun-like stars from the study of Baliunas *et al.* (1995), showing the "Ca II H index" versus time. The solar index varies as shown in the lower left hand panel, in phase with sunspots, the index is therefore a "proxy" for sunspot-like activity on sun-like stars. Roughly as many stars show "cycling" behavior as do not. Reproduced with permission © AAS.

The continued cycling behavior of the Sun through the Maunder Minimum is hinted at already by the sunspot study of Ribes and Nesme-Ribes (1993). Further characteristics of the evolving solar magnetic field on the surface and in the interior have been reviewed by Hathaway (2010), from which Figure 2.10 is taken. While an obvious point, it must always be remembered that such figures reveal magnetic field patterns evolving both through and across a 2D surface of a 3D system. Care must be taken in interpreting such data only in terms of dynamics across a 2D surface.

2.4.3 *Smaller Scales, Sunspots, Flares and CMEs*

Spruit (2011) has emphasized more local aspects of sunspots. Spots emerge first by exhibiting fragmented surface magnetic fields with mixed polarities. From this, clumps of opposite polarity then form and diverge without influence by convection. Spruit continues:

> This striking behavior is the opposite of diffusion. To force it into a diffusion picture, one would have to reverse the arrow of time. Instead of opposite polarities decaying by diffusing into each other, they segregate out from a mix. The MHD equations are completely symmetric with respect to the sign of the magnetic field, however. There are no flows (no matter how complex) that can separate fields of different signs out of a mixture. This rules out a priori all models attempting to explain the formation of sunspots and active regions by turbulent diffusion... The observations... demonstrate that the orientation and location of the polarities seen in an active region must already have been present in the initial conditions: in the layers below the surface from which the magnetic field traveled to the surface.

These and other observations of Spruit are discussed in Section 2.6.1

Spruit then argues that models of dynamo action based upon cyclonic turbulence (originating with the ideas of Parker, 1955) cannot lie at the heart of the solar dynamo. Simply put, how can small-scale turbulence cause the intense, clumped tubes of sunspot flux?

Solar flares mostly (but not always!) originate in the neighborhood of sunspots. Following the development of MHD in the 1940s and 1950s (Section 2.5), it was realized,

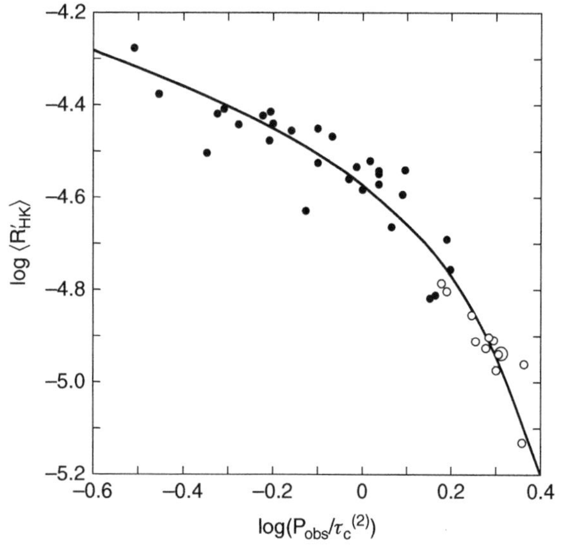

FIGURE 2.12. A scatter plot of an activity index (in this case not the S index but the more physical R'_{HK}) versus rotation period/convective turnover time, the inverse of the Rossby number, Figure 8 from Noyes *et al.* (1984). Reproduced with permission © AAS.

through a process of elimination, that the sudden release of 10^{32} ergs of energy in a few minutes implies the storage and release of magnetic free energy in tenuous regions of the Sun's atmosphere – the corona. The best introduction to this problem I have found is "The Physics of Solar Flares", proceedings of AAS-NASA Symposium, editors Wilmot N. Hess, ed., NASA SP-50 1964.

2.4.4 *Summary of Observed Properties*

Somehow, the solar plasma/magnetic field interactions:

- induce global solar variability to time scales of 11 years and (much) less, compared with a 10^5 year Kelvin–Helmholz time scale,
- introduce highly variable high energy tails to the distributions of photons and particles,
- produce *order out of chaos.*

As noted by Parker (1955), the order must arise because the Sun rotates. The way this happens remains an active research area (Parker, 2009; Spruit, 2011). Again, appealing to stellar behavior proves fruitful – Figure 2.12, from Noyes *et al.* (1984), shows that the Sun lies squarely in the activity levels expected in an ensemble of solar-like stars. This plot has on the abscissa the ratio of rotation period to convective turnover time, the inverse of the Rossby number, to be discussed further below.

2.5 Magnetohydrodynamics

The essential ideas behind MHD as a description of plasmas and the electromagnetic fields are firstly, the usual continuum approach is valid for the plasma (many particles in each phase element), imposing a lower limit to the scales at which MHD applies. Even at the low densities of the corona, $\sim 10^8$ particles per cm^3, this lower limit is less than 1 cm. Second, the plasma is quasi-neutral (electrons move quickly to eliminate any electric field on macroscopic scales). Third, we neglect Maxwell's displacement current, which implies that plasma evolution occurs slower than the light crossing time $t \sim \ell/c$ for the plasma ($R_{\odot}/c \sim 2$ s). Fourth, in the case of "single fluid" MHD, collisions relax different particles (atoms, ions, electrons) on time scales short compared with dynamics, we deal with densities ρ and pressures p that are sums over all constituents, which are all described by single fluid velocity $\mathbf{u}$ and temperature T.[5]

Another central aspect of MHD is not often appreciated (B. C. Low, personal communication, 2012). Einstein (1905) boldly relaxed the Galilean transformations at the root of Newton's laws of motion in order to reconcile the laws with Maxwell's equations and to account for the experimental results of Michelson and Morley (1887). Partly reversing Einstein's work, Alfvén in the 1940s (Alfvén, 1950) formulated MHD within the Galilean/Newtonian framework, but keeping Einstein's relativistic transformations to $\mathcal{O}(v/c)$. The Bard might have remarked

"What's done cannot be undone." – Lady Macbeth (Shakespeare, 1623)

but then Shakespeare preceded Newton and the notion of physical approximations. The electric field $\mathbf{E}$, magnetic field $\mathbf{B}$, charge density ρ and current density $\mathbf{j}$ transform from rest frame to one moving with velocity $\mathbf{u}$ to $\mathcal{O}(u/v)$ as, e.g. Ferraro and Plumpton (1966)

$$\mathbf{E}' \approx \mathbf{E} + \mathbf{u} \times \mathbf{B}; \quad \mathbf{B}' \approx \mathbf{B},$$
$$\rho' \approx \rho; \quad \mathbf{j}' \approx \mathbf{j}.$$

To $\mathcal{O}(u/c)$, $\mathbf{E}$ is frame-dependent, but we can speak of $\mathbf{B}$, $\mathbf{j}$ and ρ without specifying the frame of reference.

To arrive at the standard MHD equations we add Maxwell's equations, which have source terms ρ and $\mathbf{j}$. On scales larger than (usually small) Debye radii r_D, the charge density $n_p e - n_e e$ scales with $\exp -r/r_D$ (here n refers to particle number densities). However, we cannot neglect macroscopic electric currents $\mathbf{j}$, which arise from small (but finite) differences in flow speeds of the electron and ion fluids. Braginskii (1965) derives equations of motion for electrons and ions in plasmas leading to the current density $\mathbf{j}'$, neglecting electron inertia.[6] For simplicity assume a hydrogen plasma which consists only of protons (p) and electrons (e) with charge $+e$, $-e$. Then in terms of any residual macroscopic electric field (to be determined) the electron equation of motion reduces to

$$\mathbf{j}' = ne(\mathbf{u}_p - \mathbf{u}_e) = \sigma \mathbf{E}'; \quad \sigma \approx \frac{e^2 n_e \tau_e}{m_e}, \tag{2.16}$$

where $\mathbf{u}_p$ $\mathbf{u}_e$ are proton and electron fluid velocities, and τ_e is the "electron collision time." This equation is generically called "Ohm's law."

[5] (Parker, 2007, section 8.1) points out that continuum approximations often apply even in collisionless regimes since collisions conserve mass and momentum. Some of the fluid equations are still valid even in collisionless plasmas, when care is taken of higher order quantities such as the pressure tensor!

[6] Electrons are assumed light enough that their momentum changes essentially instantaneously, compared with other terms in the momentum equation for electrons. The inertial term then drops out of the equation.

We also eliminate $\mathbf{j}$ using $\mathbf{j} = \sigma\mathbf{E} + \mathbf{u} \times \mathbf{B}$ to arrive at equations written in terms of $\mathbf{u}$, $\mathbf{B}$, σ with no explicit appearance of $\mathbf{j}$ or $\mathbf{E}$. The equations are (Rempel, 2009)

$$\frac{\partial \rho}{\partial t} = -\nabla \cdot (\rho \mathbf{u}) \tag{2.17}$$

$$\rho \frac{\partial \mathbf{u}}{\partial t} = -\rho(\mathbf{u} \cdot \mathbf{grad}\,)\mathbf{u} - \mathbf{grad}\, p + \rho \mathbf{g} + \frac{1}{\mu_0}(\mathbf{curl}\,\mathbf{B}) \times \mathbf{B} + \nabla \cdot \tau \tag{2.18}$$

$$\rho \frac{\partial e}{\partial t} = -\rho(\mathbf{u} \cdot \mathbf{grad}\,)e - p\nabla \cdot \mathbf{u} + \nabla \cdot (\kappa \mathbf{grad}\, T) + Q_\nu + Q_\eta \tag{2.19}$$

$$\frac{\partial \mathbf{B}}{\partial t} = \mathbf{curl}\ (\mathbf{u} \times \mathbf{B} - \eta\, \mathbf{curl}\,\mathbf{B})\ . \tag{2.20}$$

MHD problems involve the solution of coupled partial differential equations for unknowns ρ, $\mathbf{u}$, T, $\mathbf{B}$, given an equation of state and various additional relations written in terms of these variables needed to close the system (expressions for pressure and heat flux tensors, conductivity, various source and sink terms such as radiative or convective gains or losses, even ionization). The equations themselves are non-linear, the equation of motion containing the advection term $\mathbf{u} \cdot \nabla \rho \mathbf{u}$ and the Lorentz force $\mathbf{curl}\,\mathbf{B} \times \mathbf{B}/\mu_0$, and the induction equation containing $\mathbf{curl}\,(\mathbf{u} \times \mathbf{B})$, from the frame transformation.

2.6 Magnetic Field (Re-)Generation

There are (at least) two major challenges in modern solar physics: (1) what causes the global regeneration of large-scale magnetic field with a period of $\approx$22 years, (2) why must the magnetic field appear most prominently in spots? In this section, I draw some essential points from the review by Rempel (2009) about the first question.

Dynamo theory studies conditions under which a velocity field of a highly conducting fluid can sustain a magnetic field against Ohmic decay. Rempel adopts "dynamo" to mean a system that has a magnetic energy that does not approach zero as time $t \to \infty$, owing to current systems occurring entirely within a finite volume. In the absence of induction effects, magnetic fields decay on a time scale $\tau_d = L^2/\eta$, for the Sun this is $\sim 10^9$ years, using kinetic values for conductivity (Equation (2.16)). The Sun is $\sim 4.5 \times 10^9$ years old, so the mere existence of magnetic field does not by itself require dynamo action, but the observed 22 year cycling behavior shows that something other than diffusion is going on. (Rempel, 2009, notes that at least one alternative mechanism has been proposed, but these are in conflict with helioseismology.) If one accepts the idea of "turbulent magnetic diffusivities" based upon turbulence, mixing length ideas or "mean field theory," which yield conductivities several orders of magnitude smaller, a dynamo is indeed required to sustain fields against this "enhanced diffusion." This issue is discussed below (Section 2.6.1).

The Sun could well have both a "large-scale" and a "small-scale" dynamo. If most magnetic energy is in scales associated with turbulence (surface granulation for example has scales of $\sim$1 Mm compared with $R_\odot = 700$ Mm) we speak of a small-scale dynamo. On the basis of Figure 2.7, 3D numerical simulations with Prandtl numbers $P_m = \nu/\eta \sim 1$ (e.g. Cattaneo, 1999), it is tempting to conclude that a small-scale dynamo is active on the Sun. Caution is needed since the convection zone has $P_m \ll 1$. Almost any chaotic velocity field with large R_m produces a small-scale dynamo (Cattaneo, 1999), but a large-scale dynamo requires new ingredients, such as a net mean helicity in the turbulence induced by, say Coriolis forces (Parker, 1955) and/or large-scale shears.

However, MHD dynamos require at least *some* diffusion to work at all. This is clearly seen by taking the limit of ideal MHD, ($\sigma \to \infty$), in which the magnetic field is frozen to

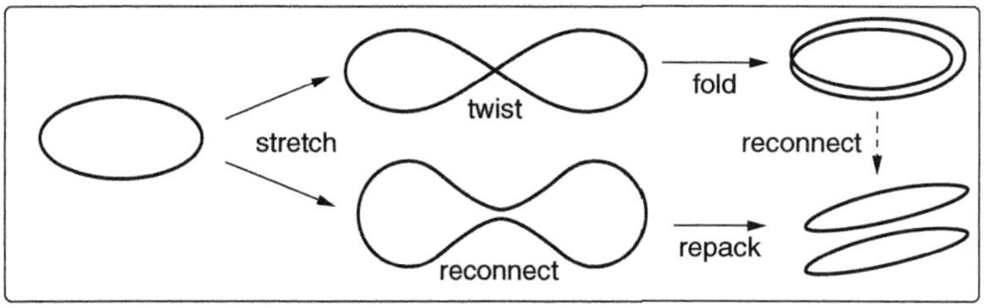

FIGURE 2.13. A figure from Rempel (2009) showing two possible dynamos. Amplification occurs during stretching, the twist–fold and reconnect–repack steps remap the amplified flux rope into the original volume element so that the process can be repeated. Both three dimensions and magnetic diffusion are *required to allow for the topology change needed to close the cycle.*

the plasma. In this case the magnetic fields entrain the same elements of plasma for ever. Thus, the *topology* of the magnetic fields is fixed, once and for all in this limit; all that can be done is to entangle tubes of plasma-entraining flux between one another, like braiding hair in which the bundles of hair are separate entities. But this can only be achieved so far as the Lorentz force ($\mathbf{j} \times \mathbf{B}$) does not act to oppose the braiding. Consider Figure 2.13 from Rempel (2009), showing the operation of "flux rope" dynamos. In the absence of diffusion, repeated stretch–twist–folds simply generate more complexity (more turns of the rope in a given volume). Eventually such a state will need such strong Lorentz forces that the limited driving energy can generate no further changes. This behavior is not capable of explaining solar magnetic behavior. With diffusion, the topology can change (the dashed "reconnect" step in the figure), allowing field lines to slip across the fluid, as is required to explain, for example, the 11 year flip in magnetic polarity of global solar magnetic fields.

The role of stretching, among other velocity gradient effects, can be seen from the ideal induction equation

$$\frac{\partial \mathbf{B}}{\partial t} = \mathbf{curl}\,(\mathbf{u} \times \mathbf{B}) = -(\mathbf{u} \cdot \mathbf{grad}\,)\mathbf{B} + (\mathbf{B} \cdot \mathbf{grad}\,)\mathbf{u} - \mathbf{B}\,\nabla \cdot \mathbf{u}. \tag{2.21}$$

The first term on the right-hand side describes advection of magnetic field, the second amplification by shear (including stretching) and the third by compression.

Helioseismology has revealed large-scale velocity shears in the solar interior. Radial shear exists near the interface of the core and convection zone ("the tachocline"), near the surface; latitudinal shear exists at intermediate latitudes throughout the convection zone (Schou *et al.*, 1998). Such sheared regions naturally lead to field amplification through Equation (2.21). Figure 1 of Spruit (2011) is reproduced in Figure 2.14. It shows the linear growth rates of toroidal field starting from a uniform poloidal field, driven by a fit to surface differential rotation (a rough approximation to interior rotation through the solar convection zone). But, as Spruit (2011) emphasizes:

> The common ingredient ... is the generation of a toroidal (azimuthally directed) field by stretching (winding-up) of the lines of a poloidal field ... This is "the easy part" ... To produce a cyclic, self-sustained field as observed there must be a second step that turns some of the toroidal field into a new poloidal component, which is again wound up, completing a field-amplification cycle that becomes independent of initial conditions.

Rempel (2009, sections 3.3.7–3.4.3) summarizes these ideas using an example of a cylindrically symmetric non-dynamo and dynamos based upon scale separation, the "mean field" dynamos. In the latter, equations for the evolution of the large-scale or "mean"

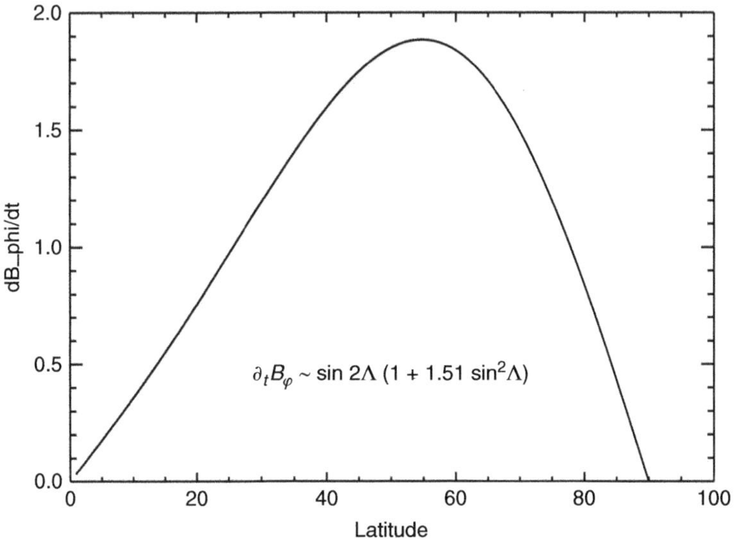

FIGURE 2.14. A figure from Spruit (2011) showing the growth rates $\frac{\partial \mathbf{B}}{\partial t}$ of a toroidal field, starting from a uniform poloidal field stretched by observed surface latitudinal differential rotation, neglecting any Lorentz force. Is this simple figure a key ingredient in the solar dynamo?

fields $\overline{\mathbf{u}}$ and $\overline{\mathbf{B}}$ are written in terms of the mean fields themselves and the correlations between the small-scale velocity and magnetic fields, $\overline{\mathbf{u}'}$, $\mathbf{B}'$. The latter are, in turn, written in terms of the large-scale fields to provide closure, based upon several strong assumptions, but drawing upon several important symmetry properties. In a nutshell, the mean field equations yield

$$\frac{\partial \overline{\mathbf{B}}}{\partial t} = \mathbf{curl}\ \left(\overline{\mathbf{u}' \times \mathbf{B}'} + \overline{\mathbf{u}} \times \overline{\mathbf{B}} - \eta\, \mathbf{curl}\, \overline{\mathbf{B}}\right), \tag{2.22}$$

where a "new" electromotive force (EMF) term arises compared with the full induction equation:

$$\overline{\mathcal{E}} = \overline{\mathbf{u}' \times \mathbf{B}'}\ . \tag{2.23}$$

Taking into consideration symmetries, the EMF can be written in terms of tensor turbulent transport coefficients $\boldsymbol{\alpha}$, $\boldsymbol{\beta}$, $\boldsymbol{\gamma}$ and $\boldsymbol{\delta}$ as

$$\overline{\mathcal{E}} = \boldsymbol{\alpha}\overline{\mathbf{B}} + \boldsymbol{\gamma} \times \overline{\mathbf{B}} - \boldsymbol{\beta}\, \mathbf{curl}\, \overline{\mathbf{B}} - \boldsymbol{\delta} \times (\mathbf{curl}\, \overline{\mathbf{B}}) + \cdots \tag{2.24}$$

Making further physical assumptions with various levels of justifiability, the tensors $\boldsymbol{\alpha}$ and $\boldsymbol{\beta}$ become diagonal. With τ_c the correlation time of components of $\mathbf{u}'$ Rempel writes (his equations (3.59), (3.60))

$$\alpha = \frac{1}{3}\alpha_{ii} = -\frac{1}{3}\tau_c\, \overline{\mathbf{u}' \cdot (\mathbf{curl}\, \mathbf{u}')}\ , \quad \eta_T = \frac{1}{3}\beta_{ii} = \frac{1}{3}\tau_c\, \overline{\mathbf{u}'^2}\ , \quad \gamma = -\frac{1}{2}\mathbf{grad}\, \eta_T\ . \tag{2.25}$$

The α effect is related to the kinetic helicity of the flow, H_k:

$$H_k = \mathbf{u} \cdot (\mathbf{curl}\, \mathbf{u})\ , \tag{2.26}$$

while the turbulent magnetic diffusivity is proportional to the intensity of the turbulence. In fact, $\eta_T \sim L u_{rms} \sim R_m \eta \gg \eta$ (Rempel, 2009, his equation (3.63)), and η_T describes the *transport* of magnetic energy via *advection* and some *reconnection* of field

(Section 2.7), the latter requiring the development of small scales to permit ultimate dissipation via Ohmic collisions.

Under the influence of rotation at angular velocity $\mathbf{\Omega}$, and when stratification exists (under a gravity vector $\mathbf{g}$), then

$$\alpha \approx \alpha_0(\mathbf{g} \cdot \mathbf{\Omega}) = \tau_c^2 v_{\mathrm{rms}}^2 \mathbf{\Omega} \cdot \mathbf{grad}\ \ln(\varrho v_{\mathrm{rms}}) . \tag{2.27}$$

Note that $\Omega\tau_c$ is the Rossby number shown on the abscissa of Figure 2.12.

In essence, the α effect "fixes" the problem quoted from Spruit above by adding the following term $\boldsymbol{\alpha}\mathbf{B}$ to the induction equation

$$\frac{\partial \overline{\mathbf{B}}}{\partial t} = \cdots + \mathbf{curl}\left(\alpha \overline{\mathbf{B}}\right) = \cdots + \alpha \mu_0 \overline{\mathbf{j}} , \tag{2.28}$$

i.e. the induced magnetic field is proportional to the mean current. The new term converts poloidal magnetic field into toroidal field and vice versa, so there exists in principle the following dynamo scenario:

$$\mathbf{B}_t \xrightarrow{\alpha} \mathbf{B}_p \xrightarrow{\alpha} \mathbf{B}_t , \tag{2.29}$$

called an "α^2-dynamo." Preferred α^2-dynamo modes are stationary solutions with poloidal and toroidal field of comparable amplitude, clearly not compatible with solar data.

Returning to the Sun, its known interior shear profiles and essential properties of time-dependent large-scale magnetic fields (Section 2.4.4), we can see that a combination of shear-driven amplification (Ω-effect) and an α effect might in principle contain essential ingredients of a solar dynamo: to deliver strong toroidal fields at the surface in the form of emerging ropes of flux generated by the Ω effect, and an α effect to complete the cycle of events needed for field reversals. Such models are epitomized by those of e.g. Dikpati and Charbonneau (1999). In this case, the α effect producing the toroidal field is assumed smaller than the Ω effect, so that we have

$$\mathbf{B}_t \xrightarrow{\alpha} \mathbf{B}_p \xrightarrow{\Omega} \mathbf{B}_t . \tag{2.30}$$

In such a dynamo, "waves" of amplified field propagate along $\mathbf{grad}\,\mathbf{\Omega}$ – to explain the latitudinal migration of spots (butterfly diagram) this requires a radial gradient in $\mathbf{\Omega}$. This basic property of $\alpha\Omega$ dynamo-waves is pitted against recent arguments by Spruit in the next section. Lastly, the "tight relationship" with Rossby number plotted in Figure 2.12 should not be regarded as supporting mean field dynamo models *per se*, given the underlying assumptions, as stressed already by Noyes *et al.* (1984).

2.6.1 *What Might be the Ingredients in the Solar Dynamo?*

In order that an $\alpha\Omega$ dynamo be consistent with the observed 22 year periodic change in global solar magnetic field, the cycle of events must take place *fast*: the speed at which the cycle equation (2.30) is limited by the slowest critical physical process. The differential rotation operates fast in the Ω part of the cycle, the ratio $\Delta\Omega/\Omega$ is ~ 0.1, so the time t where $\Delta\Omega t \sim 1$ is about 10 rotations, the best part of a year. A critical part of all dynamos is not just the existence of an α effect but the accompanying β effect which helps promote the changes of topology by generating small scales needed for kinetic diffusivities to become important. In the mean field theory this process is modeled through the "turbulent diffusivity" (Equation (2.25)), which is equivalent to an "eddy diffusivity" of $\frac{1}{3}\lambda v'$. Conveniently, the "mixing length" λ and v' values yield $\eta_T \sim 10^{12}$ cm^2 s^{-1}, orders of magnitude larger than kinetic values. Such values are precisely those needed in dynamo models to produce cycles near 11 years in length, in this formalism.

So is everything in order? Not according to Parker (2009) and Spruit (2011), who have independently re-assessed the ingredients needed to make a dynamo compatible with physical principles and observations. Both have pointed to problems with the mean field/turbulent dynamo concept. Parker points out that the concept of a turbulent diffusivity at high R_m, valid in the kinematic regime, is likely invalid in the dynamic regime. By considering the magnetic fields to be frozen up til the development of scales small enough for molecular diffusion to arise, Parker argues that the diffusion occurs when the field strength has increased by $R_m^{1/2}$. However, by the time this occurs, the magnetic stresses ($\propto B^2$) will have increased by R_m. Thus, for fields of interest, the magnetic stresses overwhelm any turbulence long before resistive diffusion can obliterate the field. In the language of mean field theory, Parker concludes that

> The idea that the azimuthal [i.e. toroidal] magnetic field is subject to ordinary turbulent diffusion, with $\eta = \frac{1}{3}u\lambda$, seems unjustified ... There is no way to account for the value $\eta_T \sim 10^{12}$ cm^2 s^{-1}, suggesting that it is necessary to re-think the $\alpha\Omega$-dynamo for the Sun.

Spruit (2011) offers a scathing critique of mean field dynamo theory, arguing instead for a return to models of the 1960s (Babcock, 1961, 1963; Leighton, 1969). In the 1960s nothing was known about interior rotation profiles of the Sun; in these early works, the surface (i.e. latitudinal) rotation profile was adopted for simplicity. In Spruit's view, latitudinal differential rotation generates toroidal field in the interior, which becomes unstable to buoyancy. The dynamic buoyant rise of these toroidal ropes of magnetic flux is subject to the Coriolis force generating automatically poloidal field. As shown in Figure 2.14 the first flux is expected to emerge near 55° latitude, and later flux emerges polewards and equatorwards. Some properties of Figure 2.10 might be considered qualitatively (if not quantitatively) compatible with this picture. Spruit cites evidence (albeit indirect) for such behavior, other support has recently been reported by S. McIntosh (private communication, 2014). Convective turbulence plays no active role in the induction equation, instead it provides stresses needed to maintain latitudinal differential rotation.

Spruit's concerns can be summarized as follows.

- The observed evolution of sunspots points to processes that are not diffusive in nature (see quote in Section 2.4.3). The observations reveal intense bundles of flux that must have been assembled by deterministic processes from below.
- Various observed properties of emerging flux appear compatible with the simple notion that the solar cycle reflects the rate of generation of toroidal field by latitudinal differential rotation (Figure 2.14). There remain important questions (why do sunspots form only at low latitudes?).
- The radial gradient ("tachocline") in the rotation profile at the base of the convection zone can do little work to stretch the magnetic field, the lower boundary being stress-free in the radiative core.
- Instead, the latitudinal gradient found in helioseismology – similar to the surface differential rotation profile known by Babcock and Leighton in the 1960s – is a more promising source for the Ω effect.
- The physical justifications underlying mean field theories are weak.
- Hale's and Joy's laws imply, albeit indirectly, that a strong (super-equipartition) field is required to survive in the convection zone.
- Such a field, roughly 10^5 G in the depths of the convection zone, is also the field strength needed for buoyant eruption (e.g. Schüssler, 1996). Several observed properties of sunspots seem compatible with such "initial conditions" for the formation of sunspots.

At least two open questions remain, one being how to maintain the toroidal field in the latitudinal shear layer as it builds up slowly to become buoyant, against the strong convective eddy motions. Spruit cites another puzzle:

> The most challenging problem may well be finding a satisfactory description for the process by which the mass of buoyant vertical flux tubes resulting from a cycle's worth of eruptions gets 'annealed' back into a simpler configuration... appeal to traditional convective 'turbulent diffusion' will not work (even if the concept itself is accepted), since the fields are now much stronger than equipartition with convection (at least near the base of the convection zone where this annealing has to take place).

I conclude that while we have different cookbooks and different recipes to make a solar cycle-like model, we still seem to be missing some key ingredients in explaining how the Sun and other stars re-generate their global magnetic fields.

2.7 Magnetic Field Dissipation, Topology, Reconnection

By "dissipation," I refer to the conversion of ordered forms of energy to disordered, thermal energy. In MHD, magnetic energy is dissipated directly through Ohmic dissipation (the ∇^2 term in the induction equation) in which the ordered differential motion of electrons and ions (or other neutral species when present), i.e. electric currents, is converted to random motion, heat, via particle collisions. The rate of conversion of this energy, $\mathbf{j} \cdot \mathbf{E}$ is merely

$$\frac{1}{2\mu_0}\frac{\partial B^2}{\partial t} = j^2/\sigma = (\mathbf{curl}\,\mathbf{B})^2/\mu_0^2\sigma\,. \tag{2.31}$$

Of course the rate of destruction of magnetic field is

$$\frac{\partial \mathbf{B}}{\partial t} = \frac{1}{\mu_0\sigma}\nabla^2\mathbf{B} = \eta\nabla^2\mathbf{B}\,. \tag{2.32}$$

Manifestly both processes involve the generation of small scales to be effective. But to dissipate energy on large scales requires a mechanism naturally generating such small scales. In the limit $R_m \to \infty$, *topology* becomes terribly important, because, owing to Alfvén's theorem, plasma is nearly frozen to specific tubes of flux. One can imagine an initial condition of a star threaded by magnetic field. Draw a closed path around some magnetic flux, the field lines on this path trace out a tube which will close back on itself, most likely after a tortuous journey. As time goes on, Alfvén's theorem implies that although plasma may be forced to move, the topology of this tube cannot change – to make knots in this requires one to cut the flux tube, but this is forbidden in ideal MHD. The topological constraints are of a global, not local, nature, in contrast with the local equation of motion.

To change magnetic topology requires diffusion, i.e. the development of smaller physical scales. *Most of the solar and solar-terrestrial phenomena of interest involve the evolving magnetism and its changing topology*, even on the largest scales. Like cutting through rubber bands, the "snip" only needs to occur *somewhere* along a tube of magnetic flux, not along the entire length. Enormous changes $\frac{\partial \mathbf{B}}{\partial t}$ can occur because of a "reconnection" (a "snip" that satisfies $\mathrm{div}\,\mathbf{B} = 0$) somewhere else along a tube, allowing the system to reach a lower energy state in a state with a quite different topology.

Magnetic reconnection in MHD is the study of changing topology of magnetic fields. Accordingly, one would think that the diffusivity η plays a central role. But in fact η plays a secondary role, something that became clear when discussing the β effect in mean field dynamos. In that case, correlations between velocity and magnetic perturbations lead to values η_T that are associated with the "turbulence" – parcels of fluid are moved bodily with speed u a mixing length λ to produce $\eta_T = \frac{1}{3}u\lambda$ to produce changes in the magnetic field orders of magnitude faster than are possible using kinetic values for η. In this picture, field lines frozen to fluid elements are advected by them without hindrance by the Lorentz force (in the kinetic regime of mean field theory), shuffling them around. No reconnection is implied in the mean field theory by η_T, this term originates from the $\mathbf{curl}\,(\mathbf{u} \times \mathbf{B})$ term. It is merely a re-shuffling, which, in MHD, can lead to reconnection if and when it generates scales small enough for genuine diffusion to occur via the kinetic value of η.

Important first steps in magnetic reconnection were taken by Dungey (1953), Sweet (1958), and Parker (1957). The famous 2D "Sweet–Parker" model, based upon a stationary flow bringing opposite polarity, collision-dominated plasmas together, yields a diffusion rate $\sim V_A/\sqrt{R_m}$, which, through R_m, depends explicitly on η. A modification by Petschek (1964) introduced slow shocks to allow the bulk of the frozen field in the Sweet–Parker picture to avoid having to flow through the diffusion region, permitting the diffusion region to be much shorter. In Petschek's scenario, much faster reconnection rates $\sim V_A/\log_e R_m$, almost independent of η, were achieved! Such reconnection scenarios are called "fast" reconnection (i.e. almost independent of R_m), they are *required* to account for the sudden release of energy seen in solar flares, coronal mass ejections, and indeed to explain the operation of the solar dynamo. For some time the ability of space plasmas like the solar corona to be able to reconnect at the Petschek rate (i.e. "fast" enough) was in doubt; it requiring a somewhat specific configuration and prescribed steady state in the velocity field. Curiously, tearing instabilities in MHD have been put forward by Bhattacharjee *et al.* (2009), which may provide MHD systems with a natural tendency to approach the Petschek rate of reconnection. Reconnection physics is a huge subject and so I simply refer readers to texts by Biskamp (2005) and Priest and Forbes (2007), and a tutorial article by Kulsrud (2011).

Below, I point out some problems I find interesting concerning the evolution and dissipation of magnetic fields under conditions where turbulent and Reynolds stresses and gas pressure dominate (the high-β regime), a regime that occurs in the solar interior and most parts of the solar photosphere. I also discuss Parker's "fundamental theorem of magnetostatics" as it pertains to the solar corona, which is mostly in the opposite, low plasma-β regime. Some surprises will arise. Figure 2.15 shows the dramatic change from hydrodynamically dominated turbulent structure characteristic of the dense photospheric plasma, to the magnetically dominated corona in the tenuous plasma there.

2.7.1 *"High-β" Plasma*

Surface magnetic fields on the Sun exhibit behavior that seems to imply enormous diffusivities at and near the visible surface. Generally speaking, direct observations of $\frac{\partial \mathbf{B}}{\partial t}$ are not made; instead "tracers" of magnetic fields are measured through "bright points" in the photosphere or overlying coronal structures, for example. An implicit assumption is that fields evolve across some two dimensional surface, there is no flux of magnetic field through the atmosphere. In one example, Abramenko *et al.* (2011) studied trajectories of tiny bright points of magnetic origin in photospheric inter-granular lanes and derived diffusion coefficients $\propto \lambda^{\gamma} u$, where $\gamma = 1$ for "normal" (thermal-like) diffusion. These concentrations of magnetic field are close to the "high-β" regime, so that the surface magnetic fields are assumed not only to be advected only across the surface but that they are unimpeded by magnetic forces. These studies typically yield $\eta_T \sim 3 \times 10^{12}$ cm^2 s^{-1}. Rather astonishingly, this surface diffusion coefficient is of the same magnitude needed to make a (3D) interior dynamo work in the kinematic regime (Parker, 2009)!

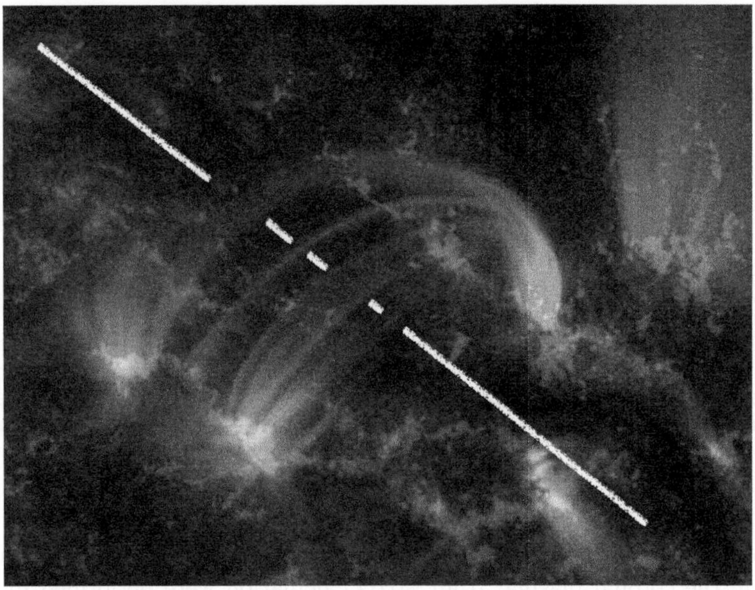

FIGURE 2.15. An image showing a longitudinal magnetogram (the slanted line roughly separates regions of opposite polarities) and some associated hot plasma in the overlying corona.The switch from dominant "turbulent" fluid motions seen in the photospheric magnetogram, an example of a "high-β plasma," to the far more ordered coronal loop structures arising from the dominant magnetic stresses, a "low-β plasma," is dramatic. But the switch occurs in a geometrically thin, poorly understood stratified layer called the "chromosphere," which is not seen in this image. The magnetograms trace out "supergranule" cells of 20–30 Mm diameter.

Does this (coincidence?) lend credibility to the idea of "turbulent diffusion"? This remains an open question. The argument of Parker (2009) says "no"; others (e.g. Lazarian, 2013) believe turbulent diffusion is a central physical process in all plasmas with small diffusive terms. The resolution of this difference will probably come from detailed studies of direct numerical simulations, to look at when and where the small-scale Lorentz forces act so as to oppose reconnection, since they can also enhance it. Rempel (2009) discusses some of these issues in the context of mean field theory in his section 3.5.2.

Curiously, it has become customary to invoke "turbulent diffusivities" to explain observations of the rotation and evolution of coronal features, such as coronal holes, in higher layers of the Sun's atmosphere where the magnetic stresses tend to dominate. It is important to understand the meaning of such work since the literature invokes diffusion and reconnection sometimes interchangably, and at various levels in the Sun's atmosphere (photosphere, corona).

On the one hand, in the work of Wang and Sheeley (1993) diffusion occurs (implicitly) in the photosphere. They solve a 2D surface induction equation using a diffusivity of 6×10^{12} cm^2 s^{-1} and macroscopic velocities based upon photospheric measurements. Consider an area originally of normal "quiet Sun" (i.e. mixed polarity) that is evolving into a coronal hole (mostly one polarity). The change in polarity must occur in surface transport models via some "cancellation" or reconnection at photospheric heights of field of opposite polarity to that of the coronal hole. Since there are no monopoles the net flux remains the same but the boundaries move. This can occur just as fast as the flows drive opposite polarities together since there is nothing to halt mutual annihilation. In this model, the corona is dealt with via a potential-field extrapolation (including a source surface at $2.5R_{\odot}$ for "open" field lines), in which the state of the corona is set

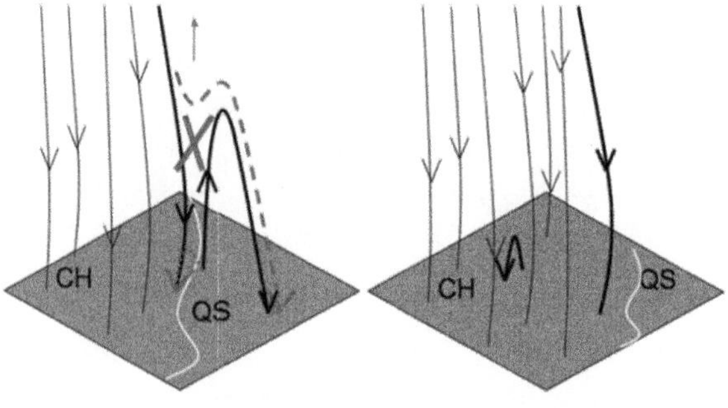

FIGURE 2.16. Interchange reconnection: the location of a coronal hole and quiet Sun boundary is indicated as a line. In the left figure, a quiet Sun loop located near a CH open field. Following the reconnection (at the "x") a small loop is created inside the CH and the open field line relocated, in the right figure. This process has been invoked to explain the rigid rotation as well as the growth of coronal holes. From Krista *et al.* (2011). Reproduced with permission © AAS.

instantaneously by the boundary conditions. By assumption no electrical currents are permitted in the coronal plasma. If an MHD calculation were performed instead using the same boundary conditions, one would impose a large value of η to dissipate all currents on time scales short compared with dynamical times, i.e. $R_m \lesssim 1$. In turn this implies very fast reconnection everywhere in the corona. Astonishingly, this simple model is found to capture important features of the evolving coronal holes under study. In the next section I discuss more recent work on effective diffusion in the corona.

2.7.2 *Low-β Plasma: Diffusion*

As well as in high-β plasmas, it has become popular to describe coronal magnetic field evolution in terms of diffusive processes. For example, superficially similar work to that of Wang and Sheeley (1993) has been presented by Fisk and Schwadron (2001), followed by later work (e.g. Krista *et al.*, 2011). In these studies, observations *in the coronal plasma* have been used to infer effective diffusivities. In their studies of coronal holes, observations[7] indicate to Fisk and Schwadron that

> Diffusion by random convective motions in supergranules is quite slow and will prove to be inadequate for the transport of open magnetic flux on the Sun. We introduce, therefore, a faster process.

The proposed process is called "interchange reconnection," where $\frac{\partial \mathbf{B}}{\partial t}$ changes locally at a point in the corona by the process illustrated in Figure 2.16 from Krista *et al.* (2011). It is quite a different process than the convective diffusion, it has a diffusion coefficient at least an order of magnitude larger. To descibe the evolution of coronal hole boundaries, Krista *et al.* (2011) find $\eta_T \lesssim 3 \times 10^{13}$ cm^2 s^{-1}. Fisk and Schwadron (2001) however found $\eta_T \sim 3.5 \times 10^{13}$ and 1.6×10^{15} ($50\times$ larger!) cm^2 s^{-1}, based upon the apparent movement of closed loops, within coronal holes and near the equator respectively. The idea is that *fast reconnection occurs somewhere well above the photosphere where the Alfvén speed is high*, between open and closed regions brought slowly together by the relatively slow photospheric footpoint motions. Instead of annihilation of fields at the photosphere, a smaller loop structure, not reaching coronal heights, remains (Figure 2.16).

[7] Simon *et al.* (1995) find $\eta_T \sim 5$–7×10^{12} cm^2 s^{-1} for supergranules.

This is an interesting departure from the "mean field" picture since magnetic forces ostensibly cannot be neglected, so this effect cannot be described by kinematics of hydrodynamic turbulence. What, then, does a "turbulent diffusivity" mean in the low-β regime? This example differs not only in plasma β from the photosphere, but also in that it implies that *reconnection is very fast.*

There are observable consequences of such large values of η occurring near the coronal base. Using $\eta \sim \frac{1}{3}L^2/t$ we can turn this around and say that such diffusion will limit the lifetime of any solar features to $t \lesssim L^2/\eta$. Modern telescopes resolve features down to about 10^7 cm on the Sun. With $\eta = 3 \times 10^{13}$ cm^2 s^{-1}, the lifetime is $\lesssim$3 s. There are many fine-scale structures living far longer than this seen near the coronal base in quiet and coronal hole regions (e.g. de Pontieu *et al.*, 2007). Additionally, if we allow that, say, 1 day is needed for a $L \sim 20$ Mm-sized active region to build up enough free energy within the corona so that a large reconnection event can trigger a flare, then $\eta \lesssim \frac{1}{3}L^2/(1 \text{ day})$, or about 10^{13} cm^2 s^{-1}. It seems that interchange reconection, if real, does not happen everywhere, all the time, in the Sun's corona. If it does happen, then we must explain how it occurs so rapidly in the low-β environment at far higher rates than can be driven by photospheric diffusion driven by turbulence there. Is there a process that can drive such dynamic changes?

2.7.3 *Low-β Plasma: Reconnection, Flares, Heating*

The "coronal heating problem" is a 70+ year old problem that is alive and well. It is a long recognized "grand challenge" for astronomy (e.g. Hoyle, 1955). Several lines of argument indicate the magnetic field as a prime source of energy (e.g., see Figure 2.15). Why is this problem so "refractory," resistant to a clean solution? After all, usually a mere 1 part in 10^6 or so of the Sun's energy flux is used to maintain a corona. Part of the problem is precisely *because* of this small energy requirement. The problem is "ill posed" in that unobservably small changes in regions of the atmosphere where we can accurately measure components of energy fluxes (via hydrodynamic or electrodynamic processes) can lead to order zero changes in the energy flux into the corona! Indeed, photospheric changes associated with large flares have only in recent years been detected through spectropolarimetry.

We have an embarrassment of riches in ideas to convert ordered magnetic energy into heat (e.g. Parnell and De Moortel, 2012). But the bulk of the mechanisms rely on the same process – the development of small-scale structures in the large coronal volume in order to permit either MHD (Ohmic heating, viscous heating) or non-MHD processes (wave–particle interactions?) to dissipate magnetic energy. Instead of focussing on individual mechanisms, I draw attention to a general property of low-β, highly conducting plasma, first discovered by Parker (1972; 1994).

In the limit of "zero β," the Lorentz force is the only game in town. Parker asks us to consider that a force-free state exists between two infinitely conducting plates at $z = 0$, $z = L$, filled with an infinitely conducting plasma, then

$$\mathbf{j} \times \mathbf{B} = 0 \quad \text{(force-free condition)}\,, \tag{2.33}$$

which is non-trivially satisfied when

$$\mathbf{curl}\,\mathbf{B} = \alpha(\mathbf{r})\mathbf{B}\,, \tag{2.34}$$

where $\alpha(\mathbf{r})$ is a measure of twist. Boundary values of B_z are specified. The ideal induction equation endows the magnetic field with a specific, fixed topology. An operational approach is taken, first to obtain the set of all *continuous* force-free fields satisfying the boundary conditions. These fields typically have different topologies, one of the fields being the unique potential field ($\alpha = 0$). The second step is to search the set $S\{\mathbf{B}_\alpha\}$ for the solution $\mathbf{B}_\alpha^{sol}$ with the correct topology τ.

Then Parker's "Fundamental Theorem of Magnetostatics" says that, in general, the solution $\mathbf{B}_\alpha^{sol}$ must be *discontinuous.* The relevance to the Sun is that, on the one hand, one can imagine the two plates being the solar "photosphere" and the intermediate plasma as the "corona," the corona living long enough for magnetostatic equilibrium to be a reasonable approximation outside of obvious dynamic events. The relevance to the heating, reconnection, flaring problems should be clear also, in that the theorem enforces the formation of magnetic discontinuities in anything other than highly tidy geometries. This is precisely the property – the systematic development of small from large scales – needed to account for dissipation and reconnection. In the physical situation on the Sun, this implies that other physics must be included – *ideal MHD causes its own demise.* In trying to become singular to satisfy the partial differential equations as well as the integral equations imposing the field-line topology, ideal MHD must break down.

On the other hand, solar plasmas are not infinitely conducting; the photosphere–corona interface involves the complications of the chromosphere, which is not force-free. Much has been discussed regarding this theorem far beyond these simple comments, but closer inspection of mathematics (Low, 2010; Janse *et al.*, 2010) and some highly non-dissipative numerical simulations (Bhattacharyya *et al.*, 2010) have revealed evidence in support of the theorem. The reader should refer to the cited papers for the formal proofs, but they can also be understood using two heuristic arguments. Using the abstract of the paper by Janse *et al.* (2010):

> A magnetic field embedded in a perfectly conducting fluid and rigidly anchored at its boundary has a specific topology invariant for all time. Subject to that topology, the force-free state of such a field generally requires the presence of tangential discontinuities (TDs). This property proposed and demonstrated by Parker [Spontaneous Current Sheets in Magnetic Fields (Oxford University Press, New York, 1994)] is explained in terms of (i) the overdetermined nature of the magnetostatic partial differential equations nonlinearly coupled to the integral equations imposing the field topology and (ii) the hyperbolic nature of the partial differential equation for the twist function α of the force-free field.

To see the origin of point (ii), simply take the divergence and curl of Equation (2.34)

$$\mathbf{B} \cdot \mathbf{grad}\, \alpha = 0 \tag{2.35}$$
$$\nabla^2\mathbf{B} + \alpha^2\mathbf{B} = \mathbf{B} \times \mathbf{grad}\, \alpha\,. \tag{2.36}$$

The second equation has complex characteristics, but the first implies that α is real along field lines, showing that along field lines the characteristics are real. There is therefore *no requirement that $\alpha(\mathbf{r})$ be continuous from one field line to the next.* Different neighboring values of α clearly must lead to some discontinuity since the twist is different at neighboring points.

In 1988, Parker proposed that this tendency for natural MHD systems near force-free equilibrium to form discontinuities is an essential ingredient in the physics of coronal heating. The consequence of this tendency is, Parker argues, a theory for coronal heating – "nanoflares" (Parker, 1988). This tendency may explain why "potential fields" (Wang and Sheeley, 1993) can often resemble the solar corona, in spite of the fact that there is no free energy in the potential field to produce the corona. The nanoflares might just release this energy on small scales, giving the appearance of a near-potential configuration on observable scales.

Lastly, Parker's work suggests the omnipresence of small-scale reconnections in untidy physical systems. As such it may provide some rationale behind at least some puzzles such as "interchange reconnection" which appear to have neither a strong physical basis nor critical observational support.

2.8 Concluding Remarks

I leave it to the reader to decide if the *history* of dynamo theory, as related by Parker (2009) and Spruit (2011), is a *tragedy*, or perhaps a *comedy*. The Parker theorem on magnetostatics reminds me personally of a classic, tragic Shakespearean hero, being defeated by its own, fundamental character flaw. What is different from Macbeth and Lady Macbeth, is that the "flaw" in ideal MHD might just lead to the correct explanation of much "coronal heating," a problem of a mere 70 years, but a significant one.

For my own part, I have learned to take seriously the weight of observational evidence and the ingenuity of theoretical scientists to understand the true implications of what the truly critical observations imply. This is one way of describing the "Scientific Method." I believe solar physics is, and will remain for some time, an "observationally driven" area of research. Either way, it is an ongoing, entertaining drama.

"I wish your horses swift and sure of foot; And so I do commend you to their backs. Farewell." – Macbeth (Shakespeare, 1623)

Acknowledgments

I am grateful to Roberto Casini, Matthias Rempel and Boon Chye Low for sharing material and discussions, and to the organizers for inviting me and supporting my participation in the school.

REFERENCES

Abramenko, V. I., Carbone, V., Yurchyshyn, V., *et al.*: 2011, *Astrophys. J.* **743**, 133

Alfvén, H.: 1950, *Cosmical Electrodynamics*, International Series of Monographs on Physics, Oxford: Clarendon Press

Babcock, H. W.: 1953, *Astrophys. J.* **118**, 387

Babcock, H. W.: 1961, *Astrophys. J.* **133**, 572

Babcock, H. W.: 1963, *Ann. Rev. Astron. Astrophys.* **1**, 41

Babcock, H. W. and Babcock, H. D.: 1952, *Publ. Astron. Soc. Pac.* **64**, 282

Baliunas, S. L., Donahue, R. A., Soon, W. H., *et al.*: 1995, *Astrophys. J.* **438**, 269, DOI:10.1086/175072

Beer, J., McCracken, K., and von Steiger, R.: 2012, *Cosmogenic Radionuclides*

Bhattacharjee, A., Huang, Y.-M., Yang, H., and Rogers, B.: 2009, *Physics of Plasmas* **16(11)**, 112102

Bhattacharyya, R., Low, B. C., and Smolarkiewicz, P. K.: 2010, *Physics of Plasmas* **17(11)**, 112901

Birkeland, K. and Muir, J. T.: 1908, *On the Cause of Magnetic Storms and the Origin of Terrestrial Magnetism*, Vol. 1 of The Norwegian Aurora Polaris Expedition, 1902–03, H. Aschehoug

Biskamp, D.: 2005, *Magnetic Reconnection in Plasmas*, Cambridge University Press, Cambridge, UK

Braginskii, S. I.: 1965, *Reviews of Plasma Physics.* **1**, 205

Carrington, R. C.: 1859, *Mon. Not. R. Astron. Soc.* **20**, 13

Casini, R. and Landi Degl'Innocenti, E.: 2008, *Plasma Polarization Spectroscopy*, Chapt. 12. Astrophysical Plasmas, 247, Springer

Cattaneo, F.: 1999, *Astrophys. J. Lett.* **515**, 39

Comte, A.: 1835, *Traité de philosophie positive,* available at gallica.bnf.fr, Vol. 2

de Pontieu, B., McIntosh, S., Hansteen, V. H., *et al.*: 2007, *Publ. Astron. Soc. Japan* **59**, 655

Dikpati, M. and Charbonneau, P.: 1999, *Astrophys. J.* **518**, 508

Dungey, J. W.: 1953, *Phil. Mag.* **44**, 725

Eddy, J. A.: 1976, *Science* **192**, 1189

Edlén, B.: 1943, *Zeitschrift für Astrofysik* **22**, 30

Einstein, A.: 1905, *Annalen der Physik* **322**, 891

Ferraro, V. C. A. and Plumpton, C.: 1966, *An Introduction to Magneto-Fluid Mechanics*, Clarendon Press, Oxford, 2nd edition

Fisk, L. A. and Schwadron, N. A.: 2001, *Astrophys. J.* **560**, 425

Gandorfer, A. M. and Povel, H. P.: 1997, *Astron. Astrophys.* **328**, 381

Grotrian, W.: 1939, *Naturwissenschaften* **27**, 214

Hale, G. E.: 1908, *Astrophys. J.* **28**, 315

Hale, G. E., Ellerman, F., Nicholson, S. B., and Joy, A. H.: 1919, *Astrophys. J.* **49**, 153

Hanle, W.: 1924, *Zeitschrift fur Physik* **30**, 93

Hathaway, D. H.: 2010, *Living Reviews in Solar Physics* **7**, 1

Hoyle, F.: 1955, *Frontiers of Astronomy*, Heinemann, London, UK

Janse, A. M., Low, B. C., and Parker, E. N.: 2010, *Physics of Plasmas* **17(9)**, 092901

Jeffries, J., Lites, B., and Skumanich, A.: 1989, *Astrophys. J.* **343**, 920

Judge, P.: 2002, in *ASP Conf. Ser. 277: Stellar Coronae in the Chandra and XMM-NEWTON Era*, 45

Judge, P. G. and Thompson, M. J.: 2012, *ArXiv e-prints*

Kirchhoff, G. and Bunsen, R.: 1860, *Annalen der Physik* **186**, 161

Kleint, L., Berdyugina, S. V., Shapiro, A. I., and Bianda, M.: 2010, *A& A* **524**, A37

Krista, L. D., Gallagher, P. T., and Bloomfield, D. S.: 2011, *Astrophys. J. Lett.* **731**, L26, DOI:10.1088/2041-8205/731/2/L26

Kulsrud, R. M.: 2011, *Physics of Plasmas* **18(11)**, 111201

Landi Degl'Innocenti, E.: 2013, *Memorie della Societa Astronomica Italiana* **84**, 391

Lazarian, A.: 2013, *Space Sci. Rev.*

Leighton, R. B.: 1969, *Astrophys. J.* **156**, 1

Lites, B. W.: 1987, *Applied Optics* **26**, 3838

Lites, B. W.: 2000, *Reviews of Geophysics* **38**, 1

Lites, B. W., Elmore, D. F., and Streander, K. V.: 2001, in M. Sigwarth (Ed.), *Advanced Solar Polarimetry – Theory, Observation, and Instrumentation*, Vol. 236 of *Astronomical Society of the Pacific Conference Series*, 33

Low, B. C.: 2010, *Astrophys. J.* **718**, 717

Maunder, E. W.: 1904, *Mon. Not. R. Astron. Soc.* **64**, 747

McCracken, K., Beer, J., Steinhilber, F., and Abreu, J.: 2013, *Space Sci. Rev.* **176**, 59

Michelson, A. A. and Morley, E. W.: 1887, *Sidereal Messenger, vol. 6, pp. 306–310* **6**, 306

Noyes, R. W., Hartmann, L. W., Baliunas, S. L., Duncan, D. K., and Vaughan, A. H.: 1984, *Astrophys. J.* **279**, 763, DOI:10.1086/161945

Parker, E. N.: 1955, *Astrophys. J.* **122**, 293

Parker, E. N.: 1957, *J. Geophys. Res.* **62**, 509

Parker, E. N.: 1965, *Plan. Space Sci.* **13**, 9

Parker, E. N.: 1972, *Astrophys. J.* **174**, 499

Parker, E. N.: 1988, *Astrophys. J.* **330**, 474

Parker, E. N.: 1994, *Spontaneous Current Sheets in Magnetic Fields with Application to Stellar X-Rays*, International Series on Astronomy and Astrophyics, Oxford University Press, Oxford

Parker, E. N.: 2007, *Conversations on Electric and Magnetic Fields in the Cosmos*, Princeton University Press, Princeton, NJ, USA

Parker, E. N.: 2009, *Space Sci. Rev.* **144**, 15

Parnell, C. E. and De Moortel, I.: 2012, *Phil. Trans. R. Soc. Lond. A* **370**, 3217

Petschek, H. E.: 1964, *NASA Special Publication* **50**, 425

Priest, E. and Forbes, T.: 2007, *Magnetic Reconnection*, Cambridge University Press, Cambridge, UK

Rempel, M.: 2009, *Heliophysics I: Plasma Physics of the Local Cosmos*, Chapt. 3. Creation and destruction of magnetic fields, 42, Cambridge University Press

Ribes, J. C. and Nesme-Ribes, E.: 1993, *Astron. Astrophys.* **276**, 549

Rosner, R., Tucker, W. H., and Vaiana, G. S.: 1978, *Astrophys. J.* **220**, 643

Saha, M. N.: 1937, *Proc. R. Soc. Lond. A.* **160**, 155

Schou, J., Antia, H. M., Basu, S., *et al.*: 1998, *Astrophys. J.* **505**, 390

Schrijver, C. J. and Harvey, K. L.: 1989, *Astrophys. J.* **343**, 481

Schüssler, M.: 1996, in K. C. Tsinganos (Ed.), *Solar and Astrophysical Magneto-hydrodynamic Flows*, Kluwer, p. 17

Schwabe, H.: 1844, *Astron. Nachr.* **21(495)**, 233

Seagraves, P. H. and Elmore, D. F.: 1994, *Proc. SPIE* **2265**, 231

Shakespeare, W.: 1623, *Macbeth*

Shibata, K., Isobe, H., Hillier, A., *et al.*: 2013, *Publ. Astron. Soc. Japan* **65**, 49

Simon, G. W., Title, A. M., and Weiss, N. O.: 1995, *ApJ* **442**, 886

Spruit, H. C.: 2011, in Miralles, M. P. & Sánchez Almeida, J. (Ed.), *The Sun, the Solar Wind, and the Heliosphere*, IAGA Special Sopron Book Series, Vol. 4. Berlin: Springer, 39

Sweet, P. A.: 1958, in B. Lehnert (Ed.), *Electromagnetic Phenomena in Cosmical Physics*, Vol. 6 of *IAU Symposium*, 123

Wang, Y.-M. and Sheeley, Jr., N. R.: 1993, *Astrophys. J.* **414**, 916

Young, C. A.: 1892, *The Sun*, Kegan Paul, Trench, Trübner & co. Ltd., London, 4th edition

3. Stellar Magnetic Fields

OLEG KOCHUKHOV

Abstract

Recent developments in astrophysical spectropolarimetry have significantly expanded our understanding of stellar magnetism. Using modern instrumentation and data analysis methods it is becoming possible to detect and characterise surface magnetic fields not just for a few peculiar objects but for essentially all types of stars, ranging from massive OB stars to brown dwarfs. The intention of this chapter is to give a general overview of the stellar magnetic field diagnostic methods and to present results of the application of these methods to different classes of stars. First, the chapter discusses the physical processes that lead to appearance of magnetic signatures in stellar spectra. Based on this discussion, the most commonly used magnetic observables are introduced for the intensity and polarisation spectra. The methods of interpretation of magnetic measurements, starting with robust field detection techniques and leading to reconstruction of the physically realistic vector maps of surface magnetic fields, are discussed. Finally, the current knowledge about the incidence, geometries and the origin of magnetic fields in stars of different masses and ages is reviewed and the astrophysical significance of magnetism in the context of stellar structure and evolution is discussed.

3.1 Introduction

This chapter presents an introduction to the studies of stellar magnetic fields. We explore the questions of how magnetic fields can be detected on the stellar surfaces, what information about magnetic field topologies can be extracted from different observations, and what are the observed properties of magnetic fields in stars of different masses and ages. The origin of magnetism in different types of stars and its relation to other physical processes operating in stellar interiors and atmospheres are briefly discussed.

The following sections start by presenting the essential theoretical foundations of the stellar magnetic field diagnostic and modelling. We consider the Zeeman effect in spectral lines and information contained in the polarisation characteristics of stellar radiation. A connection between the local magnetic field parameters and morphology of the stellar polarisation profiles is established using both detailed numerical calculations and simplified analytical considerations. Based on this discussion, we introduce different methodologies suitable for the detection and characterisation of stellar magnetic fields.

The second part of this chapter gives a concise overview of the current observational picture of stellar magnetism. We consider massive (O and early-B spectral types), intermediate-mass (mid-B to early-F), solar-type (F, G, K), and low-mass (M) stars. Magnetism in the pre-main sequence stars (Herbig Ae/Be and T Tauri stars) of different masses is also briefly considered. The key theoretical concepts and findings are summarised where appropriate, although this presentation has no intention to cover stellar magnetohydrodynamics problems in detail.

The topics discussed here have been reviewed in many recent publications. The reader is encouraged to consult these articles and books whenever more information is needed on the specific types of stars and/or specific physical processes involving magnetic fields. A useful general review of the observational studies of magnetic fields in non-degenerate stars has been published by Donati & Landstreet (2009). Further details on the observational studies of magnetic fields in low-mass stars can be found in Reiners (2012). Magnetism in isolated and binary white dwarfs (which are not considered here) has been summarised by Wickramasinghe & Ferrario (2000). A series of lectures by Landstreet (2009) provides an excellent introduction to observing and modelling of magnetic fields in the upper main sequence stars. An extensive theoretical perspective on the role of magnetic fields in the context of stellar structure and evolution can be obtained from the book by Mestel (1999). Finally, the book by Landi Degl'Innocenti & Landolfi (2004) provides a comprehensive presentation of the theoretical concepts and mathematical methods essential for interpreting magnetically induced polarisation in spectral lines.

3.2 Theoretical Basis

3.2.1 *Atom in a Magnetic Field*

The presence of an external magnetic field modifies the atomic energy levels and, as a consequence, changes the properties of spectral lines corresponding to the transitions between these levels. Generally, both the central wavelength, the number of spectral line components, and their polarisation properties are altered by the magnetic field. It is this influence of the field on the spectra of atoms (and molecules in cool stars) that enables a direct detection, measurement, and ultimately a detailed modelling of stellar magnetic fields.

Following Landstreet (2009), we consider different regimes of magnetic field interaction with an atom. The atomic Hamiltonian in a magnetic field can be expressed as

$$H = -\frac{\hbar}{2m}\nabla^2 + V(r) + \xi(r)\mathbf{L}\cdot\mathbf{S} + \frac{e\hbar}{2mc}(\mathbf{L}+2\mathbf{S})\cdot\mathbf{B} + \frac{e^2}{8mc^2}(\mathbf{B}\times\mathbf{r})^2. \tag{3.1}$$

In this equation m and e are the electron mass and charge; c and $\hbar$ are the speed of light and the Planck constant; $\mathbf{L}$ and $\mathbf{S}$ are the orbital and spin angular momentum operators, and $\mathbf{B}$ is the vector magnetic field. The first term of the Hamiltonian corresponds to the kinetic energy; the second term represents the Coulomb potential energy; the third term corresponds to the spin-orbit interaction under the L-S coupling approximation. The two final terms – the linear and quadratic field terms – describe an effect of the magnetic field $\mathbf{B}$.

Depending on the relative magnitudes of the spin-orbit interaction term and the two magnetic field terms, it is convenient to define different regimes, which can be treated using perturbation theory of the atomic Hamiltonian.

- Linear Zeeman effect: quadratic field term $\ll$ linear field term $\ll$ spin-orbit term. This typically corresponds to fields <100 kG (1 G $=10^{-4}$ T) and represents by far the most common observable manifestation of magnetic fields in the spectra of non-degenerate stars.
- Paschen–Back effect: quadratic field term $\ll$ linear field term but, at the same time, spin-orbit term $\ll$ linear field term. This regime is encountered at different field strengths, depending on the fine structure splitting. The majority of spectral lines require $B > 100$ kG to exhibit the Paschen–Back effect. However, for a number of transitions with close multiplet components (e.g. He I 5876 Å, Li I 6707 Å) significant deviations from the linear Zeeman effect can occur already for B = 1–5 kG.
- Quadratic Zeeman effect: quadratic field term $\gg$ linear field term and quadratic field term $\gg$ spin-orbit term. Requires fields >100 kG.

For field strengths exceeding $\sim$10 MG, the magnetic terms of the Hamiltonian become comparable to the Coulomb potential energy term. In that case the perturbation approach is no longer valid and the structure of an atom has to be solved in both the electrostatic and external magnetic fields. This formidable theoretical problem, relevant

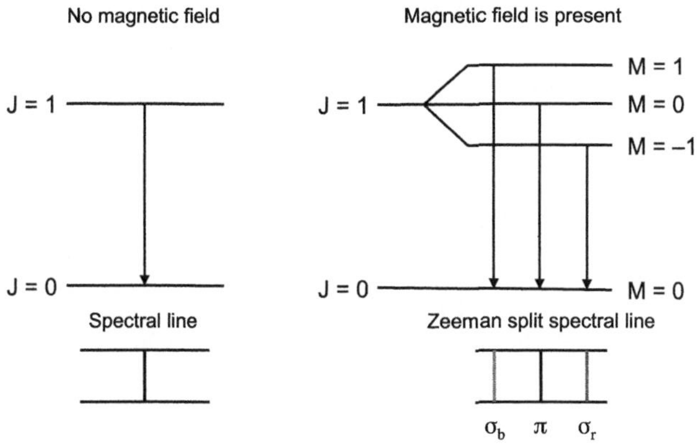

FIGURE 3.1. Zeeman splitting in a magnetic field. In the absence of the field, the transition between the upper and lower atomic levels gives a single spectral line. When an external field is present, the line splits into three (π, blue- and red-shifted σ) groups of Zeeman components.

for interpretation of the spectra of strongly magnetic white dwarfs, has been addressed so far only for a few atoms.

3.2.2 *Zeeman Splitting*

An external magnetic field removes the degeneracy of different angular momentum states, leading to line splitting. In the linear Zeeman regime the magnetic perturbation of the atomic levels is small compared with the fine structure splitting. Consequently, the magnetic splitting of spectral lines can be considered independently of each other. Each atomic level characterised by the quantum number J splits into $2J+1$ sublevels with energies

$$E = E_0 + g\frac{e\hbar}{2mc}BM, \tag{3.2}$$

corresponding to the magnetic quantum numbers $M = -J, -J+1, \ldots, J-1, J$.

The dimensionless parameter g, called the Landé factor, is a number typically lying between 0 and 3, with an average around 1.2 for any sufficiently large set of lines. The Landé factors are often determined as part of detailed quantum mechanical calculations aimed at establishing the line spectrum of a given ion.[1] Alternatively, L-S coupling can be used to determine g from the L and S quantum numbers of an energy level

$$g = \frac{3}{2} + \frac{S(S+1) - L(L+1)}{2J(J+1)}. \tag{3.3}$$

This approximation is suitable for computing the Landé factors of light elements, but its accuracy diminishes for the iron-peak group elements (see discussion in Mathys, 1990) and is largely inadequate for heavy ions.

As a consequence of the magnetic splitting of atomic levels, a single spectral line gives rise to several Zeeman components if a magnetic field is present. Figure 3.1 illustrates the simplest situation of a spectral line formed by the transition between the unsplit ($J_l = 0$) lower level and the upper level with $J_u = 1$. The selection rules allow only transitions satisfying $\Delta M = M_u - M_l = 0, \pm 1$. Accordingly, each spectral line splits into distinct groups of Zeeman components. Those corresponding to $\Delta M = 0$ are called

[1] Online data bases, such as VALD (http://vald.astro.uu.se/) provide wavelengths, the Landé factors and other line parameters resulting from such calculations.

the π components; the ones with $\Delta M = \pm 1$ are known as the blue- and red-shifted σ components (σ_b and σ_r). For the case shown in Fig. 3.1 (normal Zeeman triplet) we obtain a single component of each type.

The wavelength shifts of the Zeeman components relative to the unperturbed wavelength λ_0 are given by

$$\Delta\lambda(M_l, M_u) = (g_l M_l - g_u M_u)\Delta\lambda_B \tag{3.4}$$

with

$$\Delta\lambda_B = \frac{eB\lambda_0^2}{4\pi m_e c} = 4.67 \times 10^{-13} B\lambda_0^2 \tag{3.5}$$

for the field strength in G and wavelength in Å. The magnetic splitting of spectral lines is symmetric with respect to λ_0; it grows linearly with the field strength and quadratically with wavelength. It is therefore advantageous to observe the Zeeman effect at longer wavelengths.

The relative strengths of the components of a given type q (0 for π and ± 1 for $\sigma_{\mathrm{r,b}}$) are expressed using the 3-j symbols

$$T_q(M_l, M_u) = 3\begin{pmatrix} J_u & J_l & 1 \\ -M_u & M_l & -q \end{pmatrix}^2, \tag{3.6}$$

and can be written as simple algebraic functions of $M_{u,l}$ and $J_{u,l}$ (see Table 3.1 in Landi Degl'Innocenti & Landolfi, 2004). The strengths of the Zeeman components are normalised to unity

$$\sum_{M_l M_u} T_q(M_l, M_u) = 1 \tag{3.7}$$

for each component type.

The so-called effective Landé factor, defined by the relation

$$\bar{g} = \frac{1}{2}(g_l + g_u) + \frac{1}{4}(g_l - g_u)[J_l(J_l + 1) - J_u(J_u + 1)], \tag{3.8}$$

is a convenient characteristic of the magnetic sensitivity of a spectral line. Mathematically, it corresponds to the separation of the centres of gravity of σ components from the unperturbed wavelength of a line.

An illustration of the Zeeman splitting patterns for the lines belonging to the ^2S–^2P multiplet is presented in Fig. 3.2. These terms correspond to the lower and upper energy levels of some astrophysically interesting transitions (the Li I doublet at 6708 Å and Na I D doublet at 5889–5895 Å). Depending on the magnetic field strength and the fine structure energy separation, the magnetic splitting occurs in the linear Zeeman regime considered here, or in the Paschen–Back regime. For example, one needs fields of the order of 100 kG to reach the Paschen–Back regime for the Na D doublet. However, the lithium doublet components are much closer together in the absence of the field, leading to substantial departures from the simple Zeeman splitting for $B \geq 2$–3 kG.

In general, the transition between the Zeeman and Paschen–Back splitting (sometimes called the partial Paschen–Back regime) is characterised by the gradual loss of symmetry between the σ_b and σ_r components, weakening of some components and strengthening of others, and even by the appearance of entirely new Zeeman components absent in weak fields. For sufficiently intense magnetic fields the π and $\sigma_{\mathrm{b,r}}$ components of all lines in a multiplet merge into three groups, corresponding to the normal triplet splitting (see the right panel in Fig. 3.2).

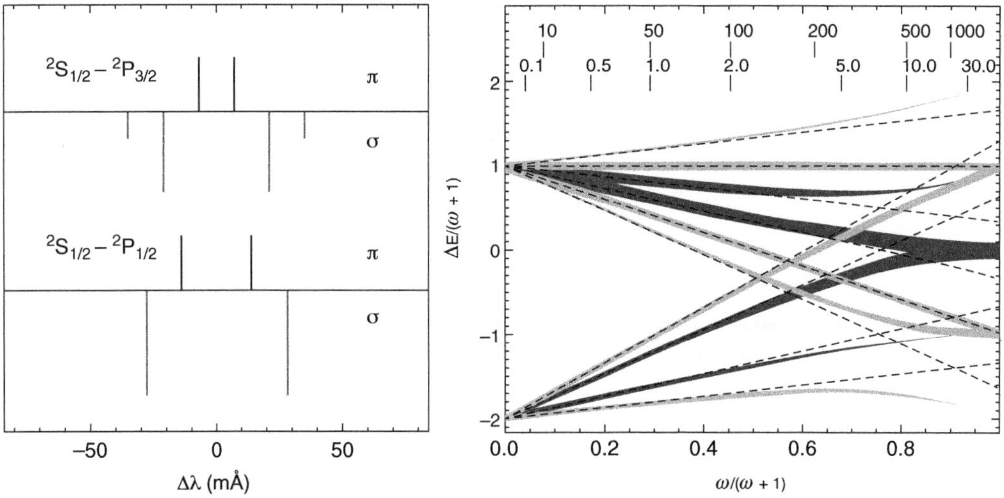

FIGURE 3.2. *Left panel:* schematic illustration of the strengths and displacements of the Zeeman components of the ${}^2S_{1/2}$–${}^2P_{1/2}$ and ${}^2S_{1/2}$–${}^2P_{3/2}$ transitions (e.g. the resonance Li I doublet at 6708 Å or the Na I D doublet at 5889–5895 Å). The π (σ) components are shown above (below) the horizontal line. The horizontal scale corresponds to $\lambda_0 = 6708$ Å and $B = 10^3$ G.
Right panel: The strength and normalised energy separation of the Zeeman components as a function of the magnetic field strength. Dashed lines show splitting for the linear Zeeman effect. Shaded curves correspond to the magnetic splitting calculated taking the Paschen–Back effect into account. The width of the curves is proportional to the strength of the respective π (darker curves) and σ (lighter curves) components. The vertical bars show the magnetic field strength (in kG) corresponding to the fine structure splitting in the Na D lines (upper row) and the Li I resonance doublet (lower row). Adapted from Kochukhov (2008). Reproduced with permission © ESO.

3.2.3 *Stokes Parameters and Polarisation of Zeeman Components*

Polarisation is a property of an electromagnetic wave to oscillate in a non-random way. If the electric field vector of the wave vibrates in a single plane as the wave propagates, such light is called linearly polarised. If the electric field vector rotates along a circle, the wave is circularly polarised. In the general case, both the linear and circular polarisation can be present, meaning that the electric vector traces an ellipse in the plane perpendicular to the propagation direction.

To describe polarisation of an electromagnetic wave we use the Stokes (column) vector $\mathbf{I} = \{I, Q, U, V\}^{\mathrm{T}}$. Its components are defined by considering measurements with ideal polarisers – devices that fully transmit one polarisation and completely block the orthogonal polarisation.

- Stokes I is the total intensity of radiation or, equivalently, the sum of any two beams with orthogonal polarisations: $I = I_0 + I_{90} = I_{45} + I_{135} = I_{\circlearrowright} + I_{\circlearrowleft}$
- Stokes Q is obtained by measuring the difference of the intensity of the beams seen through a perfect linear polariser with the transmission axis set to 0° and 90°: $Q = I_0 - I_{90}$.
- Stokes U is given by the intensity difference of the beams obtained by setting the transmission axis of an ideal linear polariser to 45° and 135°: $U = I_{45} - I_{135}$.
- Stokes V is the intensity difference of the beams transmitted by ideal polarising devices which let through either the left or the right circularly polarised light: $V = I_{\circlearrowright} - I_{\circlearrowleft}$

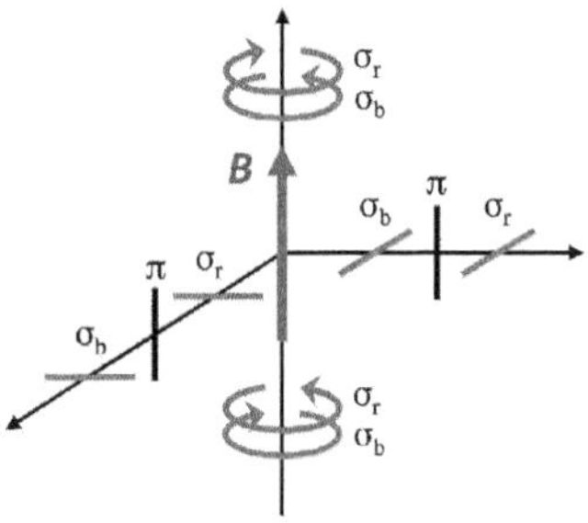

FIGURE 3.3. Polarisation properties of the radiation emitted in the π and σ components for different orientations of the magnetic field vector relative to the line of sight.

Thus, fundamentally, the polarisation diagnostic is a differential measurement of the intensity of the two beams with orthogonal polarisation states. In practice, this measurement is accomplished by recording the two beams one after another using the same detector (temporal modulation) or simultaneously on different parts of the detector (spatial modulation).

Components of a Zeeman split spectral line have distinct polarisation properties. This polarisation changes according to the angle between the magnetic field vector and the direction of the emitted light (see Fig. 3.3). For the light emitted parallel to the field vector, the π components vanish and the $\sigma_{\rm b}$ and $\sigma_{\rm r}$ components have opposite circular polarisation. On the other hand, if the line of sight is perpendicular to the field vector, the π components are linearly polarised parallel to the field and the σ components are linearly polarised perpendicular to the field. Thus, the π components can have only linear polarisation while the σ components can exhibit both circular and linear polarisation.

The sensitivity of Zeeman component polarisation to the field orientation provides a powerful method to study vector properties of magnetic field distributions on the stellar surfaces and in circumstellar environments. Polarisation signal in spectral lines is also straightforward to detect against an unpolarised background, which greatly enhances our ability to detect and characterise weak stellar magnetic fields.

3.2.4 *Polarised Radiative Transfer*

An interaction between matter and radiation is described by the radiative transfer equation. Most remote sensing methods applied in stellar physics are based, in one way or another, on the solution to this equation. Interpretation of the spectroscopic and spectropolarimetric observations aimed at extracting information about stellar magnetic fields also heavily relies on the transfer equation.

In the presence of a magnetic field a single scalar equation for the specific intensity is superseded by the analogous transfer equation for the (wavelength-dependent) Stokes vector $\mathbf{I}$

$$\frac{d\mathbf{I}}{dz} = -\mathcal{K}\mathbf{I} + \mathbf{J}, \tag{3.9}$$

where z is some geometrical coordinate, e.g. height in the stellar atmosphere. The 4×4 matrix $\mathcal{K}$ describes the absorption of light and attenuation of its polarisation characteristics. For the case of unpolarised continuum (true for all magnetic stars except MG-field white dwarfs), $\mathcal{K}$ can be expressed as

$$\mathcal{K} = \kappa_c \mathcal{I} + \sum_{\ell} \kappa_\ell \mathbf{\Phi}_\ell, \tag{3.10}$$

where κ_c is the continuum opacity coefficient, $\mathcal{I}$ is the 4×4 identity matrix and the summation is carried out over all lines with the central opacity κ_l contributing to a given wavelength. Ignoring scattering, the emission vector $\mathbf{J}$ appearing on the right hand side of Eq. (3.9) can be written as

$$\mathbf{J} = \kappa_c S_c \mathcal{J} + \sum_\ell \kappa_l S_\ell \mathbf{\Phi}_\ell \mathcal{J}, \tag{3.11}$$

where $\mathcal{J} = \{1, 0, 0, 0\}^{\mathrm{T}}$ and, under the assumption of local thermodynamic equilibrium (LTE), the line and continuum source functions are both equal to the Planck function $S_c = S_\ell = P_\lambda(T)$.

Information about the magnetic field strength and orientation is encoded in the depth-dependent components of the absorption matrix $\mathbf{\Phi}_\ell$

$$\mathbf{\Phi}_\ell = \begin{pmatrix} \eta_I & \eta_Q & \eta_U & \eta_V \\ \eta_Q & \eta_I & \rho_V & -\rho_U \\ \eta_U & -\rho_V & \eta_I & \rho_Q \\ \eta_V & \rho_U & -\rho_Q & \eta_I \end{pmatrix} \tag{3.12}$$

with the absorption η and anomalous dispersion ρ terms given by

$$\begin{aligned}
\eta_I &= \frac{1}{2}\left[\phi_{\mathrm{p}} \sin^2\theta + \frac{\phi_{\mathrm{b}} + \phi_{\mathrm{r}}}{2}(1 + \cos^2\theta)\right], \\
\eta_Q &= \frac{1}{2}\left[\phi_{\mathrm{p}} - \frac{\phi_{\mathrm{b}} + \phi_{\mathrm{r}}}{2}\right] \sin^2\theta \cos 2\chi, \\
\eta_U &= \frac{1}{2}\left[\phi_{\mathrm{p}} - \frac{\phi_{\mathrm{b}} + \phi_{\mathrm{r}}}{2}\right] \sin^2\theta \sin 2\chi, \\
\eta_V &= \frac{1}{2}\left[\phi_{\mathrm{r}} - \phi_{\mathrm{b}}\right] \cos\theta, \\
\rho_Q &= \frac{1}{2}\left[\psi_{\mathrm{p}} - \frac{\psi_{\mathrm{b}} + \psi_{\mathrm{r}}}{2}\right] \sin^2\theta \cos 2\chi, \\
\rho_U &= \frac{1}{2}\left[\psi_{\mathrm{p}} - \frac{\psi_{\mathrm{b}} + \psi_{\mathrm{r}}}{2}\right] \sin^2\theta \sin 2\chi, \\
\rho_V &= \frac{1}{2}\left[\psi_{\mathrm{r}} - \psi_{\mathrm{b}}\right] \cos\theta.
\end{aligned} \tag{3.13}$$

The angles θ and χ appearing in these expressions specify orientation of the magnetic field vector with respect to the line of sight (see Fig. 3.4). In particular, $B\cos\theta$ is the line of sight projection of the field and χ gives the orientation of the transverse field component $B\sin\theta$ in the plane perpendicular to the light propagation direction.

The quantities $\phi_{\mathrm{p,b,r}}$ and $\psi_{\mathrm{p,b,r}}$ in Eq. (3.13) represent the absorption and anomalous dispersion profiles for the three groups of Zeeman components. These profiles are obtained by adding together the Voigt H, respectively Faraday–Voigt L profiles, shifted to the positions of individual Zeeman components $\lambda_0 + \Delta\lambda(M_l, M_u)$ and weighted by their relative strengths $T_q(M_l, M_u)$

$$\begin{aligned}
\phi_q &= \sum_{M_l M_u} T_q(M_l, M_u) H\left[\frac{\lambda - \lambda_0 - \Delta\lambda(M_l, M_u)}{\Delta\lambda_D}, a\right], \\
\psi_q &= \sum_{M_l M_u} T_q(M_l, M_u) L\left[\frac{\lambda - \lambda_0 - \Delta\lambda(M_l, M_u)}{\Delta\lambda_D}, a\right].
\end{aligned} \tag{3.14}$$

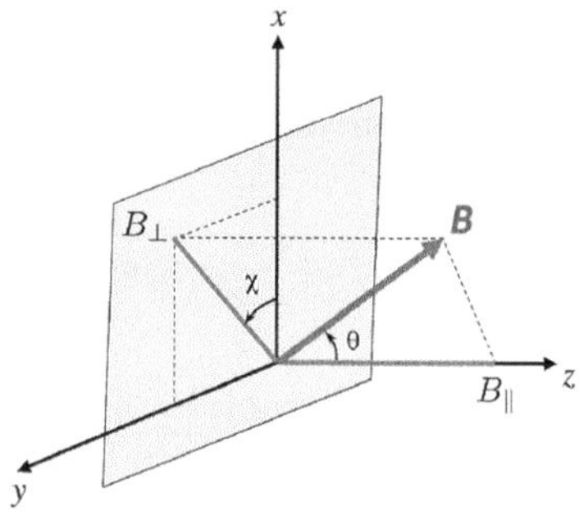

FIGURE 3.4. Definition of the angles θ and χ. The line of sight is along the z-axis. $B_{\|}$ is the line of sight component of the magnetic field vector; $B_{\perp}$ is the transverse component.

The quantities $\Delta\lambda_D$ and a, needed for calculation of the Voigt and Faraday–Voigt functions, represent the Doppler width corresponding to the thermal and turbulent broadening and the width of a Lorentzian profile describing the natural (radiative), Stark, and van der Waals broadening processes acting on a given spectral line.

3.2.5 *Numerical Calculation of Stokes Parameters*

To solve the polarised radiative transfer (PRT) equation in a stellar atmosphere we require the following input information.

(i) The strength and orientation of the magnetic field vector **B**, possibly as a function of z.

(ii) Thermodynamic information about the stellar atmosphere, i.e. the temperature and pressure as a function of z. This can be obtained from the grids of tabulated stellar atmosphere models (e.g. Kurucz, 1993; Gustafsson et al., 2008).

(iii) Information about relative concentrations of chemical elements whose lines we wish to model. These elemental abundances are normally specified in terms of the number densities relative to hydrogen. Asplund et al. (2009) provide a recent compilation of solar abundances.

(iv) A data base of the continuum opacity coefficients of relevant absorbers and a line list containing information on the position of spectral lines, their intensity (transition probability), broadening parameters and additional parameters $J_{u,l}$ and $g_{u,l}$ required for computing the Zeeman splitting patterns.

These inputs are combined to calculate arrays $\kappa_c(\lambda, z)$, $\kappa_l(\lambda, z)$, and $\mathbf{\Phi}_\ell(\lambda, z)$ for each wavelength of interest and for each depth layer of a discrete grid on which the stellar model atmosphere is tabulated (typically 50–100 points covering some 7–9 orders of magnitude in the optical depth). Using all this information, a numerical solution of the PRT equation can, in principle, be accomplished by propagating the Stokes vector **I** from the bottom (where a diffusion approximation can be used as a boundary condition) to the top of the stellar atmosphere with the help of an appropriate algorithm.

In practice, a numerical solution of the four coupled differential equations with highly variable coefficients (due to a strong depth-dependence of the line absorption coefficient κ_l and the line broadening parameters $\Delta\lambda_D$ and a) is a challenging task given relatively sparse vertical grids of the typical stellar atmosphere models. Dedicated numerical algorithms for solving the PRT equation have been discussed and compared by Landstreet (1988), Rees et al. (1989), Piskunov & Kochukhov (2002), and de la Cruz Rodríguez & Piskunov (2013). A comparison of the Stokes profile calculations with independent

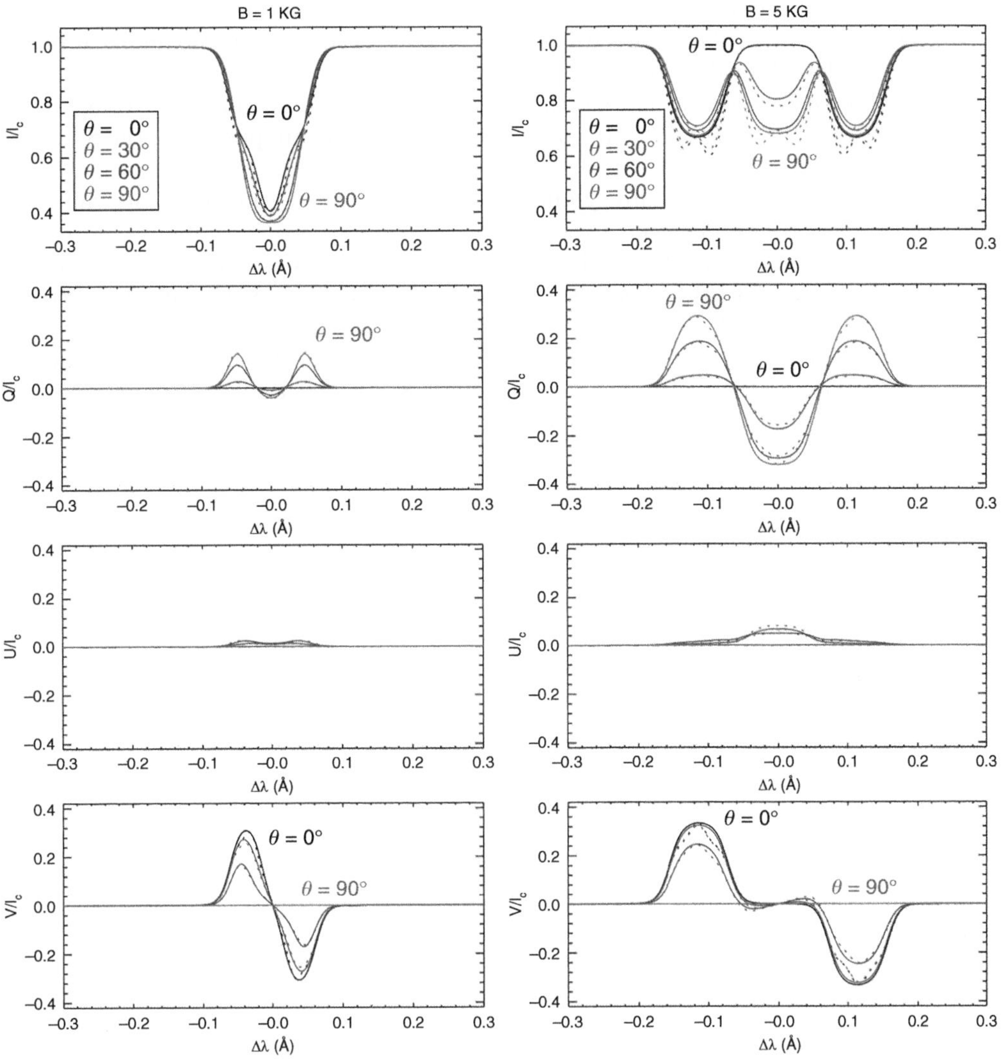

FIGURE 3.5. Stokes parameter profiles of the Fe II 5018 Å line for two values of the magnetic field strength and several orientations of the field vector with respect to the line of sight. The solid lines show results of the full numerical solution of the PRT equation in a realistic stellar model atmosphere. The dotted lines correspond to the analytical profiles computed for a Milne–Eddington atmospheric model.

PRT codes usually finds a satisfactory agreement (discrepancies below 0.02–0.05% of the continuum intensity) provided that the codes use consistent input data and the model atmosphere is tabulated on a sufficiently dense vertical grid (Wade et al., 2001; Deen, 2013).

An example of the numerical solution of the PRT equation with the code described by Kochukhov et al. (2010) is presented in Fig. 3.5. This figure shows the Stokes *IQUV* profiles of the Fe II 5018 Å line for $T_{\text{eff}} = 9000$ K, $\log g = 4.0$ model atmosphere taken from the Atlas9 grid (Kurucz, 1993). Calculations are carried for the two field strength values, 1 and 5 kG, and different orientations of the field vector relative to the line of sight. All magnetic field parameters are assumed to be constant with depth.

Several general properties of the Stokes profiles are worth noting.

- The maximum circular polarisation is observed at $\theta = 0^o$; the maximum linear polarisation is found for $\theta = 90^o$.
- The π and σ components show orthogonal linear polarisation and the π components disappear in Stokes I for $\theta = 0^o$; only the σ components are producing the Stokes V signal.
- The Stokes IQU profiles are symmetric and the Stokes V profile is antisymmetric with respect to the line centre.

All these features agree with the qualitative discussion in Section 3.2.3 and directly follow from the expressions (3.13) for the absorption components of the matrix $\mathbf{\Phi}_\ell$. A less trivial finding is the presence of a weak but non-zero signal in Stokes U, even for the field vector aligned with the reference Q direction (i.e. $\chi = 0^o$). This feature is explained by the presence of the anomalous dispersion terms ρ in $\mathbf{\Phi}_\ell$, which lead to 'magneto-optical' rotation of the linear polarisation plane as the radiation propagates through the stellar atmosphere.

3.2.6 *Approximate Solutions of the PRT Equation*

3.2.6.1 Milne–Eddington Atmosphere

The procedure of obtaining a detailed numerical solution of the PRT equations discussed above is computationally demanding and requires a large amount of input data, including information on individual spectral lines, continuous absorbers, stellar parameters, and the structure of stellar atmosphere. In many practical applications, especially in the statistical studies dealing with a large number of objects, an accurate numerical solution is impossible and probably not needed. Several useful inferences about stellar magnetic fields can be made by considering approximate analytical solutions. Among those, the one based on a so-called Milne–Eddington (ME) atmosphere is of particular interest. This solution assumes that the magnetic field, the ratio κ_l/κ_c and the absorption and anomalous dispersion profiles ϕ_q and ψ_q are all constant in the line formation region, and the source function depends linearly on the optical depth τ

$$S_c = S_\ell = P_0(1 + \beta_0 \tau). \tag{3.15}$$

Under these assumptions, a general analytical solution for the unpolarised continuum intensity I_c and the Stokes parameter profiles $X = I, Q, U, V$ emerging from the atmosphere can be derived in the form

$$I_c(\mu) = P_0(1 + \beta_0 \mu) \tag{3.16}$$

and

$$X(\mu) = P_0 F_X(\beta_0, \mu, \eta_0, \eta_I, \eta_Q, \eta_U, \eta_V, \rho_Q, \rho_U, \rho_V), \tag{3.17}$$

where μ is the cosine of the angle between the line of sight and the normal to the stellar surface and F_X are analytical functions (see Section 9.8 in Landi Degl'Innocenti & Landolfi, 2004). Thus, for a given field vector, the normalised Stokes parameters X/I_c are expressed in terms of the constant β_0, parameters $\Delta\lambda_D$ and a determining the shape of ϕ_q and ψ_q, and some line strength parameter η_0 required to specify the residual intensity of the absorption profile. All these parameters have no direct physical meaning and have to be adjusted empirically, by matching detailed calculations or observations.

With a suitable combination of free parameters, Eq. (3.17) can successfully approximate essentially any realistic Stokes vector. However, the overall accuracy of this analytical solution and hence its usefulness diminishes if one considers a sufficiently wide range of field strength and orientation. For example, Fig. 3.5 shows a nearly perfect agreement between the Stokes parameters predicted by the ME model and detailed radiative transfer calculations for $B = 1$ kG, but noticeable discrepancies occur for $B = 5$ kG. (In both cases we used the same set of η_0, β_0, $\Delta\lambda_D$, and a, determined by fitting the non-magnetic Stokes I profile.)

3.2.6.2 Weak field approximation

Another set of analytical formulas can be obtained by considering the system of PRT equations in the limit of weak magnetic field. This limit is defined by the condition $\bar{g}\Delta\lambda_B \ll \Delta\lambda_D$, i.e. any Zeeman splitting is substantially smaller than the intrinsic line width. For typical metal lines forming in the atmosphere of a solar-type star this translates to $\bar{g}B \ll 2500$ G, corresponding to the fields below 0.5–1.0 kG.[2] In that case one can re-write the system of the PRT equations for the Taylor expansion of Stokes parameters $X = X_0 + \Delta\lambda_B X_1 + (\Delta\lambda_B)^2 X_2 + \cdots$ with terms of diminishing significance (see Section 9.6 in Landi Degl'Innocenti & Landolfi, 2004). Retaining only the first-order terms, one finds that magnetic field gives rise only to circular polarisation ($I(\lambda) \approx I_0, Q \approx 0, U \approx 0$), with the shape of the Stokes V profile given by the scaled derivative of the Stokes I profile,

$$V(\lambda) \approx -\Delta\lambda_B \bar{g} \cos\theta \frac{\partial I}{\partial \lambda} = -4.67 \times 10^{-13} \bar{g} \lambda_0^2 B_{\parallel} \frac{\partial I}{\partial \lambda}, \tag{3.18}$$

for the field strength in G and wavelength in Å. This is a widely used relation, providing a simple method to estimate the longitudinal field component $B_{\parallel}$ from the Stokes I and V observations of moderate quality, independently of the shape and strength of diagnostic spectral lines. Note that there is no equivalent formula for the linear polarisation signatures applicable to lines of arbitrary strength.

Equation (3.18) requires the knowledge of the Stokes I profile, which in general has to be taken from observations or computed theoretically by solving the non-magnetic radiative transfer equation for an appropriate stellar atmosphere model. A further simplification, leading to a complete set of analytical formulas for all four Stokes parameters, is introduced by considering the weak line formation, $\kappa_l/\kappa_c \ll 1$, for which $1 - I(\lambda)/I_c \propto \eta_I$. In this case one can adopt a Gaussian function for η_I

$$\eta_I = \eta_0 \exp\left[-\frac{(\lambda - \lambda_0)^2}{2w^2}\right] \tag{3.19}$$

in combination with the expressions

$$\begin{aligned} Q(\lambda)/I_c &= -5.45 \times 10^{-26} \bar{G} (\lambda_0^2 B_{\perp})^2 \frac{\partial^2 \eta_I}{\partial \lambda^2} \cos 2\chi, \\ U(\lambda)/I_c &= -5.45 \times 10^{-26} \bar{G} (\lambda_0^2 B_{\perp})^2 \frac{\partial^2 \eta_I}{\partial \lambda^2} \sin 2\chi, \\ V(\lambda)/I_c &= -4.67 \times 10^{-13} \bar{g} \lambda_0^2 B_{\parallel} \frac{\partial \eta_I}{\partial \lambda}, \end{aligned} \tag{3.20}$$

where $\bar{G}$ is an algebraic function of $J_{u,l}$ and $g_{u,l}$. For Stokes QU $\bar{G}$ has the same role of a linear polarisation sensitivity index as $\bar{g}$ plays for Stokes V; for the spectral lines with a triplet Zeeman splitting $\bar{G} = \bar{g}^2$.

[2] For certain intrinsically very broad spectral lines, e.g. hydrogen Balmer lines, this approximation can be extended to much stronger fields.

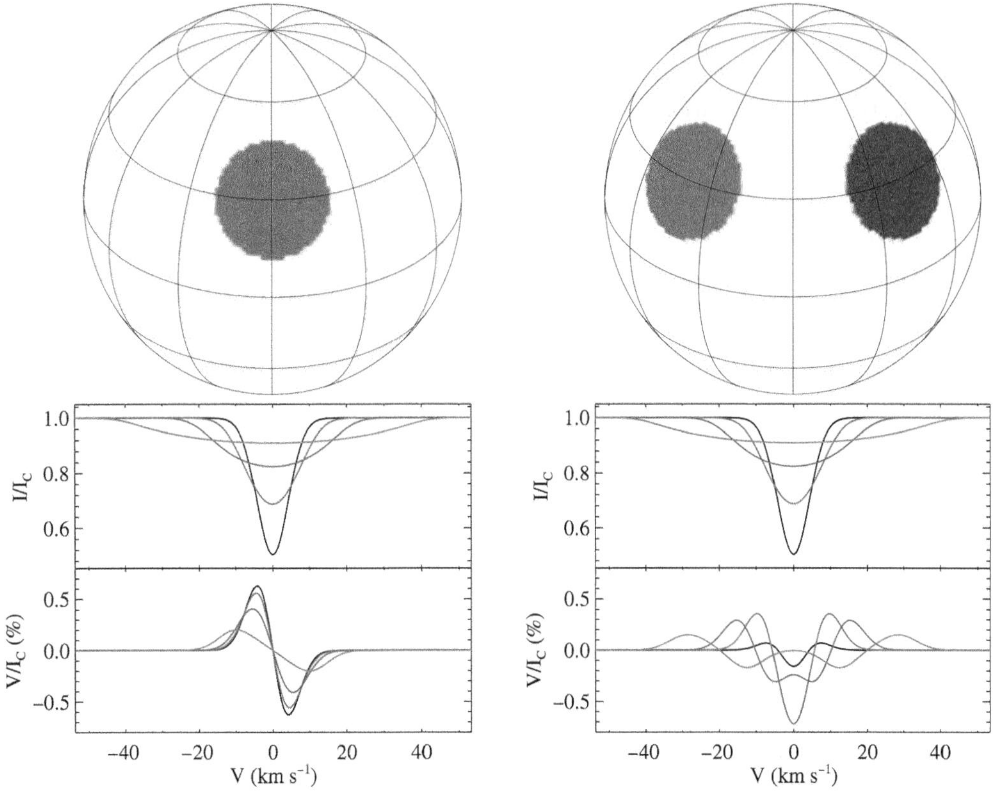

FIGURE 3.6. Stokes I and V disk integrated profiles for two radial magnetic field distributions. Calculations for a unipolar spot (left) are compared to the profiles corresponding to the bipolar magnetic map containing two spots of opposite polarity (right). The intensity and circular polarisation profiles are shown for $v_{\rm e} \sin i = 5$, 10, 20, and 40 $\rm km\,s^{-1}$.

3.2.7 *Disk Integrated Stokes Profiles*

Calculating the Stokes parameter profiles with the methods outlined above is sufficient for interpreting observations of the spatially resolved magnetic structures, e.g. sunspots. However, studies of stellar magnetic fields have to deal with a fundamentally different and considerably more complex situation. With a few exceptions (none corresponding to an interesting magnetised object), stars are seen as unresolved objects. For any realistic stellar magnetic field topology, magnetic field vector changes considerably from one region of the surface to another. Each surface zone produces its own Stokes vector, which is Doppler-shifted due to the stellar rotation and weighted according to the local brightness and projected surface area. Contributions from all surface zones on the visible stellar hemisphere add together to form the disk integrated Stokes profiles. Adding to the complexity, the result of this disk integration is time-dependent, even for stable magnetic configurations, since the field structure is observed from different aspect angles as the star rotates.

Disk integrated (flux) Stokes spectra may be morphologically very complex and do not resemble any of the local Stokes profiles. In particular, the symmetry properties of Stokes QUV discussed in Section 3.2.5 do not apply to the flux profiles. Figure 3.6 gives a feeling of the complexity introduced by the surface integration. We consider magnetic distributions consisting of a single large radial field spot and two spots with opposite polarity. In the former case, the Stokes V flux profiles have familiar S-shape structure independently of the stellar rotational velocity. But in the latter case, spectral contributions from the two spots add up non-trivially, producing symmetric Stokes V profile with an amplitude and morphology strongly dependent on the projected stellar rotational velocity $v_e \sin i$.

Evidently, special diagnostic methods have to be applied to interpret stellar Stokes flux profiles to infer a meaningful information about the surface magnetic field distribution from disk integrated observables. An assessment of the information content, reliability, and applicability of these techniques to different classes of magnetic stars is one of the central topics of the observational research in stellar magnetism.

3.3 Methods of Stellar Magnetic Field Diagnostic

3.3.1 *Stellar Magnetic Field Observables*

3.3.1.1 Mean Longitudinal Magnetic Field

Historically, the vast majority of stellar magnetic field studies, including the very first attempts to diagnose magnetic fields in stars other than the Sun (Babcock, 1947), relied on detecting the Zeeman-induced circular polarisation in spectral lines. This diagnostic approach can be applied to spectra of a moderate signal-to-noise ratio ($S/N \sim 100$) and resolution ($\lambda/\Delta\lambda = 5000$–$20\,000$) obtained with the help of a circular polarisation analyser. As was shown above, in the presence of a magnetic field the σ components of spectral lines exhibit non-zero Stokes V signal. Therefore, the two spectra with orthogonal circular polarisation recorded on the detector – RCP $(I+V)$ and LCP $(I-V)$ spectra – show a displacement depending on the wavelength and magnetic sensitivity of a spectral line. Using the weak line approximation and assuming that the parameters governing the line formation (e.g. P_0, β_0, η_0 in the case of Milne–Eddington atmosphere) are not changing across the stellar surface, one can relate the shift between the centres of gravity of the RCP and LCP line profiles to the mean longitudinal magnetic field $\langle B_z \rangle$ – the line of sight magnetic field component averaged over the visible stellar surface –

$$(\lambda_R - \lambda_L)/2 = 4.67 \times 10^{-13} \bar{g} \lambda_0^2 \langle B_z \rangle. \tag{3.21}$$

Here the centres of gravities are defined in the usual way

$$\lambda_{R,L} = \frac{\int \lambda \mathcal{F}_{R,L} d\lambda}{\int \mathcal{F}_{R,L} d\lambda}, \tag{3.22}$$

where the quantities $\mathcal{F}$ represent the relative line depressions $\mathcal{F}_X = (I_c - X)/I_c$ in the RCP and LCP spectra ($X = I_{R,L}$) or in any of the Stokes parameters ($X = I, Q, U, V$). Note that with these definitions the left-hand side of Eq. (3.21) is mathematically equivalent to the first moment of the Stokes V profile normalised to the equivalent width of a spectral line (Borra & Vaughan, 1977)

$$(\lambda_R - \lambda_L)/2 = -\frac{\int \lambda \mathcal{F}_V d\lambda}{\int \mathcal{F}_I d\lambda}. \tag{3.23}$$

The mean longitudinal field, derived either from the RCP and LCP spectra with Eq. (3.21) or using directly the Stokes I and V spectra with Eq. (3.23), represents most of the catalogued stellar magnetic field measurements (e.g. Bychkov et al., 2003). In practice, longitudinal field can be determined from isolated metal lines taking into account their individual Landé factors (Mathys, 1991; Leone et al., 2000) or by cross-correlating large spectral intervals assuming a single mean effective Landé factor (Kudryavtsev et al.,

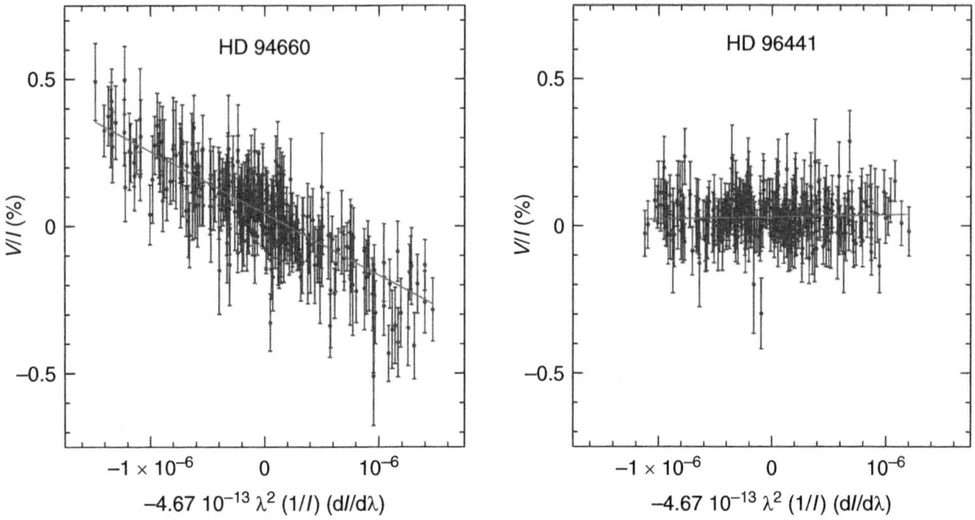

FIGURE 3.7. Evidence of a strong field in the magnetic A-type star HD 94660 (left) from the slope of V/I vs $(1/I)(dI/d\lambda)$ regression. For this particular observation, the straight line fit indicates $\langle B_z \rangle = -2085 \pm 85$ G. In comparison, no correlation is evident for the non-magnetic star HD 96441 (right). From Bagnulo et al. (2002b). Reproduced with permission © ESO.

2006). A variety of spectropolarimeters at small (1–2 m) and medium size (3–6 m) telescopes around the world are capable of recording the data suitable for such $\langle B_z \rangle$ measurements (Plachinda & Tarasova, 1999; Kim et al., 2007; Monin et al., 2012).

Another approach to deriving the mean longitudinal magnetic field was developed by Landstreet et al. (1975) for application to early-type stars with broad spectral lines. This technique relies on the photopolarimetric measurements using narrow-band interference filters centred on the wings of one of the hydrogen Balmer lines. The photoelectric circular polarisation measurements are related to $\langle B_z \rangle$ using Eq. (3.18) applicable in the weak field limit. Subsequently, Bagnulo et al. (2002b) extended this method to low-resolution ($\lambda/\Delta\lambda = 1000$–$5000$) CCD circularly polarised spectra obtained with the FORS1/2 instrument at the 8 m ESO VLT. In this implementation, the mean longitudinal field is deduced from the slope of the linear regression

$$\frac{V}{I} = -4.67 \times 10^{-13} \bar{g} \lambda^2 \langle B_z \rangle \frac{1}{I} \frac{dI}{d\lambda} \tag{3.24}$$

applied to every spectrum pixel assuming $\bar{g} = 1.0$ for hydrogen lines and $\bar{g} = 1.2$ for the regions dominated by metal lines. Figure 3.7 shows an example of using this method for determination of the longitudinal field in a magnetic A-type star. Since currently FORS is the only instrument suitable for stellar magnetometry at 8–10 m telescopes, it is widely used for magnetic field surveys of different, mostly early-type, stars (e.g. Bagnulo et al., 2006; Kochukhov & Bagnulo, 2006; Wade et al., 2007).

Both the integral method based on the analysis of medium-resolution spectra and the differential method relying on low-resolution data allow to retrieve $\langle B_z \rangle$ with a precision of 20–50 G for bright stars. Accuracies of $\sim$10 G can nominally be achieved with FORS thanks to the large collecting area of VLT. However, at this level of precision systematic instrumental artefacts, typical of non-stabilised Cassegrain-mounted instruments, are likely to dominate over the photon noise errors (Bagnulo et al., 2012).

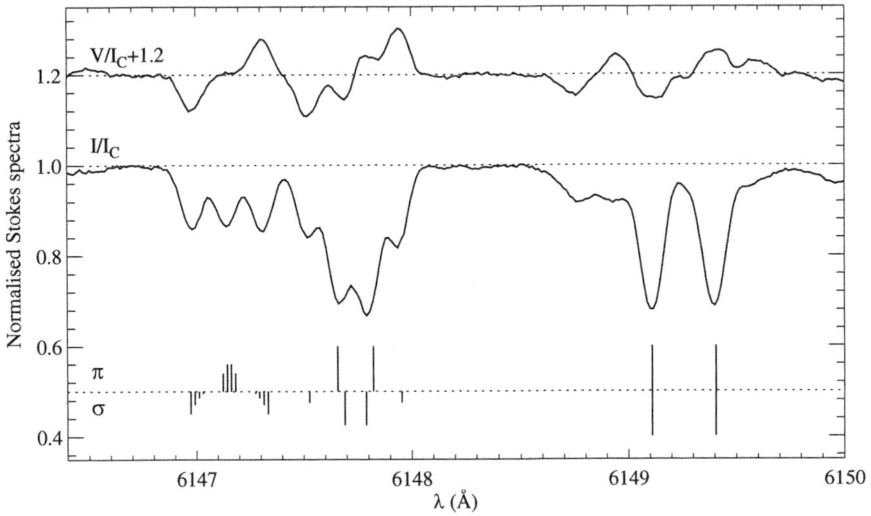

FIGURE 3.8. High-resolution intensity (middle) and circular polarisation (top) spectrum of the strongly magnetic A star HD 94660. This star has $\langle B \rangle = 6.2$ kG, leading to a strong polarisation signal and magnetic splitting clearly visible in many spectral lines. The bottom plot schematically shows the Zeeman splitting patterns of the main spectral features contributing to this wavelength region.

3.3.1.2 Mean Field Modulus

Detecting the Zeeman broadening and splitting of lines in the Stokes I spectrum is another direct way of ascertaining the existence of a magnetic field on the stellar surface. However, unlike the circular polarisation field detections, for which any Stokes V line signal constitutes an evidence of the field, disentangling the magnetic effects from other processes broadening the intensity line profiles (e.g. the thermal and rotational Doppler broadening, instrumental smearing) is highly challenging and requires spectra with $S/N \geq 200$ and resolution $\geq 10^5$.

Typically, resolved Zeeman split spectral lines are visible in stellar spectra only if the field strength exceeds ~ 2 kG and the rotational broadening is negligible (see example in Fig. 3.8). These conditions are met for a few dozen strongly magnetic, slowly rotating peculiar A-type stars (Mathys et al., 1997; Freyhammer et al., 2008). For these objects the separation of the σ and π components of the lines with a triplet splitting pattern enables essentially a model-independent way to measure the field strength averaged over the stellar surface, also known as the mean field modulus $\langle B \rangle$,

$$\lambda_\sigma - \lambda_\pi = \pm 4.67 \times 10^{-13} \bar{g} \lambda_0^2 \langle B \rangle. \qquad (3.25)$$

In addition, the field inclination with respect to the line of sight can be constrained to some extent by considering the relative intensities of the π and σ components (see expression for η_I in Eq. (3.13)). The Zeeman doublets (lines with coinciding π and σ components), such as Fe II 6149 Å, are even more useful since their components show separation of twice the value given by Eq. (3.25).

For late-type active stars, magnetic broadening studies based on the Stokes I spectra use near-IR observations to take advantage of the λ^2 dependence of Zeeman splitting (Valenti et al., 1995; Johns-Krull, 2007). But rather than dealing with a distinct Zeeman splitting corresponding to a single, well-defined value of $\langle B \rangle$, as is often the case for early-type stars, analyses of late-type objects have to deal with intrinsically complex, multi-component field geometries and consequently have to retrieve a set of field strengths and corresponding filling factors. This is typically accomplished using an elaborate spectrum synthesis modelling (Johns-Krull et al., 1999; Shulyak et al., 2014). In the best circumstances of $v_e \sin i \leq 1$–$2\ \mathrm{km\,s^{-1}}$ and an exceptionally high S/N ratio these methods are able to detect Zeeman broadening corresponding to $\langle B \rangle \geq 200$–$500$ G (Anderson et al., 2010; Kochukhov et al., 2013a).

3.3.1.3 Moment Technique

The line profile information contained in the high-resolution ($\lambda/\Delta\lambda \geq 30\,000$) Stokes parameter spectra can be exploited beyond the $\langle B_z \rangle$ measurements by considering different moments of line profiles and their relation to the stellar magnetic field topologies. Mathys (1988, 1989) showed that, under the same assumptions as used in Section 3.3.1.1 (weak line limit, no surface inhomogeneities), the moment of order n of the Stokes parameter X about the line's centre of gravity λ_I

$$R_X^{(n)}(\lambda_I) = \frac{1}{W_\lambda} \int r_{\mathcal{F}_X}(\lambda - \lambda_I)(\lambda - \lambda_I)^n d\lambda \tag{3.26}$$

can be expressed as a linear combination of moments of different orders m about the plane defined by the stellar rotational axis and the line of sight of products of the local field modulus $B(x, y)$ and the field components $B_j(x, y)$ parallel and perpendicular to the line of sight (z-axis)

$$\langle x^m B^\ell B_j^k \rangle = \frac{3}{2\pi} \iint x^m B^\ell(x, y) B_j^k(x - y) \sqrt{1 - x^2 - y^2} dx dy. \tag{3.27}$$

Although formulated for an arbitrary moment of any Stokes parameter, practical applications of the moment technique are limited to the following.

(i) The first moment of Stokes V, $R_V^{(1)}(\lambda_I)$, which gives the mean longitudinal magnetic field $\langle B_z \rangle$ as discussed above.

(ii) The second moment of Stokes V, $R_V^{(2)}(\lambda_I)$, which characterises the difference in width of the RCP and LCP spectra arising due to a combination of the rotational Doppler shifts and an asymmetric distribution of the B_z field component relative to the central stellar meridian. This so-called crossover effect gives a measure of the moment $\langle x B_z \rangle$ (Mathys, 1995a) and is particularly useful for quantifying symmetric Stokes V profiles similar to those shown in the left panel of Fig. 3.6.

(iii) The second moment of Stokes I, $R_I^{(2)}(\lambda_I)$, which characterises the Zeeman broadening of spectral lines in the intensity spectra and allows us to derive the mean quadratic field $\langle B_z^2 \rangle + \langle B^2 \rangle$ (Mathys, 1995b).

The moment measurements are accomplished by determining $R_X^{(n)}(\lambda_I)$ for a set of unblended spectral lines and performing multiple linear regressions of these measurements against atomic line parameters. Of the three moments, the mean quadratic field is the least reliable as it requires disentangling magnetic effects from several other broadening mechanisms using a procedure that has to be fine-tuned for individual stars and ions (Mathys & Hubrig, 2006).

In general, the moment technique is only effective in the situations when the stellar surface magnetic field distribution is dominated by a large-scale component, yielding informative low-order moments according to Eq. (3.27). Accordingly, this magnetic diagnostic method is primarily used for early-type stars with globally organised dipolar-like fields.

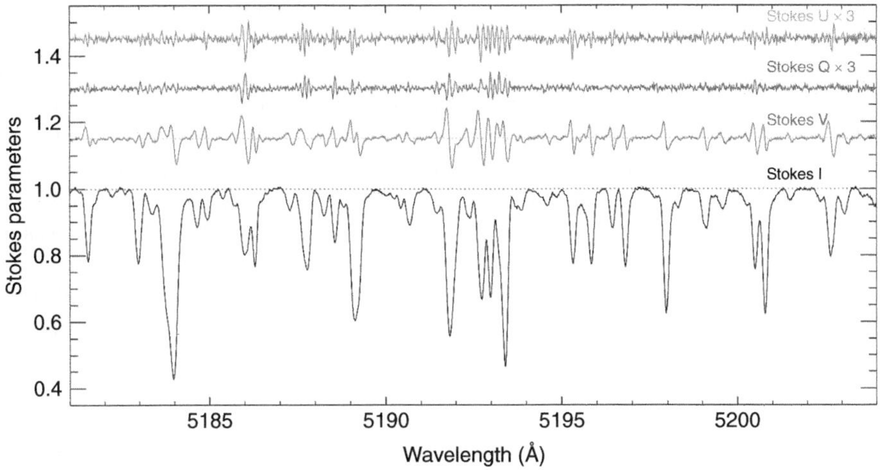

FIGURE 3.9. Normalised four Stokes parameter spectra of the cool magnetic A star HD 24712 (an object with a moderately strong dipolar magnetic field with $\langle B \rangle \approx 3.5$ kG). In this plot the polarisation profiles are offset vertically; the Stokes Q and U spectra are enhanced by a factor of 3 relative to Stokes I and V. These spectra, collected with the HARPS polarimeter, represent the highest quality full Stokes vector data that is currently possible to obtain for stars other than the Sun. Adapted from Rusomarov et al. (2013). Reproduced with permission © ESO.

3.3.1.4 Broad-Band Linear Polarisation

Compared with a multitude of techniques of extracting information about stellar magnetic fields from the Stokes V data at different resolving powers, few methods, short of direct Stokes profile modelling, exist for the analysis of linear polarisation. In a very special physical setting, requiring strong magnetic fields and deep spectral lines, a differential saturation of the π and σ components can lead to a non-zero net linear polarisation signal when integrating over individual spectral lines or over a wide wavelength region (Landi Deglinnocenti et al., 1981). A number of cool magnetic A stars exhibit this broad-band linear polarisation phenomenon (BBLP) that is amenable to photopolarimetric measurements in broad-band filters (Leroy, 1995). The stellar broad-band Q/I and U/I signals are of the order of 0.1%. The periodic variability of the BBLP observables can provide complementary information about the transverse components of stellar magnetic field, helping to constrain the large-scale field topology models (Landolfi et al., 1993).

3.3.1.5 High-Resolution Stokes Parameter Spectra

Direct observation and interpretation of the spectrally resolved circular and linear polarisation signatures in spectral lines is the ultimate methodology of studying stellar magnetic fields. Several wide wavelength coverage, thermally stabilised high-dispersion spectropolarimeters (Narval at 2 m Pic du Midi telescope, ESPaDOnS at 3.6 m CFHT and HARPSpol at 3.6 m ESO telescope) currently in operation can record spectra in all four Stokes parameters with a S/N and resolution sufficient to examine the Stokes signatures of individual spectral lines (see Fig. 3.9). Such observations are now routinely possible for intermediate mass stars with kG-strength fields down to magnitudes 7–8 mag for the full Stokes vector magnetometry (Wade et al., 2000; Silvester et al., 2012) and down to 11–12 mag for the spectra limited to Stokes I and V (Wade et al., 2012b).

However, such strongly magnetic stars are rather uncommon and there is a considerable interest in studying less extreme objects with much weaker fields, e.g. field strengths of ~1 G comparable to the global field of the Sun. Such fields do not produce detectable polarisation signatures in individual spectral lines, even for observations reaching $S/N \sim 10^3$. To overcome this limitation, one can take advantage of a wide wavelength coverage of typical night-time spectropolarimeters and try to combine information from many spectral lines. Indeed, according to the discussion in Section 3.2.6.2, in the weak line and weak field limits all intensity and polarisation line profiles should have approximately the same shape and only differ in strength according to the line's central intensity, wavelength, and the magnetic sensitivity parameter $\bar{g}$ or $\bar{G}$. The process of disk integration, being a linear addition of the Doppler-shifted Stokes profiles corresponding to zones with different magnetic field parameters, acts in the same way on all the lines and thus does not hinder the line averaging. Following this line of reasoning, Donati et al. (1997) developed the least squares deconvolution (LSD) technique for extracting highly precise mean Stokes V signatures from polarisation observations with a moderate S/N. This technique was subsequently extended to all four Stokes parameters by Wade et al. (2000) and Kochukhov et al. (2010), and characterised in detail using numerical simulations in the latter study.

LSD aims to represent the entire stellar spectrum as a linear superposition of scaled mean profiles

$$\begin{aligned} I(v) &= 1 - \sum_i w_I^i Z_I(v - v^i), & w_I^i &= d_i, \\ V(v) &= \sum_i w_V^i Z_V(v - v^i), & w_V^i &= \bar{g}\lambda_i d_i, \\ Q(v) &= \sum_i w_Q^i Z_Q(v - v^i), & w_Q^i &= \bar{G}\lambda_i^2 d_i, \end{aligned} \tag{3.28}$$

where v^i are positions of individual spectral lines in the velocity space and w_X^i are weights appropriate in the weak field and weak line limits. Each of Eqs. (3.28) can be equivalently formulated as a convolution of the mean profile Z_X and the line pattern consisting of δ-functions centred at individual lines, or as a matrix multiplication

$$\mathbf{X} = \mathbf{M} \cdot \mathbf{Z}_X, \tag{3.29}$$

where $\mathbf{M}$ is the line pattern matrix. When working with actual observations, the left-hand side of this equation is the observed intensity or polarisation spectrum $X_{\rm obs}$ with associated errors $\sigma_{\rm obs}$ and $\mathbf{Z}_X$ is the mean profile that is sought for. The corresponding least squares problem

$$\chi^2 = (\mathbf{X}_{\rm obs} - \mathbf{M} \cdot \mathbf{Z}_X)^{\rm T} \cdot \mathbf{E}_{\rm obs}^2 \cdot (\mathbf{X}_{\rm obs} - \mathbf{M} \cdot \mathbf{Z}_X) \rightarrow \min \tag{3.30}$$

has a solution

$$\mathbf{Z}_X = (\mathbf{M}^{\rm T} \cdot \mathbf{E}_{\rm obs}^2 \cdot \mathbf{M})^{-1} \cdot \mathbf{M}^{\rm T} \cdot \mathbf{E}_{\rm obs}^2 \cdot \mathbf{X}_{\rm obs}, \tag{3.31}$$

where $\mathbf{E}_{\rm obs}$ is a diagonal matrix containing the inverse of observational errors, $1/\sigma_{\rm obs}$.

LSD proved to be a conceptually simple but remarkably powerful method of detecting stellar magnetic fields. Owing to a large number (from a few hundred to several thousand) of spectral lines recorded simultaneously by modern optical spectropolarimeters, the line-addition method can achieve the S/N gain of 10–50 and yield secure detections of polarisation signatures with amplitudes as low as $\sim 10^{-5}$ (Lignières et al., 2009; Petit et al., 2011). Unlike the longitudinal field diagnostic, which cannot recognise the presence of complex fields that give no net longitudinal field component, the LSD profiles are sensitive to fields at different spatial scales. Consequently, this method can be used both

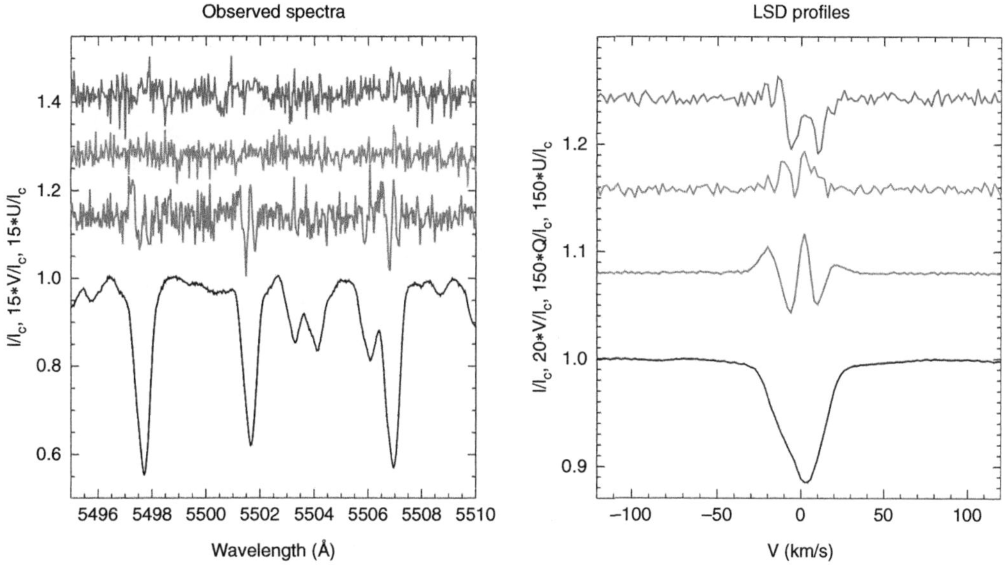

FIGURE 3.10. Example of applying the LSD line-addition procedure to the four Stokes parameter spectra of the active cool star II Peg. *Left panel:* a segment of spectra centred at three Fe I lines with large Landé factors. Polarisation signatures are marginally detected in Stokes V. *Right panel:* LSD profiles extracted from the entire optical spectrum. Both circular and linear polarisation is clearly detected. Adapted from Rosén et al. (2013).

for early-type stars with globally organised fields and for late-type stars with much more complex field topologies. For example, recently LSD was used to obtain the first full Stokes vector observations of late-type active stars with relatively weak and complex fields (see Fig. 3.10; Kochukhov et al., 2011b; Rosén et al., 2013).

Beyond a mere detection of the Zeeman polarisation signatures, the LSD profiles can be used for derivation of the mean longitudinal magnetic field, as explained in Section 3.3.1.1, and net linear polarisation equivalent to the BBLP measurements (e.g. Silvester et al., 2012). $\langle B_z \rangle$ sensitivity of 0.5–1.0 G can be easily achieved with LSD for slowly rotating bright stars with rich absorption spectra. However, detailed quantitative modelling of the LSD profile shapes is less straightforward because it is not immediately obvious that LSD spectra respond to magnetic field and other local parameters in the same way as a single spectral line with mean parameters. This problem was addressed by Kochukhov et al. (2010), who showed that the single-line approximation of the LSD profiles is reasonably accurate for Stokes I and V up to the field strength of $\approx$2 kG, but is not applicable to Stokes QU at all. Nevertheless, any LSD profile can in principle be modelled with a direct, albeit computationally expensive, approach by calculating the full synthetic Stokes spectra and applying to them the same line-averaging procedure as used to treat the observations (Kochukhov & Sudnik, 2013).

3.3.2 *Techniques to Reconstruct Stellar Magnetic Field Topologies*

3.3.2.1 Multipolar Field Modelling

Fitting the phase curves of integral magnetic observables (the mean field modulus and the three quantities provided by the moment technique) with a low-order multipolar field geometry is the basic approach to characterising globally organised stellar magnetic

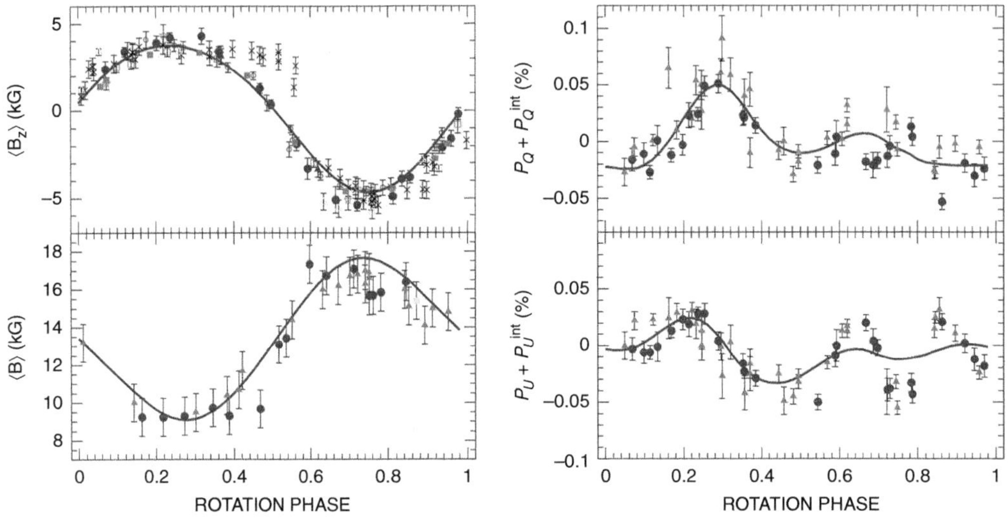

FIGURE 3.11. Multipolar fits to the integral magnetic observables for the A-type star 53 Cam. In this case, a non-axisymmetric dipole-quadrupole model is fitted to the phase curves of $\langle B \rangle$, $\langle B_z \rangle$, and BBLP observations. From Bagnulo et al. (2001). Reproduced with permission © ESO.

fields. Assuming a linear limb-darkening law

$$I(\mu)/I_0 = (1 - u + u\mu) \tag{3.32}$$

with the limb-darkening coefficient u, it is possible to establish (e.g. Leroy et al., 1994) that, for rotating star with an oblique dipolar field with the polar strength $B_{\rm p}$, variability of the mean longitudinal field and the mean field modulus is given by

$$\langle B_z \rangle = B_{\rm p} \frac{15 + u}{20(3 - u)} \cos\alpha \tag{3.33}$$

and

$$\langle B \rangle = B_{\rm p} \frac{3}{3 - u} \left[(0.778 - 0.226u) \cos^2\alpha + (0.648 - 0.234u) \sin^2\alpha \right]. \tag{3.34}$$

Here the angle α between the dipolar axis and the line of sight changes with phase φ according to

$$\cos\alpha = \cos i \cos\beta + \sin i \sin\beta \cos\varphi, \tag{3.35}$$

where β is the magnetic field obliquity with respect to the stellar rotational axis and i is the angle between the rotational axis and the line of sight.

These equations can be used to establish $B_{\rm p}$, β, and i from a time-series of $\langle B \rangle$ and $\langle B_{\rm z} \rangle$ measurements. More complex dipole–quadrupole–octupole field configurations can be explored using simultaneous fits to $\langle B \rangle$, $\langle B_{\rm z} \rangle$, $\langle xB_{\rm z} \rangle$, and $\langle B_{\rm z}^2 \rangle + \langle B^2 \rangle$ curves (Landstreet & Mathys, 2000; Bagnulo et al., 2002a). Some of these analyses also included fits to the BBLP measurements (see Fig. 3.11 and Bagnulo et al., 2001). A number of recent studies (Alecian et al., 2008; Grunhut et al., 2012) applied dipolar fits directly to the LSD Stokes V profiles instead of using the intermediate integral magnetic observables.

Multipolar fitting is a convenient method to study magnetic stars using a few simple magnetic observables. It is indispensable for large-scale statistical studies and is robust when a stellar field is topologically similar to a dipolar configuration. However, it does not allow characterising deviations from the assumed field geometry and becomes increasingly unreliable and meaningless for the fields structured on smaller scales. A diversity of the low-order multipolar field parameterisations suggested in the literature adds to the confusion. There are no objective criteria to decide when to employ, for example, an offset dipole vs an aligned axisymmetric dipole–quadrupole–octupole vs non-axisymmetric combination of dipole and quadrupole. Using different model geometries to interpret the same observations often leads to discrepant surface magnetic maps (Kochukhov, 2006) and generally does not guarantee that a model fitting the phase curves of integral observables reproduces the Stokes profiles themselves (Bagnulo et al., 2001; Kochukhov et al., 2011a).

3.3.2.2 Zeeman Doppler Imaging

In general, interpretation of the Stokes parameter signatures in stellar spectral lines requires modelling unresolved surface magnetic field distributions. For stars with intrinsically simple and stable field topologies this can be accomplished by using the moment technique and multipolar model fitting, as discussed above. However, these methods are ineffective for stars with complex fields, especially for late-type active stars for which most of the field energy is concentrated on smaller scales. In that case an indirect imaging of stellar surface, also known as Doppler imaging (DI) and Zeeman/Magnetic Doppler imaging (ZDI), is the only technique capable of extracting quantitative information about a stellar magnetic field.

Doppler imaging takes advantage of the spatial resolution of the stellar surface provided by the Doppler effect. If the rotational Doppler broadening dominates line profiles, a contribution from the zone at specific longitude at the stellar surface is limited to a certain range of velocities within the disk integrated line profile. Any inhomogeneity at the stellar surface, e.g. a cool spot, produces a distortion that moves across the line profile as the star rotates (see Fig. 3.12a). Using a time series of the line profile recordings, it is possible to reconstruct a two-dimensional map of the stellar surface in terms of some parameter (brightness, temperature, chemical abundance, magnetic field).

An essential aspect of DI is that it is formally an ill-posed mathematical problem, i.e. an infinite number of solutions can provide an acceptable fit to a given data set. One has to introduce an additional constraint, regularisation, to ensure uniqueness of the solution. The two most popular regularisation strategies are the maximum entropy (selects a solution with the least deviation from the mean) and Tikhonov regularisation (favours solutions with the maximum local smoothness). Both methods are implemented by adding a penalty function to the χ^2 of the fit to observations.

Doppler imaging was originally introduced in the context of studies of chemical spots in A-type stars (e.g. Khokhlova et al., 1986). The technique was subsequently extended to reconstruction of the temperature inhomogeneities on late-type stars (Vogt et al., 1987). A procedure to map the vector magnetic field using circular polarisation was developed by Brown et al. (1991). It was generalised to all four Stokes parameters by Donati & Brown (1997) using simplified analytical treatment of the Stokes spectra and by Piskunov & Kochukhov (2002) using numerical line profile calculations with realistic model atmospheres.

Figure 3.12b,c shows examples of the Stokes V signatures of a magnetic spot with the radial and meridional magnetic field. It is evident that, similar to temperature inhomogeneities, the velocity shift relative to the line centre encodes the longitude of a spot. The latitude position of a surface feature and orientation of the field within it can be discerned

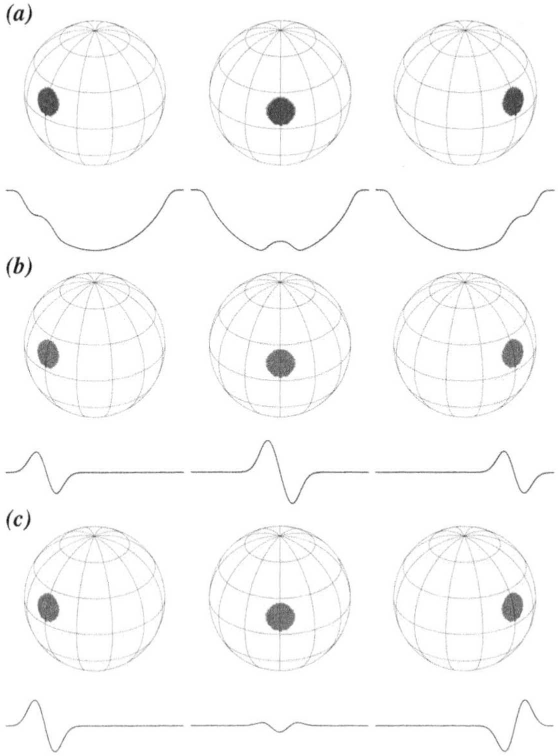

FIGURE 3.12. Signatures of starspots in the intensity and circular polarisation line profiles of a rotating star. (*a*) Variability of Stokes I due to a cool spot; (*b*) radial field spot signature in Stokes V; (*c*) azimuthal field spot signature in Stokes V. In each case the star is shown at three rotational phases, separated by 0.125 of the rotational period.

from the temporal variation of the polarisation signatures. For example, Fig. 3.12 shows that the Stokes V signature corresponding to a radial field spot maintains its sign and reaches maximum amplitude when the spot is in the centre of the stellar disk. In contrast, the signature of an azimuthal field spot changes sign when the spot crosses the disk centre.

It is worthwhile to note that much of the information extracted by ZDI comes from the rotational modulation of the Stokes signatures. This effect leads to a significant variability of the polarisation spectra even when rotational Doppler broadening is negligible. Therefore, it is possible to apply the vector ZDI to much slower rotators than the stars that can be studied with the conventional temperature/abundance scalar DI methods.

A reliable reconstruction of the full vector magnetic field map requires observations in all four Stokes parameters (Kochukhov & Piskunov, 2002; Rosén & Kochukhov, 2012). With the circular polarisation data alone the radial and meridional field components cannot be distinguished for a large fraction of the stellar surface. It is also essential to account for the influence of temperature/chemical composition starspots when reconstructing the field topology and vice versa. In other words, DI with Stokes I and ZDI with polarisation spectra should be carried out simultaneously and in a self-consistent manner – an aspect still neglected by many ZDI studies.

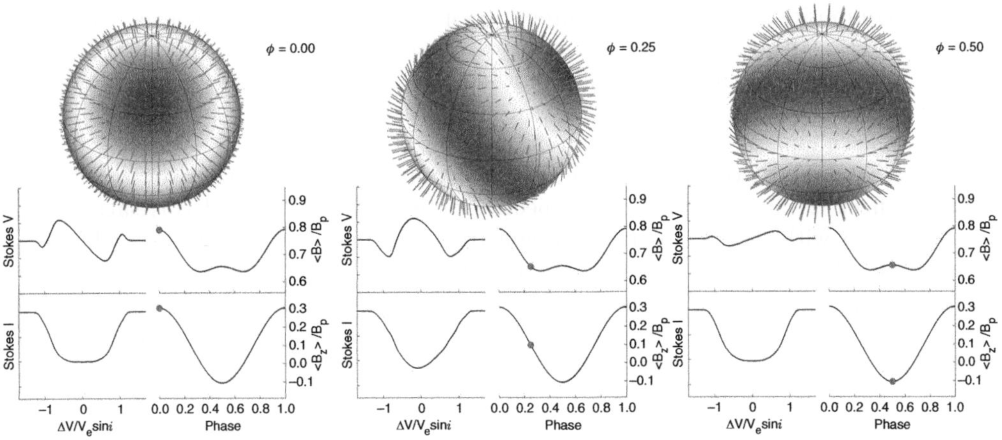

FIGURE 3.13. Oblique rotator variation of a star with dipolar field. The spherical plots show magnetic field distribution for three different rotational phases, with the underlying image corresponding to the field modulus and the vector map showing the field orientation. The Stokes I and V profiles and the corresponding $\langle B \rangle$, $\langle B_z \rangle$ parameters are shown below. These calculations are for 4 kG dipolar field inclined by $\beta = 50^{\circ}$ with respect to the stellar rotational axis.

3.4 Stellar Magnetism Across the Hertzsprung–Russell Diagram

3.4.1 *Intermediate-Mass Stars*

Intermediate-mass stars, defined here as B, A and F stars in the mass range 1.5–8 $M_{\odot}$, were the first objects outside the solar system for which evidence of magnetic fields was found from the circular polarisation observations (Babcock, 1958) and from the Zeeman splitting of absorption lines in the intensity spectra (Babcock, 1960). These observations, initially limited to $\langle B_z \rangle$ and $\langle B \rangle$ measurements, indicated the presence of fields with strengths from a few hundred G up to 34 kG covering the entire stellar surface. It was also recognised that these surprisingly strong fields always occur in the A and B stars with anomalously intense spectral line of heavy elements, known as the upper main sequence chemically peculiar (CP) stars or Ap/Bp stars.

Magnetic CP stars comprise a small fraction of all A and B stars; depending on the spectral type, they represent from a few percent up to 10–15% (Catalano & Leone, 1994) of all intermediate-mass stars. Besides having abnormal absorption spectra, these stars are characterised by a slow rotation compared to normal A and B stars. Ap stars also often show a remarkable periodic variability of brightness, line profile intensities, spectral energy distribution, and magnetic field with a common period. This periodicity typically lies in the range of 1–10 d, although some objects exhibit periods of the order of years and decades. The variability period clearly anti-correlates with the rotational Doppler broadening of spectral lines, suggesting that the observed variation is caused by an inhomogeneous surface structure (in this case, chemical spots) of a rotating star. In this context, the magnetic variability, typically seen as smooth, sinusoidal $\langle B_z \rangle$ changes, is ascribed to a non-axisymmetric, roughly dipolar magnetic field frozen in the stellar plasma and rotating with the star (see Fig. 3.13).

This "oblique rotator model" (Stibbs, 1950) provided a successful empirical framework for the interpretation of variability of magnetic Ap stars. Later, development of the radiative diffusion theory established a plausible connection between a non-uniform surface distribution of chemical elements and the presence of a magnetic field (Michaud, 1970; Michaud et al., 1981). It was understood that, on average, Ap stars have roughly solar chemical composition. However, the outer envelopes of these stars lack convective mixing and are additionally stabilised by the strong magnetic field. These conditions facilitate a vertical segregation of chemical elements owing to an interplay of the radiative pressure and gravitational settling. Additionally, magnetic field breaks the spherical symmetry, leading to accumulation of elements in different regions at the stellar surface.

A classical analysis of the magnetic field structure in Ap stars consists in detecting Zeeman effect in circularly polarised spectra, determining $\langle B_z \rangle$, and correlating these measurements with the variability of the spectroscopic and photometric observables (e.g. Leone et al., 2000). Parameters of the dipolar field geometry, B_p and β, are determined by fitting Eq. (3.33) to $\langle B_z \rangle$ measurements. An example of such analysis can be found in Borra & Landstreet (1980). Attempts to interpret simultaneously several integral magnetic observables, e.g. those provided by the moment technique, typically require introducing some deviation from the purely dipolar field geometry (Landstreet & Mathys, 2000; Bagnulo et al., 2002a). However, with a few exceptions, these studies still find that the dipolar component dominates the field geometry. Interestingly, there appears to be a statistically significant relation between the dipolar field obliquity β and the stellar rotation period: the fast rotating stars ($P_{\rm rot} < 25$ d) tend to have large βs while the majority of slow rotators show dipolar fields nearly aligned with the stellar rotational axes.

Recently, studies using the LSD technique have reached sensitivity to the magnetic fields significantly weaker than 100 G, finding that dipolar field components of all known magnetic Ap stars are stronger than $\approx$300 G (Aurière et al., 2007). It was suggested that fields weaker than this limit cannot withstand shearing by the differential rotation in the stellar interiors. Nevertheless, very high precision spectropolarimetric observations of bright normal stars, Vega and Sirius, allowed to discover sub-G magnetic fields on the surfaces of these stars (Lignières et al., 2009; Petit et al., 2011). Follow-up observations did not reveal variability of Vega's field topology over a time span of one year (Petit et al., 2010). Thus, A and B stars may possess two types of stable magnetic fields with very different typical strengths and probably different physical origins.

Application of the abundance DI to Ap stars, initiated in the 1970s (Goncharskii et al., 1977), nowadays includes reconstruction of multiple abundance maps from high-resolution spectra (Lueftinger et al., 2003; Kochukhov et al., 2004b; Nesvacil et al., 2012). Somewhat surprisingly, these studies uncovered considerably more complex and diverse abundance maps than anticipated by the theoretical diffusion calculations (e.g. Alecian & Stift, 2010). The latter predict accumulation of elements primarily in the horizontal field regions, i.e. at the equator of a dipolar field. Contrary to this hypothesis, some elements are observed to accumulate at the magnetic poles, others at the magnetic equators, and some elements exhibit no apparent relation to the field topology whatsoever (see Fig. 3.14). Evidently, some key physical ingredients are missing in the current theoretical diffusion models. The discovery of chemical spots in non-magnetic HgMn stars (Adelman et al., 2002; Kochukhov et al., 2005) points in the same direction.

An extension of the Ap-star Doppler imaging to magnetic field mapping with high-resolution polarisation spectra (Piskunov & Kochukhov, 2002) opened new possibilities for styling magnetic field topologies of these stars. First, ZDI enables physically realistic modelling of the Stokes profiles fully taking into account all aspects of the Zeeman effect and the presence of chemical inhomogeneities. Second, Kochukhov & Piskunov (2002) showed that ZDI inversions based on the spectra in all four Stokes parameters are able to recover the field distribution making no a priori assumptions about the global magnetic topology. Taking advantage of these key improvements of the magnetic

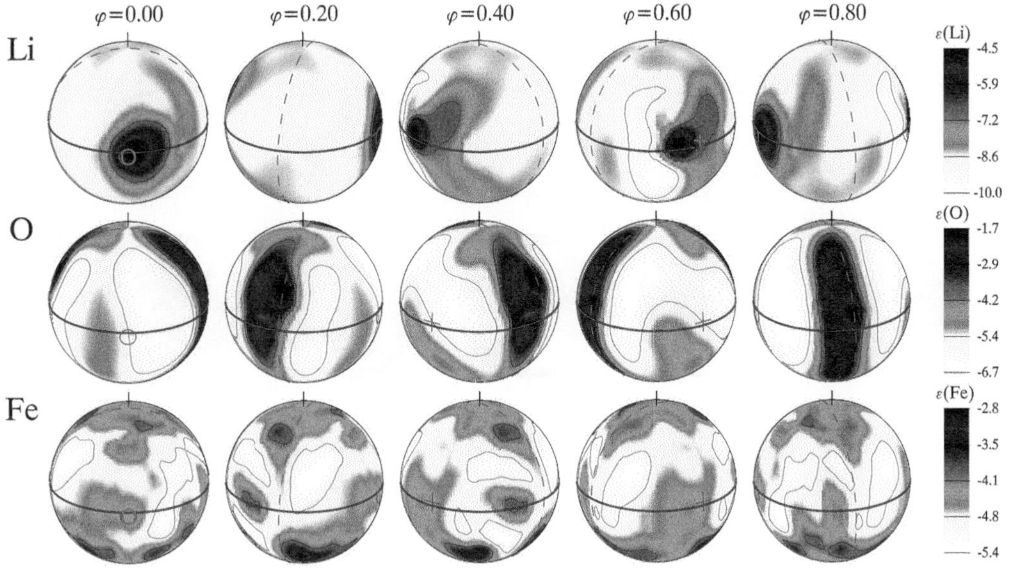

FIGURE 3.14. Typical chemical abundance distributions reconstructed for a magnetic Ap star using Doppler imaging. In this case, the maps of Li, O and Fe are shown for the cool Ap star HD 83368. The scale bars at the right give abundance values in logarithmic units $\log(N_{\rm el}/N_{\rm tot})$. The magnetic field structure of this star is approximately dipolar, with $\beta \approx 90^{\circ}$. The '+' and 'o' symbols indicate positions of the positive and negative magnetic poles, while the dashed line corresponds to the magnetic equator. The thick line and the vertical bar show the rotational equator and pole, respectively. Adapted from Kochukhov et al. (2004b). Reproduced with permission © ESO.

modelling methodology and using the first high-resolution full Stokes vector stellar observations by Wade et al. (2000), Kochukhov et al. (2004a) and Kochukhov & Wade (2010) carried out the first four Stokes parameter inversions for a couple of bright Ap stars. These studies revealed that, when the Stokes Q and U data are included in the reconstruction, significant deviations from a dipolar field topology are necessary to fit the observations. As illustrated by Fig. 3.15, these deviations do not resemble quadrupole or octupole components invoked by multipolar fits to the phase curves of integral magnetic observables. Instead, the overall radial field distribution remains roughly dipolar, but some small-scale magnetic features appear in the horizontal field maps. A new generation of the full Stokes vector observations of Ap stars currently underway (Silvester et al., 2012; Rusomarov et al., 2013) will be used to elucidate how widespread are these small-scale magnetic structures and how their incidence changes with stellar parameters.

Given the stability of the magnetic fields in Ap stars, as judged by repeatability of $\langle B_{\rm z} \rangle$ measurements, the lack of a correlation between the field strength and the stellar rotation rate, and the presence of the field in only a small fraction of stars, it is rather unlikely that the field is generated by a contemporary dynamo mechanism in the stellar interior, similar to the one operating in the Sun and other cool stars. Instead, the magnetism of early-type stars is attributed to the so-called 'fossil' magnetic field, i.e. the field acquired by the star at a previous evolutionary stage. The absence of an appreciable field evolution at the main sequence (Kochukhov & Bagnulo, 2006) indicates that Ap stars become magnetic early in their life. The exact origin of the fossil fields is unclear; it may be related to a dynamo action at the pre-main sequence Hyashi stage, produced by the stellar mergers,

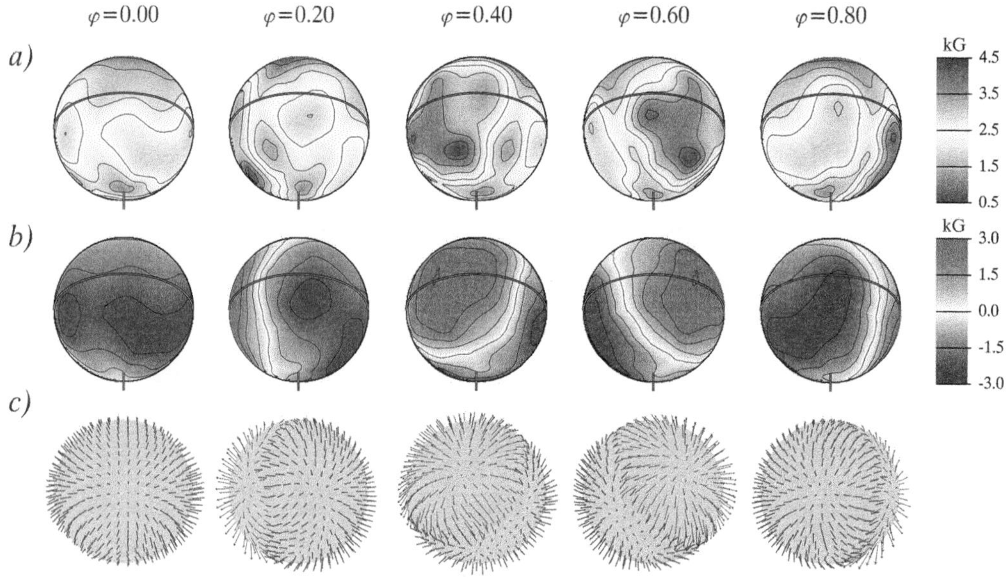

FIGURE 3.15. Magnetic field map of the Ap star α^2 CVn reconstructed with the ZDI analysis of the four Stokes parameter spectra. The spherical plots show distributions of (*a*) the field modulus, (*b*) radial field, and (*c*) field orientation. Adapted from Kochukhov & Wade (2010). Reproduced with permission © ESO.

or may simply represent a strong-field tail of the magnetic field distribution existing in proto-stellar clouds. Detailed hydrodynamic calculations and observations of magnetism in young early-type stars are required to distinguish between these different theoretical possibilities.

Despite the continuing uncertainty about the origin of fossil fields, the problem of their long-term stability has been successfully addressed by the 3-D magneto-hydrodynamical simulations (Braithwaite & Nordlund, 2006) and subsequent analytical studies (Duez & Mathis, 2010). It has been shown that, to retain dynamical stability, the global field must have a mixed poloidal–toroidal configuration in the stellar interior and that stable configurations more complex than a dipolar geometry are possible (Braithwaite, 2008). These calculations, however, have so far neglected stellar rotation and used computational grids that were too coarse to resolve the outer stellar layers where the magnetic field properties are actually constrained by observations.

Another interesting manifestation of the magnetism in intermediate-mass stars is the presence of global magneto-acoustic pulsations. The rapidly oscillating Ap (roAp) stars pulsate in high-overtone non-radial *p*-modes with periods 6–24 min, which are excited and shaped by the processes closely related to the presence of global field (Bigot et al., 2000; Saio & Gautschy, 2004; Kochukhov, 2004). Differently from all other pulsating stars, non-radial modes in roAp stars are aligned not with the stellar rotational axis, but with the magnetic field axis, leading to a characteristic rotational modulation of the phase and amplitude of oscillations (Kurtz, 1982). Asteroseismology of the magnetic A-type stars supplies rich information about the fundamental stellar parameters and atmospheric structure (Ryabchikova et al., 2007; Saio et al., 2012) and so far provides the only possibility to observationally constrain the interior field characteristics (Cunha et al., 2003).

3.4.2 *Massive Stars*

It was know for a long time that the phenomenon of fossil stellar magnetism extends at least to B2–B3 spectral types (Bohlender et al., 1987). The presence of fields in these stars typically correlates with an abnormal He abundance (He-weak and He-rich stars). At the same time, until a few years ago, little was known about the magnetic properties of OB stars with $M > 8M_{\odot}$ – the stars that end their life as core-collapse supernovae. A series of individual studies (e.g. Donati et al., 2006a,b) and, especially, the recently completed MiMeS ('Magnetism in Massive Stars') survey (Wade et al., 2012a) have significantly expanded our knowledge about magnetic fields of massive stars. These investigations demonstrated that about 7% of early-B and O-type stars have magnetic fields and that, for the majority of these objects, the observed field characteristics, including typical strengths, geometries, and stability, are very similar to those of the intermediate-mass Ap/Bp stars. Studies of large stellar samples also ascertained that pulsating SPB and β Cep stars are not distinguished by their magnetic properties (Silvester et al., 2009). None of more than 50 studied B stars with emission lines (classical Be stars) was found to be magnetic. On the other hand, a certain rare class of O stars with peculiar emission lines (Of?p stars) all turned out to be magnetic (Donati et al., 2006a; Wade et al., 2012b).

Interestingly, several early-B stars were found to possess profoundly non-dipolar field topologies (Donati et al., 2006b; Kochukhov et al., 2011a), such as the one illustrated in Fig. 3.16 for the He-peculiar B2 star HD 37776. The complex nature of these fields, not seen in lower mass B and A stars, is evident without any modelling from a non-sinusoidal longitudinal field curve and a complex morphology of the Stokes V profiles.

In addition to strong global magnetic fields in a handful of OB stars, weak complex magnetic fields could potentially exist on the surfaces of many more objects. Stellar evolutionary calculations with updated opacities suggested the presence of a sub-surface iron convection zone in massive stars (Cantiello et al., 2009). It is reasonable to suppose that convectively driven waves overshoot into the optically thin layers, leading to surface inhomogeneities and turbulent line broadening. The resulting non-uniform surface structure offers an attractive explanation of the stochastic photometric variability of massive stars (Belkacem et al., 2009) and ubiquitous observations of the discrete absorption components (DACs) in their winds (Cranmer & Owocki, 1996). The turbulent dynamo mechanism naturally generates magnetic fields in the sub-surface convection zone. These fields may appear at the surface (Cantiello et al., 2009) and could potentially be detected using modern spectropolarimetric data (Kochukhov & Sudnik, 2013).

Mass loss is the dominant process in the outer envelopes of massive stars. The wind outflow quickly removes any chemically segregated material, rendering radiative diffusion ineffective. As a result, no chemical anomalies or starspots are found on the surfaces of magnetic massive stars. However, the stellar wind itself interacts strongly with the magnetic field. Close to the stellar surface the field directs the outflow from the polar regions towards the magnetic equator; here the flows from the two magnetic hemispheres collide, producing an X-ray emission. Further from the star the mass loss takes over the magnetic field, opening the field lines.

There are many observational manifestations of this magnetically confined mass loss scenario (Babel & Montmerle, 1997). Besides explaining some features of the short-wavelength emission of massive magnetic stars, it predicts the existence of torus-like (for small magnetic obliquities) or cloud-like (large magnetic obliquities) regions where the material lost in the wind can accumulate (e.g. Townsend & Owocki, 2005), forming stable or dynamically evolving magnetospheric structure. Ud-Doula & Owocki (2002) suggested a useful classification of the massive star magnetospheres according to the

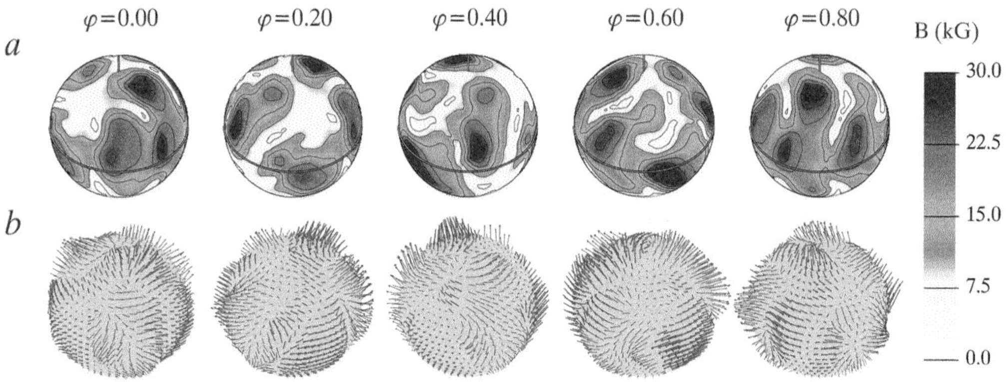

FIGURE 3.16. Unusually complex magnetic field of the B-type star HD 37776 reconstructed with ZDI. The spherical maps show (*a*) surface distributions of the field modulus and (*b*) field orientation. From Kochukhov et al. (2011a). Reproduced with permission © AAS.

balance between the magnetic field energy and kinetic energy of the wind. They defined the so-called magnetic confinement parameter

$$\eta_\star = \frac{B_{\rm eq}^2 R^2}{\dot{M} V_\infty}, \tag{3.36}$$

where $B_{\rm eq}$ is the field strength at the magnetic equator, R is the stellar radius, $\dot{M}$ is the mass-loss rate, and V_∞ is the terminal wind speed. The confinement parameter is related to the Alfvén radius R_A

$$R_A = \eta_\star^{1/4} R. \tag{3.37}$$

On the other hand, the centrifugal support parameter (the ratio of the rotational speed $V_{\rm rot}$ to the orbital speed at the surface of radius R)

$$W = \frac{V_{\rm rot}}{\sqrt{GM/R}} \tag{3.38}$$

is related to the Keplerian radius

$$R_K = W^{-2/3} R. \tag{3.39}$$

Thus, the two parameters $\eta_\star$ and W define the relative location of the Alfvén and Keplerian radii. As schematically illustrated in the left panel of Fig. 3.17, the wind material accumulates in a stable structure ('centrifugal magnetosphere') in the region between R_K and R_A if $R_K < R_A$. Closer to the star and for the entire magnetosphere with $R_K > R_A$ no stable structures can occur, leading to the 'dynamical magnetosphere' behaviour. Petit et al. (2013) classified all known magnetic massive stars in terms of R_A and R_K.

Magnetic channelling of the stellar wind explains periodic variability of the wind-sensitive UV emission lines (Henrichs et al., 2013) – an observable now understood to be a useful proxy of the presence of a global field in a massive star. Furthermore, the presence of the magnetospheric material can be recognised by the variable Doppler-shifted emission in the strong optical lines, such as the hydrogen Hα and He I (see example in the right panel of Fig. 3.17). The density and spatial distribution of these clouds provide a novel diagnostics of the stellar wind and circumstellar magnetic field geometry (Townsend et al., 2005; Sundqvist et al., 2012). Another observational signature of the

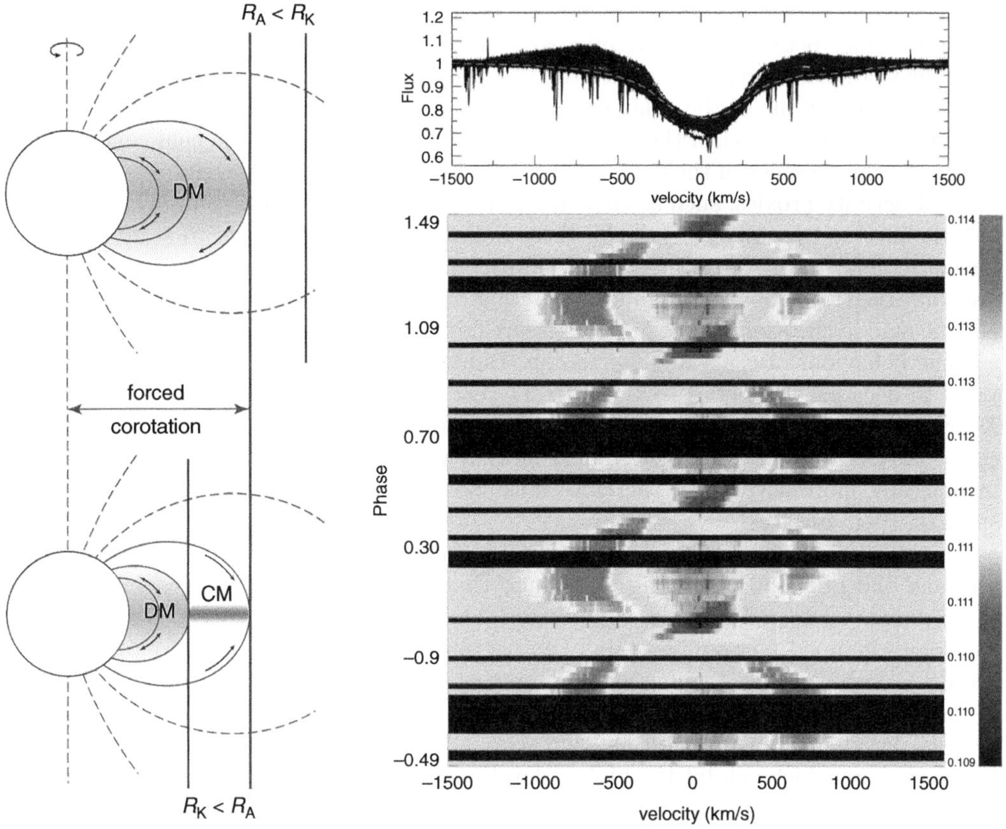

FIGURE 3.17. *Left panel:* schematic illustration of the dynamic (DM) and centrifugal regimes (CM) of massive star magnetospheres. From Petit et al. (2013).
Right panel: Hα profile variation of the magnetic star HR 7355 showing evidence of corotating magnetospheric clouds. Adapted from Rivinius et al. (2013).

magnetospheric material is the photometric variability caused by transits of corotating clouds across the stellar disk (Townsend, 2008).

Yet another consequence of the interaction between magnetic field and mass loss is a modification of the stellar angular momentum (Ud-Doula et al., 2009). Many magnetic massive stars have short rotation periods and very stable photometric light curves. This facilitates measuring rotation periods with a precision of better than 1 s over a time span of a few years and thus directly probing the rotation period evolution. Spin-down rates measured for several massive stars are reasonably consistent with theoretical predictions (Townsend et al., 2010), although the late-B rapidly rotating Bp star CU Vir shows enigmatic rotational period variations, which cannot be attributed to a simple linear period increase (Mikulášek et al., 2011).

3.4.3 *Solar-Type Stars*

Here the solar-type stars are defined very broadly as objects with sizeable outer convection zones in the mass range $0.6M_\odot \leq M \leq 1.5M_\odot$. This corresponds to main sequence and evolved (sub-giants and giants) stars with the spectral classes from mid-F to late-K. All these stars show qualitatively similar manifestations of the 'magnetic activity', believed to be a result of the dynamo process operating in the stellar interiors, either at the interface between the convective and radiative regions (tachocline) or throughout the convection zone itself. Thanks to the vast observational data available from spatially resolved observations of the solar surface, magnetic activity of solar-type stars is often considered in the context of the solar paradigm. That is, the magnetic field generation is cyclic and the presence of the field at the stellar surface invariably correlates with various indirect activity proxies, such as cool starspots, flares, enhanced emission in the chromospheric lines and X-ray emission. All these indicators are much easier to study than the magnetic field itself. Accordingly, even now nearly all 'magnetic activity' investigations are concerned with observing and interpreting one or another magnetic field proxy, but not the field itself. Along these lines of research, it was found that many cool stars exhibit a cyclic behaviour of e.g. Ca HK emission (Baliunas et al., 1995), showing periods from years to decades, analogous to the solar 11-year cycle. The average chromospheric and X-ray emission clearly anti-correlates with the stellar age and rotation period (Noyes et al., 1984; Wright et al., 2011), indicating that young and/or rapidly rotating stars are intrinsically more active and therefore must possess stronger fields or larger magnetic spots on their surfaces.

It was not until the end of 1990s, after the introduction of echelle spectropolarimeters and the LSD line-averaging technique, that the magnetic field signatures started to be directly detected for solar-type stars (Donati et al., 1997) and the magnetic field topologies mapped for some especially active objects, including primary components of the RS CVn binaries and young rapidly rotating dwarfs (Donati et al., 2003; Kochukhov et al., 2013b). These analyses have confirmed some expectations of the solar magnetic activity paradigm: the strongest fields, reaching 200–1000 G locally, were found precisely for those stars which exhibit the highest levels of the X-ray and chromospheric emission. At the same time, the field topology revealed by the magnetic inversions turned out to be quite different from what is observed for the Sun. Instead of bipolar spot groups with predominantly radial fields, the majority of solar-type active stars show strong azimuthal magnetic features, often arranged into rings encircling the star at different latitudes. These structures evolve on a time scale of a few months, sometimes changing sign. Figure 3.18 shows typical ZDI mapping results, in this case corresponding to the RS CVn variable star II Peg studied by Kochukhov et al. (2013b).

A connection between cool spots and magnetic fields is yet to be established observationally for cool stars other than the Sun. From solar observations we know that the regions with stronger surface magnetic fields have a lower temperature with respect to the surrounding photosphere due to magnetic inhibition of the convective energy transport. But this correlation is generally not seen in the ZDI maps of solar-type stars (Donati & Collier Cameron, 1997; Kochukhov et al., 2013b), even though simultaneous magnetic and temperature mapping should be sensitive to kG fields inside large monolithic cool spots (Rosén & Kochukhov, 2012).

During the past 10 years the stellar spectropolarimetric observations reached the sensitivity sufficient to probe not only the field topologies of very active rapidly rotating stars, but also to study global fields of the Sun-like dwarfs. In particular, systematic analyses of bright objects with typical local field strengths as low as 2–10 G are becoming possible. A series of studies of stars only slightly more active than the Sun detected polarity reversals of the global field (Fares et al., 2009; Morgenthaler et al., 2012). It remains unclear how these observations relate to the activity cycles measured using proxy indicators. Petit et al. (2008) have reconstructed magnetic field topologies for several G dwarfs with parameters very similar to the Sun. They found that the field topology changes drastically with the stellar rotation period even within this small and carefully defined

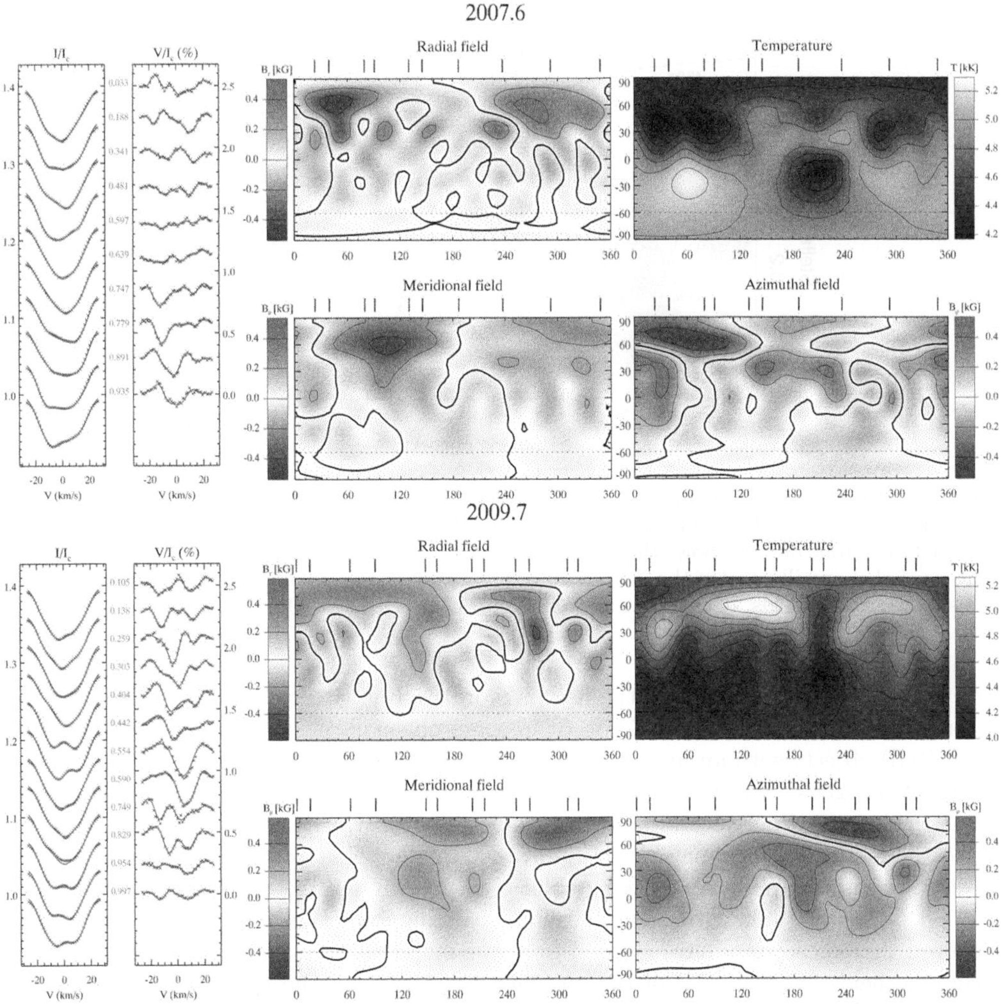

FIGURE 3.18. Simultaneous ZDI reconstruction of the vector magnetic field and temperature distribution for the RS CVn star II Peg at two different epochs. For each epoch the two columns on the left compare the observed (symbols) and theoretical (lines) LSD Stokes *I* and *V* profiles. The four rectangular maps illustrate distributions of the radial, meridional, and azimuthal field components, and temperature for each epoch. Significant evolution of the surface structure over a time span of two years is evident. From Kochukhov et al. (2013b). Reproduced with permission © ESO.

group of solar twins. The stars with periods above ≈15 d exhibit a dominant poloidal field while for the objects with shorter periods toroidal field components are stronger.

The overall statistics of the ZDI studies of cool active stars, including the results obtained for solar-type stars and for low-mass dwarfs discussed in the next section, is presented in Fig. 3.19. This plot clearly demonstrates the transition between different types of magnetic field geometries as a function of stellar mass and rotation period. These observations are yet to be interpreted by theoretical dynamo models.

Regarding evolved stars, magnetic fields at the level of 0.5–10 G have been detected for the descendants of solar-type stars (Aurière et al., 2009) and for more massive giants (Grunhut et al., 2010). The general pattern of the magnetic activity in giants follows the

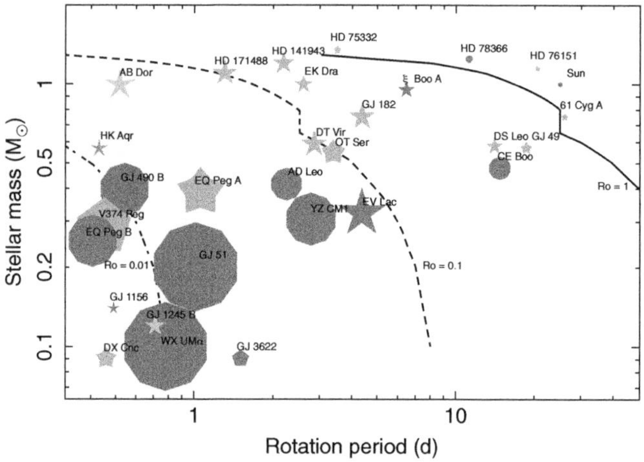

FIGURE 3.19. Properties of the magnetic topologies of cool stars, including solar-type and low-mass stars. The symbol size indicates magnetic field energy, ranging from a few G to 1.5 kG in this plot. The tone of grey shows the field configuration (dark grey: poloidal, light grey: toroidal). The symbol shape illustrates the degree of axisymmetry of the poloidal field component (stars: non-axisymmetric fields, decagons: axisymmetric). From Donati (2011).

one of the main sequence stars: the field is more readily detected in faster rotators. The strongest local fields (100–500 G) occur in some unusually fast rotating (periods of less than 10 d) objects with an uncertain evolutionary origin, e.g. FK Com and related stars (Petit et al., 2004; Konstantinova-Antova et al., 2012). Other giants have much longer rotation periods (from several months to a few years), which complicates ZDI mapping since the field structure may change during a single stellar rotation. We also know of several examples of giants with long rotational periods but exceptionally strong (up to 100 G) magnetic fields with a significant global component (Aurière et al., 2011). It is believed that these stars have evolved from the magnetic Ap stars, retaining their fossil magnetic fields. The physics of the envelope convection in the presence of such strong organised fields is yet to be addressed by theoretical calculations.

To summarise, direct systematic studies of the magnetic fields in solar-type stars are becoming possible but, to a large extent, are still in their infancy. ZDI analyses of the surface magnetic field topologies using polarisation data yield unexpected and interesting results, illustrating the importance of complementarity between the traditional activity proxy studies and characterising the magnetic field itself.

3.4.4 *Low-Mass Stars*

The observational signatures of stellar magnetism and probably its physical origin undergo a significant change in low-mass ($M \leq 0.6 M_\odot$) stars. The interior structure of these stars – M dwarfs and brown dwarfs – changes from partially convective to fully convective around mass 0.35 $M_\odot$. No tachocline dynamo can operate in the latter stars and yet M dwarfs exhibit copious signs of magnetic activity independently of their mass. Zeeman splitting of atomic lines, corresponding to the mean field modulus of 2–4 kG, has been recognised in the spectra of some M dwarfs (Johns-Krull & Valenti, 1996; Kochukhov et al., 2009). Many more objects, including brown dwarfs, are amenable to the Zeeman broadening measurements with magnetically sensitive FeH lines (Reiners & Basri, 2007), which retrieve comparable field strengths.

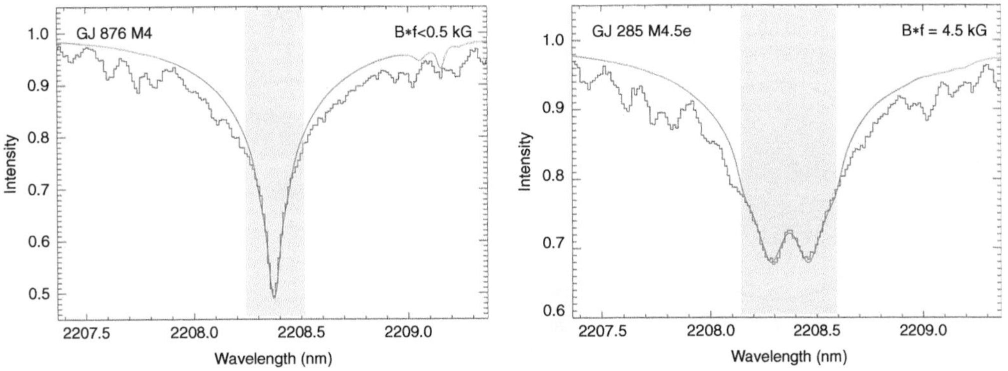

FIGURE 3.20. Spectrum synthesis analysis of the near-IR Na I 2208.4 nm line in the spectra of two M dwarf stars. For the inactive dwarf GJ 876 (left panel) only an upper limit of $Bf < 0.5$ kG can be determined. On the other hand, the spectrum of the active M dwarf YZ CMi (GJ 285, right panel) shows a clear Zeeman splitting, corresponding to $Bf = 4.5$ kG. Adapted from Kochukhov et al. (2009).

Unlike early-type magnetic stars, which show similar kG fields with little scatter of the field modulus across the stellar surface, M dwarfs clearly have multiple field components spanning a range from 1–2 kG up to 7–8 kG (Shulyak et al., 2014). To this end, the 'typical' field intensity reported by the Stokes I Zeeman broadening analyses corresponds to the product of the field strengths and filling factors of individual field components, $\sum B \cdot f$ (see Fig. 3.20).

A different picture emerges from the circular polarisation ZDI studies of M dwarfs (Donati et al., 2008; Morin et al., 2008). Interpretation of the Stokes V observations shows the presence of stable global poloidal fields, often aligned with the stellar rotational axis, in fully convective stars and somewhat weaker non-axisymmetric toroidal fields in more massive M dwarfs. There are exceptions to this rule as some low-mass M dwarfs also show weak toroidal fields (see Fig. 3.19). Evidently, the dynamo process operating in these stars can produce very different surface field topologies for the same stellar mass.

It should be noted that the fields recovered from polarisation are generally much weaker and contain a negligible magnetic energy compared to the fields measured using Stokes I (Reiners & Basri, 2009). To some extent, this discrepancy is expected since polarisation studies are only sensitive to the global, organised component of the field and miss a significant fraction of small-scale complex fields, but the degree of disagreement of the field strengths retrieved by different techniques is disconcerting. Incidentally, all ZDI studies of M dwarfs are based on the analytical profile calculations with Milne–Eddington atmospheres whereas the Stokes I Zeeman broadening analyses employ detailed spectrum synthesis.

Several theoretical studies have addressed the problem of the magnetic field generation in fully convective stars (Dobler et al., 2006; Browning, 2008; Gastine et al., 2013). These investigations confirmed that the fields reaching equipartition (magnetic energy comparable to the kinetic energy of convective flows) can be generated in the M dwarf interiors. However, at the moment it is unclear if any of these models predict a reasonable distribution of the magnetic field strength or can explain the observed co-existence of the strong small-scale fields and weaker global magnetic topologies. Interestingly, the observed field strength in giant planets, brown dwarfs and low-mass stars appears to

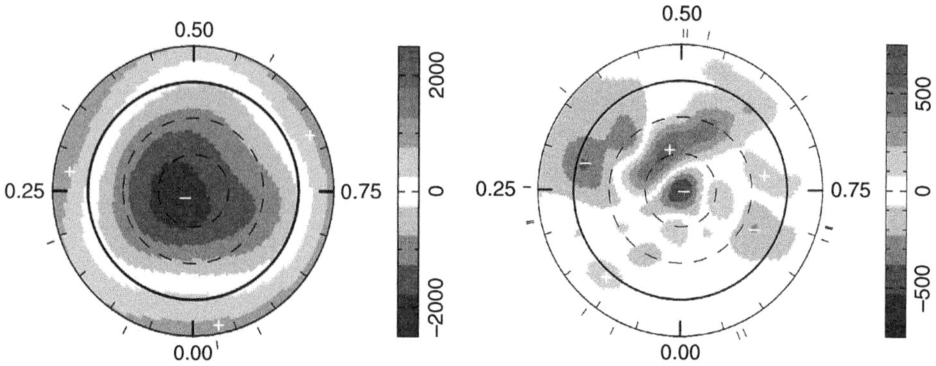

FIGURE 3.21. Representative ZDI results for T Tau stars with fully convective interior (left panel, AA Tau) and with radiative core (right panel, V2247 Oph). Both panels present the flattened polar projection of the radial field component, with the grey scale giving the field strength in G. The thick circle indicates the stellar equator. From Donati et al. (2010a,b).

follow the same scaling relation deduced by geodynamo models (Christensen et al., 2009), hinting at a universal dynamo mechanism relevant for all these vastly different objects.

3.4.5 *Pre-Main Sequence Stars*

Stellar magnetic fields are believed to play a key role in stellar formation and early stellar evolution. According to widely accepted theoretical frameworks, the field truncates accretion disks at a certain radius and governs the accretion flows towards the stellar surface, effectively controlling the accretion rate. Interaction of the stellar field and the disk also leads to an angular momentum loss and thus determines the rotation period with which a star settles on the main sequence. Furthermore, stellar magnetism affects planetary system formation since the disk truncation stops the inward migration of giant planets.

Until recently all these theoretical ideas lacked solid footing in observations. Development of high-resolution near-infrared spectroscopy and optical spectropolarimetry led to a significant progress in diagnosing magnetic fields of young stars. In particular, the mean field modulus was estimated for several dozen young solar-type stars (T Tauri stars) using the splitting of the magnetically sensitive Ti I lines at 2.2 μm (Johns-Krull, 2007). These measurements reveal 2–3 kG fields with multiple components. However, the observed field strengths do not correlate with stellar parameters or accretion rates.

On the other hand, circular spectropolarimetry demonstrated the presence of a variety of global field geometries, which do correlate with the stellar mass in a way resembling the situation for the main sequence M dwarf stars. Low-mass fully convective T Tau stars have strong (1–2 kG) global fields dominated by a dipolar or octupolar component (Donati et al., 2010a). These fields are stable and are typically aligned with the stellar rotational axis. More massive T Tau stars with radiative cores show more complex fields with local strengths of a few hundred G (Donati et al., 2010b). Figure 3.21 shows examples of these two field geometry types. Similar to the magnetic mapping of M dwarfs, ZDI inversions applied to T Tau stars have so far relied on approximate analytical treatment of the Stokes V profiles.

Thus, the global photospheric field topologies have a clear connection to the interior structure of T Tau stars (Gregory et al., 2012). At the same time, spectropolarimetric investigations (e.g. Johns-Krull et al., 2013) also discovered remarkably strong circular polarisation signals in the He I emission lines, which originate in the accretion funnels close to the stellar surface. It is unclear how 3–6 kG longitudinal fields measured in these lines are related to much weaker fields inferred from the photospheric lines.

Systematic magnetic surveys were also carried out for young intermediate-mass stars (Wade et al., 2007; Alecian et al., 2013), known as Herbig Ae/Be stars. Magnetic fields were detected for a small fraction of these stars. All objects analysed in detail exhibit stable, oblique dipole-like field configurations (Alecian et al., 2008, 2009). These stars have diverse atmospheric chemical compositions and do not show evidence of spots (Folsom et al., 2012). It is thus plausible that magnetic Herbig Ae/Be stars are the progenitors of the main sequence Ap/Bp stars. This implies that intermediate-mass stars become magnetic very early in their evolution and before any substantial surface abundance peculiarities have time to develop.

Acknowledgments

I acknowledge research funding from the Knut and Alice Wallenberg Foundation, the Swedish Research Council, and the Göran Gustafsson Foundation.

REFERENCES

Adelman, S. J., Gulliver, A. F., Kochukhov, O. P., & Ryabchikova, T. A. 2002, *ApJ*, **575**, 449

Alecian, G. & Stift, M. J. 2010, *A&A*, **516**, A53

Alecian, E., Catala, C., Wade, G. A., et al. 2008, *MNRAS*, **385**, 391

Alecian, E., Wade, G. A., Catala, C., et al. 2009, *MNRAS*, **400**, 354

Alecian, E., Wade, G. A., Catala, C., et al. 2013, *MNRAS*, **429**, 1001

Anderson, R. I., Reiners, A., & Solanki, S. K. 2010, *A&A*, **522**, A81

Asplund, M., Grevesse, N., Sauval, A. J., & Scott, P. 2009, *ARA&A*, **47**, 481

Aurière, M., Wade, G. A., Silvester, J., et al. 2007, *A&A*, **475**, 1053

Aurière, M., Wade, G. A., Konstantinova-Antova, R., et al. 2009, *A&A*, **504**, 231

Aurière, M., Konstantinova-Antova, R., Petit, P., et al. 2011, *A&A*, **534**, A139

Babcock, H. W. 1947, *ApJ*, **105**, 105

Babcock, H. W. 1958, *ApJ*, **128**, 228

Babcock, H. W. 1960, *ApJ*, **132**, 521

Babel, J. & Montmerle, T. 1997, *A&A*, **323**, 121

Bagnulo, S., Wade, G. A., Donati, J.-F., et al. 2001, *A&A*, **369**, 889

Bagnulo, S., Landi Degl'Innocenti, M., Landolfi, M., & Mathys, G. 2002a, *A&A*, **394**, 1023

Bagnulo, S., Szeifert, T., Wade, G. A., Landstreet, J. D., & Mathys, G. 2002b, *A&A*, **389**, 191

Bagnulo, S., Landstreet, J. D., Mason, E., et al. 2006, *A&A*, **450**, 777

Bagnulo, S., Landstreet, J. D., Fossati, L., & Kochukhov, O. 2012, *A&A*, **538**, A129

Baliunas, S. L., Donahue, R. A., Soon, W. H., et al. 1995, *ApJ*, **438**, 269

Belkacem, K., Samadi, R., Goupil, M.-J., et al. 2009, *Science*, **324**, 1540

Bigot, L., Provost, J., Berthomieu, G., Dziembowski, W. A., & Goode, P. R. 2000, *A&A*, **356**, 218

Bohlender, D. A., Landstreet, J. D., Brown, D. N., & Thompson, I. B. 1987, *ApJ*, **323**, 325

Borra, E. F. & Landstreet, J. D. 1980, *Astrophys. J. Suppl.*, **42**, 421

Borra, E. F. & Vaughan, A. H. 1977, *ApJ*, **216**, 462

Braithwaite, J. 2008, *MNRAS*, **386**, 1947

Braithwaite, J. & Nordlund, Å. 2006, *A&A*, **450**, 1077

Brown, S. F., Donati, J.-F., Rees, D. E., & Semel, M. 1991, *A&A*, **250**, 463

Browning, M. K. 2008, *ApJ*, **676**, 1262

Bychkov, V. D., Bychkova, L. V., & Madej, J. 2003, *A&A*, **407**, 631

Cantiello, M., Langer, N., Brott, I., et al. 2009, *A&A*, **499**, 279

Catalano, F. A. & Leone, F. 1994, in *Astronomical Society of the Pacific Conference Series, Vol. 60, The MK Process at 50 Years: A Powerful Tool for Astrophysical Insight*, ed. C. J. Corbally, R. O. Gray, & R. F. Garrison, 110

Christensen, U. R., Holzwarth, V., & Reiners, A. 2009, *Nat.*, **457**, 167

Cranmer, S. R. & Owocki, S. P. 1996, *ApJ*, **462**, 469

Cunha, M. S., Fernandes, J. M. M. B., & Monteiro, M. J. P. F. G. 2003, *MNRAS*, **343**, 831

de la Cruz Rodríguez, J. & Piskunov, N. 2013, *ApJ*, **764**, 33

Deen, C. P. 2013, *AJ*, **146**, 51

Dobler, W., Stix, M., & Brandenburg, A. 2006, *ApJ*, **638**, 336

Donati, J.-F. 2011, in *IAU Symposium, Vol. 271, IAU Symposium*, ed. N. H. Brummell, A. S. Brun, M. S. Miesch, & Y. Ponty, 23–31

Donati, J.-F. & Brown, S. F. 1997, *A&A*, **326**, 1135

Donati, J. & Collier Cameron, A. 1997, *MNRAS*, **291**, 1

Donati, J.-F. & Landstreet, J. D. 2009, *ARA&A*, **47**, 333

Donati, J.-F., Semel, M., Carter, B. D., Rees, D. E., & Collier Cameron, A. 1997, *MNRAS*, **291**, 658

Donati, J.-F., Cameron, A. C., Semel, M., et al. 2003, *MNRAS*, **345**, 1145

Donati, J.-F., Howarth, I. D., Bouret, J.-C., et al. 2006a, *MNRAS*, **365**, L6

Donati, J.-F., Howarth, I. D., Jardine, M. M., et al. 2006b, *MNRAS*, **370**, 629

Donati, J.-F., Morin, J., Petit, P., et al. 2008, *MNRAS*, **390**, 545

Donati, J.-F., Skelly, M. B., Bouvier, J., et al. 2010a, *MNRAS*, **409**, 1347

Donati, J.-F., Skelly, M. B., Bouvier, J., et al. 2010b, *MNRAS*, **402**, 1426

Duez, V. & Mathis, S. 2010, *A&A*, **517**, A58

Fares, R., Donati, J.-F., Moutou, C., et al. 2009, *MNRAS*, **398**, 1383

Folsom, C. P., Bagnulo, S., Wade, G. A., et al. 2012, *MNRAS*, **422**, 2072

Freyhammer, L. M., Elkin, V. G., Kurtz, D. W., Mathys, G., & Martinez, P. 2008, *MNRAS*, **389**, 441

Gastine, T., Morin, J., Duarte, L., et al. 2013, *A&A*, **549**, L5

Goncharskii, A. V., Stepanov, V. V., Kokhlova, V. L., & Yagola, A. G. 1977, *Soviet Astronomy Letters*, **3**, 147

Gregory, S. G., Donati, J.-F., Morin, J., et al. 2012, *ApJ*, **755**, 97

Grunhut, J. H., Wade, G. A., Hanes, D. A., & Alecian, E. 2010, *MNRAS*, **408**, 2290

Grunhut, J. H., Wade, G. A., Sundqvist, J. O., et al. 2012, *MNRAS*, **426**, 2208

Gustafsson, B., Edvardsson, B., Eriksson, K., et al. 2008, *A&A*, **486**, 951

Henrichs, H. F., de Jong, J. A., Verdugo, E., et al. 2013, *A&A*, **555**, A46

Johns-Krull, C. M. 2007, *ApJ*, **664**, 975

Johns-Krull, C. M. & Valenti, J. A. 1996, *ApJL*, **459**, L95

Johns-Krull, C. M., Valenti, J. A., & Koresko, C. 1999, *ApJ*, **516**, 900

Johns-Krull, C. M., Chen, W., Valenti, J. A., et al. 2013, *ApJ*, **765**, 11

Khokhlova, V. L., Rice, J. B., & Wehlau, W. H. 1986, *ApJ*, **307**, 768

Kim, K.-M., Han, I., Valyavin, G. G., et al. 2007, *PASP*, **119**, 1052

Kochukhov, O. 2004, *ApJL*, **615**, L149

Kochukhov, O. 2006, *A&A*, **454**, 321

Kochukhov, O. 2008, *A&A*, **483**, 557

Kochukhov, O. & Bagnulo, S. 2006, *A&A*, **450**, 763

Kochukhov, O. & Piskunov, N. 2002, *A&A*, **388**, 868

Kochukhov, O. & Sudnik, N. 2013, *A&A*, **554**, A93

Kochukhov, O. & Wade, G. A. 2010, *A&A*, **513**, A13

Kochukhov, O., Bagnulo, S., Wade, G. A., et al. 2004a, *A&A*, **414**, 613

Kochukhov, O., Drake, N. A., Piskunov, N., & de la Reza, R. 2004b, *A&A*, **424**, 935

Kochukhov, O., Piskunov, N., Sachkov, M., & Kudryavtsev, D. 2005, *A&A*, **439**, 1093

Kochukhov, O., Heiter, U., Piskunov, N., et al. 2009, in *American Institute of Physics Conference Series, Vol. 1094, American Institute of Physics Conference Series*, ed. E. Stempels, 124–129

Kochukhov, O., Makaganiuk, V., & Piskunov, N. 2010, *A&A*, **524**, A5

Kochukhov, O., Lundin, A., Romanyuk, I., & Kudryavtsev, D. 2011a, *ApJ*, **726**, 24

Kochukhov, O., Makaganiuk, V., Piskunov, N., et al. 2011b, *ApJL*, **732**, L19

Kochukhov, O., Makaganiuk, V., Piskunov, N., et al. 2013a, *A&A*, **554**, A61

Kochukhov, O., Mantere, M. J., Hackman, T., & Ilyin, I. 2013b, *A&A*, **550**, A84

Konstantinova-Antova, R., Aurière, M., Petit, P., et al. 2012, *A&A*, **541**, A44

Kudryavtsev, D. O., Romanyuk, I. I., Elkin, V. G., & Paunzen, E. 2006, *MNRAS*, **372**, 1804

Kurtz, D. W. 1982, *MNRAS*, **200**, 807

Kurucz, R. 1993, ATLAS9 Stellar Atmosphere Programs and 2 km/s grid. Kurucz CD-ROM No. 13. Cambridge, Mass.: Smithsonian Astrophysical Observatory.

Landi Degl'Innocenti, E. & Landolfi, M. 2004, *Astrophysics and Space Science Library, Vol. 307, Polarization in Spectral Lines* (Kluwer Academic Publishers)

Landi Deglinnocenti, M., Calamai, G., Landi Deglinnocenti, E., & Patriarchi, P. 1981, *ApJ*, **249**, 228

Landolfi, M., Landi degl'Innocenti, E., Landi degl'Innocenti, M., & Leroy, J. L. 1993, *A&A*, **272**, 285

Landstreet, J. D. 1988, *ApJ*, **326**, 967

Landstreet, J. D. 2009, in *EAS Publications Series*, Vol. 39, EAS Publications Series, ed. C. Neiner & J.-P. Zahn, 1–20

Landstreet, J. D. & Mathys, G. 2000, *A&A*, **359**, 213

Landstreet, J. D., Borra, E. F., Angel, J. R. P., & Illing, R. M. E. 1975, *ApJ*, **201**, 624

Leone, F., Catanzaro, G., & Catalano, S. 2000, *A&A*, **355**, 315

Leroy, J. L. 1995, *A&A*, **114**, 79

Leroy, J. L., Bagnulo, S., Landolfi, M., & Degl'Innocenti, E. L. 1994, *A&A*, **284**, 174

Lignières, F., Petit, P., Böhm, T., & Aurière, M. 2009, *A&A*, **500**, L41

Lueftinger, T., Kuschnig, R., Piskunov, N. E., & Weiss, W. W. 2003, *A&A*, **406**, 1033

Mathys, G. 1988, *A&A*, **189**, 179

Mathys, G. 1989, *Fundamentals of Cosmic Physics*, **13**, 143

Mathys, G. 1990, *A&A*, **236**, 527

Mathys, G. 1991, *A&A*, **89**, 121

Mathys, G. 1995a, *A&A*, **293**, 733

Mathys, G. 1995b, *A&A*, **293**, 746

Mathys, G. & Hubrig, S. 2006, *A&A*, **453**, 699

Mathys, G., Hubrig, S., Landstreet, J. D., Lanz, T., & Manfroid, J. 1997, *A&A*, **123**, 353

Mestel, L. 1999, *Stellar Magnetism* (Oxford Science Publications)

Michaud, G. 1970, *ApJ*, **160**, 641

Michaud, G., Charland, Y., & Megessier, C. 1981, *A&A*, **103**, 244

Mikulášek, Z., Krtička, J., Henry, G. W., et al. 2011, *A&A*, **534**, L5

Monin, D., Bohlender, D., Hardy, T., Saddlemyer, L., & Fletcher, M. 2012, *PASP*, **124**, 329

Morgenthaler, A., Petit, P., Saar, S., et al. 2012, *A&A*, **540**, A138

Morin, J., Donati, J.-F., Petit, P., et al. 2008, *MNRAS*, **390**, 567

Nesvacil, N., Lüftinger, T., Shulyak, D., et al. 2012, *A&A*, **537**, A151

Noyes, R. W., Hartmann, L. W., Baliunas, S. L., Duncan, D. K., & Vaughan, A. H. 1984, *ApJ*, **279**, 763

Petit, P., Donati, J.-F., Oliveira, J. M., et al. 2004, *MNRAS*, **351**, 826

Petit, P., Dintrans, B., Solanki, S. K., et al. 2008, *MNRAS*, **388**, 80

Petit, P., Lignières, F., Wade, G. A., et al. 2010, *A&A*, **523**, A41

Petit, P., Lignières, F., Aurière, M., et al. 2011, *A&A*, **532**, L13

Petit, V., Owocki, S. P., Wade, G. A., et al. 2013, *MNRAS*, **429**, 398

Piskunov, N. & Kochukhov, O. 2002, *A&A*, **381**, 736

Plachinda, S. I. & Tarasova, T. N. 1999, *ApJ*, **514**, 402

Rees, D. E., Durrant, C. J., & Murphy, G. A. 1989, *ApJ*, **339**, 1093

Reiners, A. 2012, *Living Reviews in Solar Physics*, **9**, 1

Reiners, A. & Basri, G. 2007, *ApJ*, **656**, 1121

Reiners, A. & Basri, G. 2009, *A&A*, **496**, 787

Rivinius, T., Townsend, R. H. D., Kochukhov, O., et al. 2013, *MNRAS*, **429**, 177

Rosén, L. & Kochukhov, O. 2012, *A&A*, **548**, A8

Rosén, L., Kochukhov, O., & Wade, G. A. 2013, *MNRAS*, **436**, L10

Rusomarov, N., Kochukhov, O., Piskunov, N., et al. 2013, *A&A*, **558**, A8

Ryabchikova, T., Sachkov, M., Kochukhov, O., & Lyashko, D. 2007, *A&A*, **473**, 907

Saio, H. & Gautschy, A. 2004, *MNRAS*, **350**, 485

Saio, H., Gruberbauer, M., Weiss, W. W., Matthews, J. M., & Ryabchikova, T. 2012, *MNRAS*, **420**, 283

Shulyak, D., Reiners, A., Seemann, U., Kochukhov, O., & Piskunov, N. 2014, *A&A*, **563**, A35

Silvester, J., Neiner, C., Henrichs, H. F., et al. 2009, *MNRAS*, **398**, 1505

Silvester, J., Wade, G. A., Kochukhov, O., et al. 2012, *MNRAS*, **426**, 1003

Stibbs, D. W. N. 1950, *MNRAS*, **110**, 395

Sundqvist, J. O., ud-Doula, A., Owocki, S. P., et al. 2012, *MNRAS*, **423**, L21

Townsend, R. H. D. 2008, *MNRAS*, **389**, 559

Townsend, R. H. D. & Owocki, S. P. 2005, *MNRAS*, **357**, 251

Townsend, R. H. D., Owocki, S. P., & Groote, D. 2005, *ApJL*, **630**, L81

Townsend, R. H. D., Oksala, M. E., Cohen, D. H., Owocki, S. P., & ud-Doula, A. 2010, *ApJL*, **714**, L318

Ud-Doula, A. & Owocki, S. P. 2002, *ApJ*, **576**, 413

Ud-Doula, A., Owocki, S. P., & Townsend, R. H. D. 2009, *MNRAS*, **392**, 1022

Valenti, J. A., Marcy, G. W., & Basri, G. 1995, *ApJ*, **439**, 939

Vogt, S. S., Penrod, G. D., & Hatzes, A. P. 1987, *ApJ*, **321**, 496

Wade, G. A., Donati, J.-F., Landstreet, J. D., & Shorlin, S. L. S. 2000, *MNRAS*, **313**, 823

Wade, G. A., Bagnulo, S., Kochukhov, O., et al. 2001, *A&A*, **374**, 265

Wade, G. A., Bagnulo, S., Drouin, D., Landstreet, J. D., & Monin, D. 2007, *MNRAS*, **376**, 1145

Wade, G. A., Grunhut, J. H., & MiMeS Collaboration. 2012a, in *Astronomical Society of the Pacific Conference Series, Vol. 464, Circumstellar Dynamics at High Resolution*, ed. A. C. Carciofi & T. Rivinius, 405

Wade, G. A., Maíz Apellániz, J., Martins, F., et al. 2012b, *MNRAS*, **425**, 1278

Wickramasinghe, D. T. & Ferrario, L. 2000, *PASP*, **112**, 873

Wright, N. J., Drake, J. J., Mamajek, E. E., & Henry, G. W. 2011, *ApJ*, **743**, 48

4. The Role of Magnetic Fields in AGN Activity and Feedback

RONY KEPPENS, OLIVER PORTH AND
HANS (J. P.) GOEDBLOED

Abstract

Active galactic nuclei (AGNs), the luminous, compact core regions of galaxies where accretion occurs onto supermassive black holes, can dramatically influence their entire host galaxy evolution by a process referred to as AGN feedback. Energy feedback to the galaxy is the result of combined radiation fields and directed outflows, and especially radio-loud active galaxies show pronounced jets and lobes. Their synchrotron radio emission indicates that dynamically important magnetic fields are at play in AGN jet collimation, stability, energy transfer to the intergalactic medium and their overall morphological appearance. Current knowledge on the launching mechanisms for such highly energetic relativistic jets, as well as the near black-hole accretion processes themselves, all invoke magnetic fields as active agents in angular momentum, mass and energy redistributions. In this review, we cover aspects of AGN feedback and the role played by magnetic fields, almost necessarily studied at vastly different length and timescales. We emphasize how typical large-scale galaxy interaction studies rely on parametric prescriptions for feedback, while detailed dedicated studies for near black-hole dynamics and relativistic jet propagation exist which take full account of magnetic field influences. We discuss representative hydro to magnetohydrodynamic (MHD) numerical simulations that exploit analogies with less energetic X-ray binary sources or even protostellar accretion-ejection systems, emphasize relativistic MHD descriptions, and point out that magnetic fields in accretion disks yield many linear instability routes to turbulence that have scarcely been recognized in the astrophysical community. In combination, they serve to show that magnetic field influences in AGN accretion, jet launch, energy feedback, and overall evolution are still far from completely understood, although many aspects have been disclosed by advanced analytical and numerical relativistic MHD studies.

4.1 Motivation: Astrophysical Jets

Radio galaxies confront us with dramatic views on energy redistributions at all scales, as mediated by central massive black holes lurking in their nucleus. A clear example is provided by the elliptical galaxy NGC5532, a nearby (red shift $z = 0.0237$, type S0) galaxy where the stellar distribution is in sharp contrast with its double-jetted appearance in radio images. In VLA observations at 1.45GHz frequencies (wavelength ~20 cm) of the same radio galaxy also known as 3C296, prominent jets stretch over half a million light years, clearly emerging from the central core. The basic ingredients return in many other radio-loud AGNs, and involve black hole masses in a range of $10^6 - 10^{10} M_\odot$, and flow speeds throughout the jets that have Lorentz factors of order 10 (e.g., Marscher, 2006). The nonthermal radio emission is due to relativistic electrons spiraling about magnetic field lines, emitting the characteristic synchrotron radiation. As early as 1974, Fanaroff and Riley performed a statistical survey with close to 60 well-resolved radio galaxies mapped at 1.4 GHz or 5 GHz with the Cambridge One-Mile telescope, and found a morphological distinction at a fairly specific luminosity value $L_{178\mathrm{MHz}}$ (taken at 178 MHz). This FR classification divided all sources with $L_{178\mathrm{MHz}} < 2 \times 10^{25}$ W Hz^{-1}sr^{-1} in a first

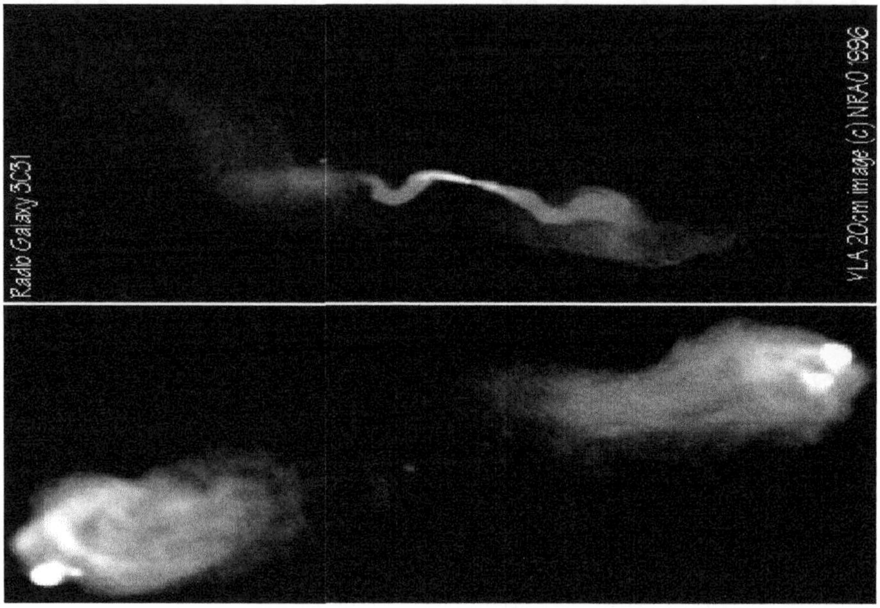

FIGURE 4.1. 3C31 (top), a typical FR I source, contrasted with Cygnus A (bottom), as prototype FR II object, both as seen in radio observations.

category with their brightest regions near the core, while the more energetic FR II had lobes and hot spots further away from the center. Although 3C296 was not in the original sample, it is nowadays known to be of FR I type (like 3C31 or 3C449, to name but two of the sample), which typically are accompanied with jets that are relativistic at parsec scale, becoming more diffuse and sub-relativistic at kiloparsec distances. The FR II sources have narrow, highly relativistic jets ending in extended lobes and hot spots (e.g. Cygnus A also known as 3C405, or 3C47 from the original 1974 survey). Figure 4.1 shows some typical radio images.

The fact that radio morphologies correlate with overall energy content suggests that depending on their absolute energy content, jets transport, mix and deposit energy differently in their intergalactic medium (IGM). The energy content of the jets is likely determined by the near black-hole launch and acceleration conditions, which one may expect to be similar on both sides of their accretion disk. Meanwhile, Hybrid Morphology Radio Sources (HYMORS; Gopal-Krishna and Wiita, 2002) such as NGC612 have been identified, where an FR II lobe with hot spot is seen on one side, while a diffuse jet at the other side suggests an FR I structure. They are indicative for ambient medium influences on the radio morphologies, so that both jet launch and internal structure, as well as the traversed IGM stratification, should be taken into account when trying to quantify mixing efficiencies, energy transport and jet deceleration. The workings of powerful AGN jets can also be contrasted with their far less energetic counterparts in X-ray binary (XRB) sources, with the archetypical SS433 system at 5.5 kiloparsec as a prime example. The microquasar[1] system SS433 demonstrates a clearly helical jet path at the $\mathcal{O}(0.1)$ parsec

[1] The term microquasar was coined by Mirabel and Rodríguez (1998) to emphasize the downscaled nature of these galactic sources of relativistic jets. Their close proximity is not sufficient to compensate for the correspondingly smaller gravitational radius of microquasars, thus the spatial resolution in terms of the physical length scale is, in fact, worse compared with the

views provided by the VLA, while VLBA studies have imaged the precessing, time-variable jet emission as seen closer to its source, with individual blobs of typical sizes of a few hundred astronomical units. SS433 has a source geometry involving an accretion disk around the 9 $M_\odot$ black hole as formed by Roche lobe overflow from a 30 $M_\odot$ companion. As an AGN jet analog, its mildly relativistic, but helical, jet calls for full 3D studies to quantify its impact on the environment. Meanwhile, numerical modeling of AGN or XRB jets focusing on jet morphologies has exploited Newtonian to relativistic, 2D to 3D, and hydro to magnetohydrodynamic models, with and without internal jets, as well as external medium stratification being incorporated. The picture that emerged is that all these aspects can play a key role in the way energy deposition is realized by mixing, defying simple parametric prescriptions. In a recent example of one such study adressing the FRI/II variation, Krause *et al.* (2012) used conically injected, axisymmetric hydro jets traversing a homogeneous medium. Exploiting the (grid-adaptive) FLASH community code, six hydro jets with opening angles from 5−30° at Mach numbers 5 up to 500 were intercompared in their shock-governed progression. Using classical force balance arguments, such conical jets can recollimate at distances where the sideways ram pressure of the jet matches the external medium pressure. One also expects dynamical changes at those distances from the inlet where the jet density drops below its ambient value, and/or where the ram pressure in the further jet path falls below the external value. Together, these translate into different jet morphologies as seen in density or pressure maps. The study classified the jet morphologies for all six models throughout the entire simulated time intervals, using a local instantaneous compression measure as a mock-up emissivity, to generate line-of-sight integrated sideviews. These maps then quantify an FR index using the extent and location of the brightest feature along the jet, much akin to the original FR classification. It was found that while all jets started at an FR II-like index, only jets at large opening angles (30°) could eventually turn into more FR I type. Despite its 2D Newtonian hydro approach to AGN jet aspects, the study highlighted that simple geometric aspects prevailing near the jet base influence its ultimate observational classification. Later in this review, we provide many more examples where gradually more advanced models are used in AGN jet studies, but first discuss their connection with the AGN feedback scenarios.

4.2 AGN Feedback

The AGN jets in essence redirect and redistribute matter, angular momentum and energy in preferred directions to the large-scale surroundings, with FR II jets pumping of the order of $\mathcal{O}(10^{46})\,\mathrm{ergs\,s^{-1}}$ in their IGM. Thereby, they shock-reheat the IGM, and observations reveal large-scale bubbles, blown by probably successive jet launch events. A clear example (Fabian *et al.*, 2003) is provided by observations using the Chandra capabilities for imaging the entire Perseus cluster at energy bands 0.3−0.7 keV. Such instantaneous maps already display how the active galaxy NGC 1275 interacts on a large scale with the intra-cluster medium (ICM) by blowing bubbles, which (when subtracting images) clearly display pressure ripples reminiscent of sound waves (up to weak shocks). One estimates that they transfer up to four times their product of pressure and volume, which is related to their mechanical power. That jets are capable of inflating bubbles, which, when jet activity ceases, behave as buoyant regions in the ICM, has

nearest AGN sources. However, in terms of temporal scales, their smaller gravitational radius works to the favor of the observer: processes that take millions of years in close proximity to an AGN will advance within a matter of weeks in microquasars.

been investigated numerically in idealized 2D hydro conical jet studies (opening angles of 15°), injected into externally stratified density distributions (Reynolds *et al.*, 2002). These Newtonian simulations followed the changeover from an active to a passive (jet ceases) phase, to address jet-related morphologies in dead radio galaxies. When the jet stops feeding energy into its self-inflated cocoon, pressure balance with the surroundings is restored by cocoon expansion, followed by a buoyant stage. Shear-flow Kelvin–Helmholtz and bouyancy-related Rayleigh–Taylor instabilities can destroy the cocoon integrity, and the rising plumes cause an up to 20-fold increase in the entropy of the surroundings.

4.2.1 *The Black Hole Mass–Velocity Dispersion Relation*

While the jetted outflows associated with massive black holes from AGNs are providing so-called kinetic feedback, where their energy injection affects the central cooling rates, feedback can also operate through more isotropically distributed winds and/or radiation fields from the core regions. AGN feedback appears to operate over an enormous range of central black hole masses, with observationally established correlations between central black hole masses $M_{\rm bh}$ from 10^6 to $10^{10}\,M_\odot$ and the stellar velocity dispersion σ in the interior galaxy bulge region.[2] An example study by Graham *et al.* (2011) refined the $M_{\rm bh}$–σ correlation by including only galaxies for which the central black hole mass measurement has been well established (Fig. 4.2). Using 64 galaxies, with types ranging from elliptical, lenticular, spiral to barred, they find the relation

$$\log(M_{\rm bh}/M_\odot) = 8.13 + 5.13\log[\sigma/200\ {\rm km\,s^{-1}}]\,. \tag{4.1}$$

The fact that only galaxies with known black hole mass are included is important, as the relation has, since its discovery in 2000 by Ferrarese and Merritt, also been used to quantify black hole masses for galaxies with known velocity dispersions only. The exact slope, here suggesting a $\bar{M}_{\rm bh} \propto \bar{\sigma}^{5.13}$ proportionality, has gradually been refined, and, depending on the galaxy type, can be closer to a fourth power, e.g. the 20 barred galaxies alone provided $\bar{M}_{\rm bh} \propto \bar{\sigma}^{4.34}$. That a fourth or fifth power-law dependence prevails is expected on theoretical grounds, which use simple order of magnitude estimates. Indeed, suppose that radiation pressure balances inward gravity; a momentum balance demands that the Eddington luminosity, which scales with black hole mass $M_{\rm bh}$ appears as $L_{\rm edd}/c = G\,f\,M_{\rm gal}^2/r^2$. The right-hand side quantifies the gravitational force at distance r, where the gas content of a galaxy with mass $M_{\rm gal}$ is taken as a fraction f of the galactic mass. When we assume an isothermal bulge with constant temperature $T = \sigma^2 m_p/3k$ (with proton mass m_p and Boltzmann constant k) and density fall-off as $\rho \propto \sigma^2/r^2$, the galaxy mass contained up to radius r scales with $\sigma^2 r$. The above relation then instantly gives $\bar{M}_{\rm bh} \propto \bar{\sigma}^4$ (Fabian, 1999). If we rather consider a galactic wind, its constant mass flux $\rho r^2 v_{\rm w} \propto \sigma^2 v_{\rm w}$ gives a power to the wind going as $v_{\rm w}^3\sigma^2$, which can again be estimated to be a fraction of the Eddington luminosity $L_{\rm edd} \propto M_{\rm bh}$. The resulting wind speed $v_{\rm w} \propto (M_{\rm bh}/\sigma^2)^{1/3}$ exceeds the escape speed found from $\sqrt{2GM_{\rm gal}/r}$ if $\bar{M}_{\rm bh} > \bar{\sigma}^5$. The latter argument, due to Silk and Rees (1998), uses power balance, while the former works with momentum balance. In any case, these arguments prove it likely

[2] This dispersion σ is generally a function of radial distance to the core, and quantifies the spread in stellar velocities at that distance. In practice, it can be determined from spectroscopic data, as follows (Tonry and Davis, 1979). Having the observed intensity versus wavelength spectrum $g(n)$, this needs to be correlated to a template spectrum at zero red shift with the known instrumental broadening $t(n)$, to quantify both the velocity dispersion σ and the red shift δ. In formulae, $g(n) \approx \alpha t(n) * b(n-\delta)$ is written as a convolution of a multiple (α) of the template spectrum with a symmetric (Gaussian bell) function $b(n)$ with dispersion σ. A least square fit analysis is used to quantify these distances and stellar velocity dispersions.

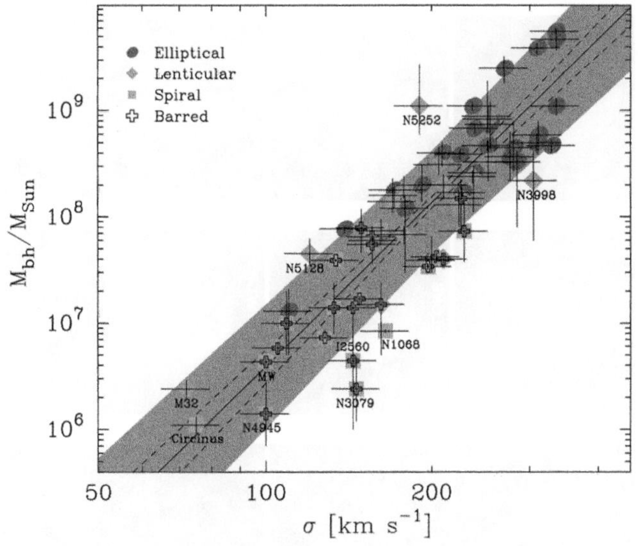

FIGURE 4.2. The observationally established $M_{\rm bh}-\sigma$ correlation, from the updated survey by Graham *et al.* (2011).

that the central black hole mass is regulated by the feedback process, since, when it exceeds the critical σ^5 related limit, either radiation and/or strong winds would sweep away the central matter, preventing further accretion. When the feedback happens more through such isotropic radiative or wind influences, one talks about radiative or wind mode feedback, distinguishing it from the kinetic mode more associated with directed outflows (see also the review by Fabian, 2012).

4.2.2 *Modeling Efforts at Large Scales and AGN Feedback*

The observationally established correlations have also been integrated in cosmological simulations, which follow galaxy formation and evolution from "first principles," by evolving dark matter, gas, and stellar dynamics over extreme length and time scales. Smooth Particle Hydrodynamics (SPH) has matured to a popular means of handling the gas dynamical evolution, with many impressive results. At the same time, these large-scale simulations almost necessarily need to handle many important aspects at sub-grid scale resolution (a term that is somewhat confusing as SPH does not employ an actual grid), because the size of the central black hole regulated region differs up to nine orders of magnitude with the spatial extent of its host galaxy. Modern simulations employing N-body (for the stellar gravitational interaction) plus SPH descriptions for the gas + dark matter therefore rely on numerous parameterizations, for aspects going from the star formation rate (SFR), over supernovae regulated feedback, to black-hole formation, accretion onto the black hole, mergers of black holes, and the feedback process just described. In particular for the feedback parameterization, in practice one redistributes a parameterized fraction of the black-hole accretion rate over all nearby stars (particles within a few SPH smoothing lengths of the black hole), raising their internal energy or momentum. The latter choice then reflects whether thermal or kinetic feedback is at play. The fact that different parameterizations can lead to dramatically different evolutionary scenarios has been illustrated well in a recent comparative study of AGN feedback algorithms by Wurster and Thacker (2013). Using their N-body + SPH Hydra

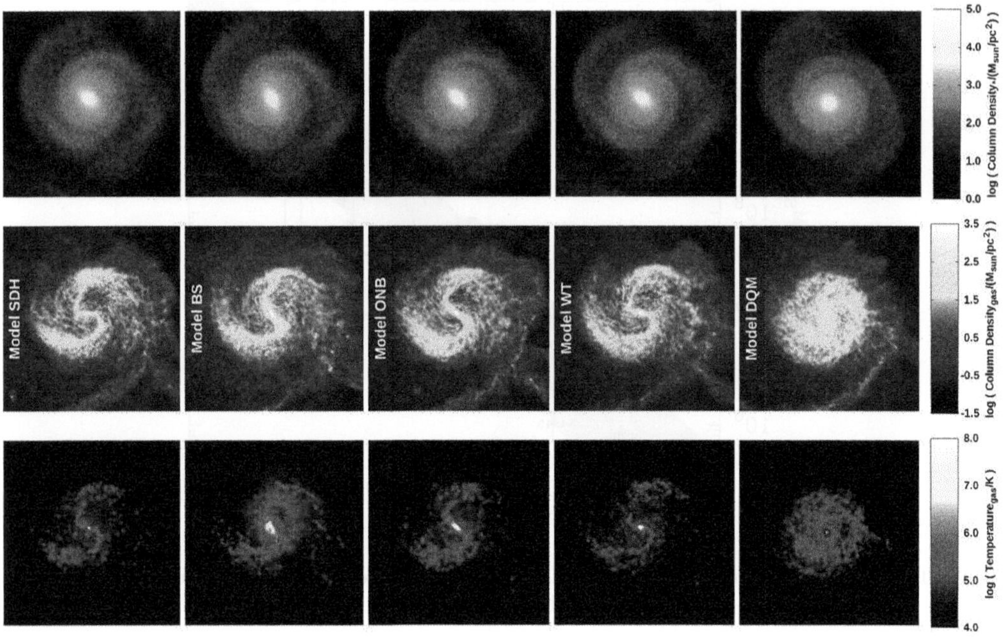

FIGURE 4.3. Zoomed view on a simulated galaxy undergoing a merger event, at the time of first maximal separation (apoapsis). The five columns adopt different parameterizations for the feedback process, resulting in clear morphological differences as seen in stellar density (top), gas density (middle) to gas temperature (bottom) instantaneous distributions. From Wurster and Thacker (2013).

code, they performed five simulations of the same merger event between two Milky Way sized galaxies, each containing a $10^5\, M_\odot$ black hole, spanning a 1.5 Gyr time frame. The initial conditions as well as the SFR parameterization employed during the simulation were identical, and only the feedback parameterization differed from case to case. This feedback incorporates a heuristic way to quantify black-hole accretion, black-hole growth (involving near black-hole particle removal), feedback mode (thermal or kinetic energy dumped on nearby particles), and black-hole movement and merger prescriptions. While the overall behavior of the merger is qualitatively alike, the star, gas, and temperature distribution differs markedly, shown visually at apoapsis in Fig. 4.3 (time of first furthest separation between central black holes, at 480 Myr). The five prescriptions lead to dramatic differences in black-hole mass growth and associated accretion rate (one model does not even show a black-hole merger event within the simulated time frame). While the SFR prescription is parameterized identically, its dependence on the local thermodynamic properties then also shows order of magnitude variations between the models, and between runs at varying resolutions (increased particle numbers) of the same model. In the end state obtained, one can extract the actual $M_{\rm bh}$–σ relation for each model run, and verify whether it obeys the observationally established correlation. It was found that only three model runs comply within a meaningful uncertainty. Also, instantaneous differences in thermodynamic properties near the black hole would vary from model to model over six orders of magnitude. It is therefore obvious that feedback processes are still characterized by many unsolved aspects. In particular, magnetic fields are known to be a key player in virtually all processes affecting the near black-hole accretion and wind or jet outflows, while many modern cosmological simulations do not yet consider any dynamically important magnetic field evolution at all.

4.2.3 *Hints for Magnetic Field Influences on Feedback*

Incorporating magnetic fields in galaxy-cluster size simulations is a modern research effort, and Sutter *et al.* (2012) used the publicly available 3D MHD-AMR code Flash to follow the evolution of an isolated galaxy cluster in a box of (2000 kpc)^3 over 6 Gyr. A central $3 \times 10^9\, M_\odot$ black hole regulates the cluster evolution of total mass $10^{14}\, M_\odot$, with a parameterized accretion rate, and feedback enters as parameterized mass, momentum and magnetic field source terms. When also magnetic field is injected, it was found that the unwinding of the injected field (an analytically prescribed magnetic tower topology was used) can reduce the accretion onto the black hole over several Gyr, as it drives plasma away from the center through the Lorentz force. Varying the magnetic topology injection from a single directed jet, to several interacting bubbles at fixed locations, to randomly superposed bubbles, also dramatically influences the later accretion rates onto the black hole. Not surprisingly, the case with randomly superposed bubbles ultimately causes near hydrodynamic behavior, due to its chaotic field entanglement and the ensuing (numerical) magnetic reconnection. Another noteworthy effort to include magnetic field influences in galaxy cluster evolutions is the work by Mendygral *et al.* (2012), which starts from an earlier SPH cosmological study performed by the Gadget code (in its SPMHD variant), extracting a cluster of size (1008 kpc)^3 from it. Then, cylindrical jets with purely toroidal fields, parameterized by their overall luminosity of $6 \times 10^{44}\, \mathrm{erg\, s^{-1}}$ were repeatedly turned on and off with a 26 Myr period. When turning the evolution in synthetic X-ray views (ray-traced images of $\rho^2/\sqrt{T}$), a rippled structure marked by jet-blown cavities and bright rims is detected after several jet cycles, very reminiscent of the Perseus cluster observations.

While these simulations give the first hints of how magnetic fields can mitigate the far-field cluster conditions through jet activity, it is fair to state that the jet models employed are toy models, with e.g. simplifications in field topology, or reducing its strength to effectively turn it into a merely passively advected quantity. Certainly in the near black-hole jet launch and acceleration region, as well as for its overall stability, propagation and mixing efficiency, the details in the field variation matter. The multi-scale aspect of the feedback process will have to face these different scales and how they interact in future multi-scale modeling. For models dealing with the dynamical AGN jet details, relativistic hydro to MHD models have meanwhile matured to tell us about precisely the processes parametrically handled in the large-scale setups. In what follows, the emphasis will be on such models, where relativistic effects, dynamically important magnetic fields, more realistic field topologies, and near black-hole launch conditions are gradually incorporated.

4.3 Relativistic Hydro Jet Models

4.3.1 *Relativistic Hydro Preliminaries*

As XRB and AGN jets demonstrate bulk velocities at significant fractions of the speed of light, dynamical models where propagation, stability, and mixing are studied must at least incorporate special relativity in their prescription. In a pre-chosen Lorentzian lab frame, the spatial three-velocity $\mathbf{v}$ then enters the Lorentz factor $\Gamma = 1/\sqrt{1 - v^2/c^2}$, and the governing particle number conservation and energy–momentum conservation laws in the four-dimensional Minkowski space-time can be formulated as straightforward generalizations of the familiar Euler equations for Newtonian gas dynamics. As such,

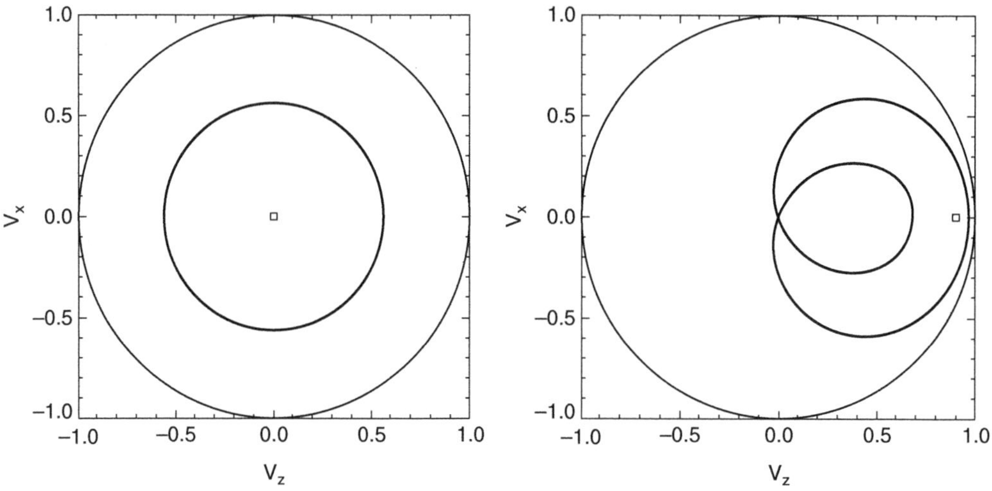

FIGURE 4.4. Phase speed diagram for plane sound waves, as seen from the rest frame (left) versus a moving vantage point (right). From Keppens and Meliani (2008).

they also allow for linear sound wave solutions, traversing a uniform, static gas (rest frame pressure p, density ρ, and enthalpy h given by $\rho h = \rho c^2 + \gamma p/(\gamma - 1)$, with γ the polytropic index) as plane waves behaving as $\exp(-i\omega t + i\mathbf{k}\cdot\mathbf{x})$ adopting time-space coordinates $(t, \mathbf{x})$. The dispersion relation for such plane sound waves quantifies their isotropic propagation at phase speed $\omega/kc = \sqrt{\gamma p/\rho h}$ in the chosen rest frame. In this rest frame, the group speed that dictates how a local perturbation redistributes its energy is also a spherically symmetric front traveling away at the same speed. However, for moving sources, the Lorentz transform already confronts us with a clear relativistic effect for such simple sound wave perturbations. Indeed, the same plane wave in a frame moving at speed $\mathbf{v}$ will have a changed frequency ω' and altered wave vector direction $\mathbf{k}'$, combining the relativistic Doppler effect with relativistic wave aberration. These are given by

$$\omega = \Gamma\left(\omega' + \mathbf{k}'\cdot\mathbf{v}\right),$$

$$\mathbf{k} = \mathbf{k}' + \mathbf{v}\left[\frac{\omega'\Gamma}{c^2} + (\mathbf{k}'\cdot\mathbf{v})\frac{\Gamma - 1}{v^2}\right].$$

These relations allow us to quantify the phase speed and group speed diagrams also for a moving source. The phase speed diagram shows an intricate double-loop structure (Keppens and Meliani, 2008), as shown in Fig. 4.4, which can be turned into the group speed diagram by performing a standard Huygens construction. What emerges is the fact that the wave front emanating from a point perturbation, when seen from a moving vantage point, will no longer appear spherically symmetric. These basic facts are already true for simple linear wave pertubations, while nonlinear effects can cause sound wave steepening into shock structures, for which relativistic effects now complicate the basic Rankine–Hugoniot shock relations. Taking proper account of such discontinuous, shock-dominated, relativistic flow conditions is done by computationally aided research. Since the chosen lab Lorentz frame allows a prescription using hyperbolic partial differential equations, expressing conservation laws, one can turn a standard Euler solver into a relativistic gas dynamics code. To allow proper shock-capturing, the employed discretizations use the conservative form of the governing equations, where conservative variables are updated by fluxes. The phase speed relation shown in Fig. 4.4 as the double loop structure then typically enters the numerical flux expressions, determining the local characteristic speed.

4.3.2 *Selected Relativistic Jet Modeling Efforts*

One such code is the open-source MPI-AMRVAC code (Keppens *et al.*, 2012), designed to handle hyperbolic PDEs with shock-capturing techniques while employing a fully parallel algorithm to time-advance the block-adaptive grid. With this code, several relativistic hydro studies in astrophysical jet conditions have been undertaken, with special attention to the following research questions.

- How do highly energetic jets (especially FR II jets) decelerate? Where is the energy primarily deposited, and what is the role of external medium variations?
- What is the role of geometric effects dictated closer to the source, and how do cylindrical to conically injected jets differ in their mixing efficiencies?
- How does jet precession alter mixing and propagation characteristics?

The numerical approach always prescribes jet properties from observationally well-established parameters, such as the bulk Lorentz factor, the overall kinetic luminosity of the jet beam, and geometric factors relating to jet opening angle or jet precession characteristics. These parameterized jets then enter the domain filled with IGM material, which can be varied from uniform to (dis)continuously changing. By exploiting grid-adaptive capabilities, highly detailed simulations by Meliani *et al.* (2008) have shown that the density contrast between the jet and IGM controls the overall morphology, with underdense jets leading to turbulent cocoons in between the internally shocked jet beams and the overarching bow shock front. For highly relativistic jets ($\Gamma \approx 20$) with FR II energy content, decelerations to FR I type propagation speeds can be achieved when the jet path encounters a dramatic density discontinuity in the IGM stratification. Discontinuous density jumps varying from 10 to 1000 were adopted, and only the larger contrasts can effectively slow down both FR I and II jet setups, where ultimately underdense jet propagation triggers effective mixing. Part of the directed jet energy can be reflected back upon encountering such density changeovers, while the shocked contact interface can itself become liable to Richtmeyer–Meshkov instability development.

In a follow-up study (Monceau-Baroux *et al.*, 2012), similarly highly relativistic jets, now with significant opening angles ($5-10^\circ$), were simulated as they traversed a realistically decreasing density IGM atmosphere. As in the non-relativistic study mentioned earlier, recollimation aspects could then be quantified. After the first recollimation shock, the jet beam gets into a relativistically hot regime (where the effective polytropic index γ drops to $4/3$, changing shock compression ratios) due to a rarefaction pattern enclosing the jet recollimation zone. Further shocks down the jet beam path result from interactions with vortices shed from the working surface separating jet material from shocked, swept-up, IGM matter. The near-static recollimation shock patterns could act as efficient Fermi acceleration sites for charged particles. Using masks to instantaneously differentiate the jet beam, instability mixing zone, and shocked ISM cocoon region, quantitative aspects of energy transfer could be established. As summarized for several jet setups in Fig. 4.5, it was found that (i) wider opening angle jets decelerate faster, and are accompanied by wider mixing zones; (ii) energy transfer mainly happens in the shocked IGM region, by cocoon traversing waves/shocks and at the frontal bow shock; and (iii) finite opening angle jets can feed up to 70% of their energy into the shocked IGM regions.

The third research topic listed above (jet precession) forces full 3D relativistic studies, which Monceau-Baroux *et al.* (2014) investigated specifically for the well-studied case of

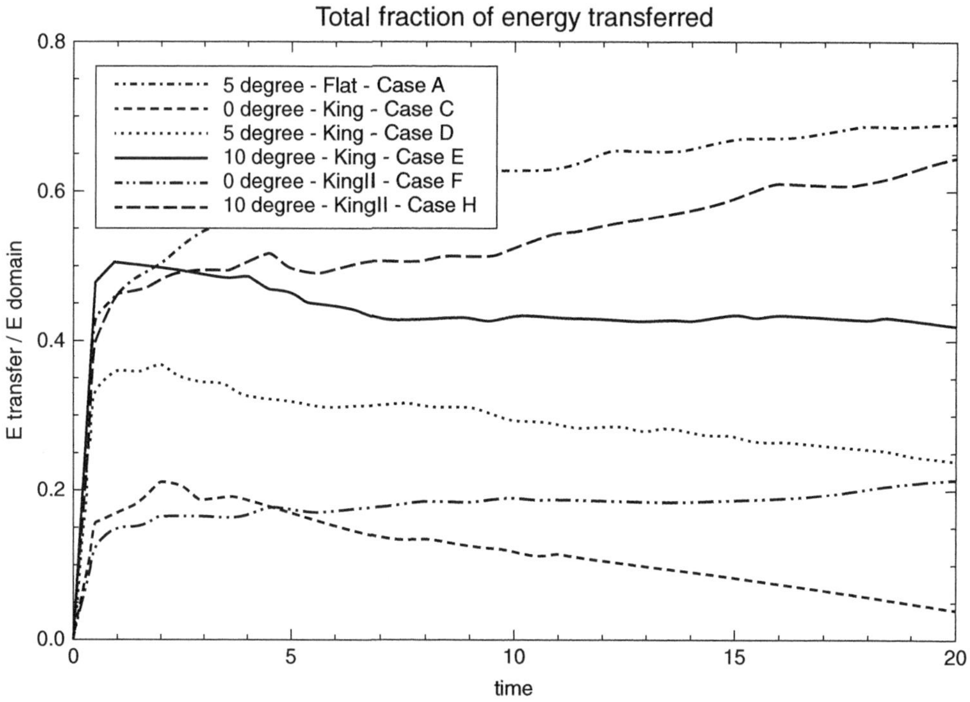

FIGURE 4.5. The fraction of energy transferred to the mixing zone and shocked ISM medium, for different relativistic hydro, cylindrical to conical jet setups. From Monceau-Baroux *et al.* (2012). Reproduced with permission © ESO.

the SS433 jet. As mentioned before, the mildly relativistic conditions in the jets of this XRB system have been observed at sub-parsec scales, and detailed radio maps using VLA have earlier only been confronted with a purely kinematic model of a precessing jet at its canonical 0.26c propagation speed. By performing simulations where the precessing jet is treated as an inner boundary prescription, its interaction with the ISM is self-consistently incorporated. Turning the 3D data into a virtual radio map provides a convincing match with the (details in) the VLA findings (see Fig. 4.6), improving the purely kinematic model. By numerically exploring cases with higher than the canonical 0.26c propagation speed, it was found that overdense, slow jets gradually build up a full helical jet pattern. In contrast, underdense, faster jets first inflate their cocoon surroundings, and undergo effective slow-down, much akin to their more energetic AGN counterparts.

4.4 Relativistic Magnetohydrodynamic Models

4.4.1 *Relativistic MHD Basics*

To do justice to magnetic field mediated influences in AGN jet dynamics, one must progress to relativistic magnetohydrodynamical descriptions. In terms of the four-velocity U^α, which has a temporal ($\alpha = 0$) and spatial ($\alpha = 1, 2, 3$) component given by $(c\Gamma, \Gamma \mathbf{v})$ in a fixed Lorentzian lab frame, particle conservation becomes $\partial_\alpha (\rho U^\alpha) = 0$, as in the hydro case. Energy–momentum conservation writes as

$$\partial_\beta \left(T^{\alpha\beta} + T^{\alpha\beta}_{\mathrm{em}} \right) = 0 \,, \tag{4.2}$$

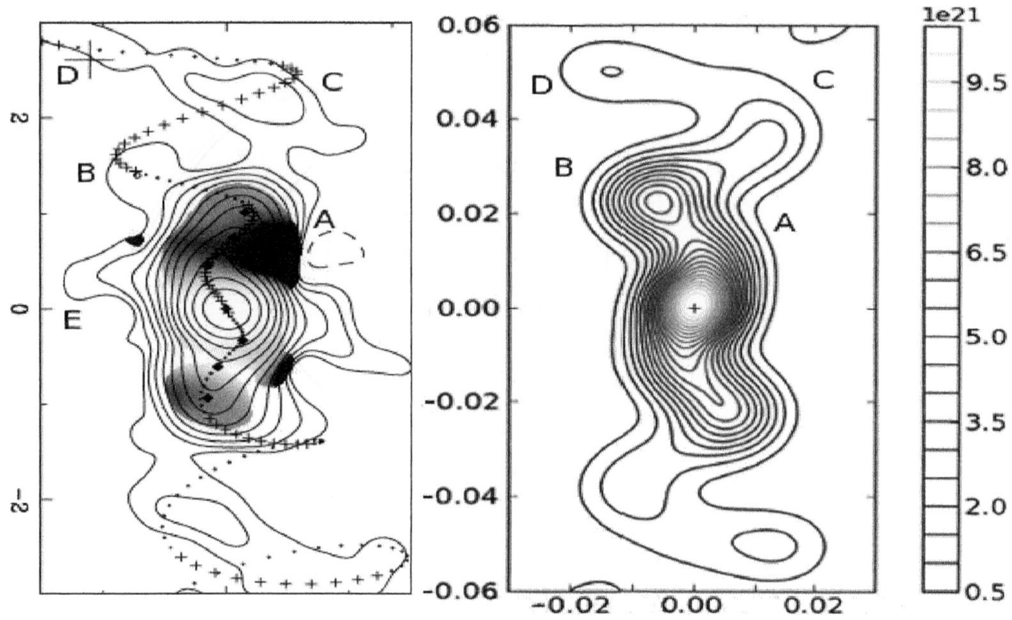

FIGURE 4.6. VLA observed (left panel, kinematic model overplotted) versus simulated radio observation (right) for the 3D precessing relativistic jet study in Monceau-Baroux *et al.* (2014). Spatial coordinates are given in pc. Both observed and simulated radio maps have contours with steps of factors of $\sqrt{2}$. Reproduced with permission © ESO.

where we deliberately separate the stress–energy tensor in a part already present for gas dynamics $T^{\alpha\beta}$ and the electromagnetic energy–stress tensor $T^{\alpha\beta}_{\rm em}$. These partial stress–energy tensor expressions contain (electromagnetic) energy density, energy flux, momentum flux, and stress contributions. For the electromagnetic part, these are readily written in terms of the lab frame specific electric $\mathbf{E}$ and magnetic $\mathbf{B}$ three-vectors, with e.g. the fluxes given by the Poynting flux $\mathbf{E} \times \mathbf{B}/\mu_0$. Adopting ideal MHD conditions, where $\mathbf{E} = -\mathbf{v} \times \mathbf{B}$, the system can be closed with the homogeneous Maxwell equations, appearing in the time-space split fashion as the usual $\nabla \cdot \mathbf{B} = 0$ and induction equation

$$\frac{\partial \mathbf{B}}{\partial t} - \nabla \times (\mathbf{v} \times \mathbf{B}) = \mathbf{0} \,. \tag{4.3}$$

The latter equation is identical to the Newtonian MHD equations, but it should be kept in mind that the relativistic MHD equations adopt the (Lorentz-invariant) full set of Maxwell equations, including displacement current contributions. The space-time split way of writing again ensures that traditional conservative, shock-capturing discretizations exploited for Newtonian MHD simulations, can relatively straightforwardly be carried over to special relativistic MHD settings (see e.g. Goedbloed *et al.*, 2010). The main difficulty is to properly handle extreme contrasts (static plasma to media moving at ultra-relativistic speeds, low to high plasma beta), ensuring positive pressures, energies, etc. The flux expressions needed to update the conservative variables now typically require an iterative nonlinear procedure to compute primitive variables like proper density, pressure, and velocity from them, while this was a mere algebraic process in the Newtonian case.

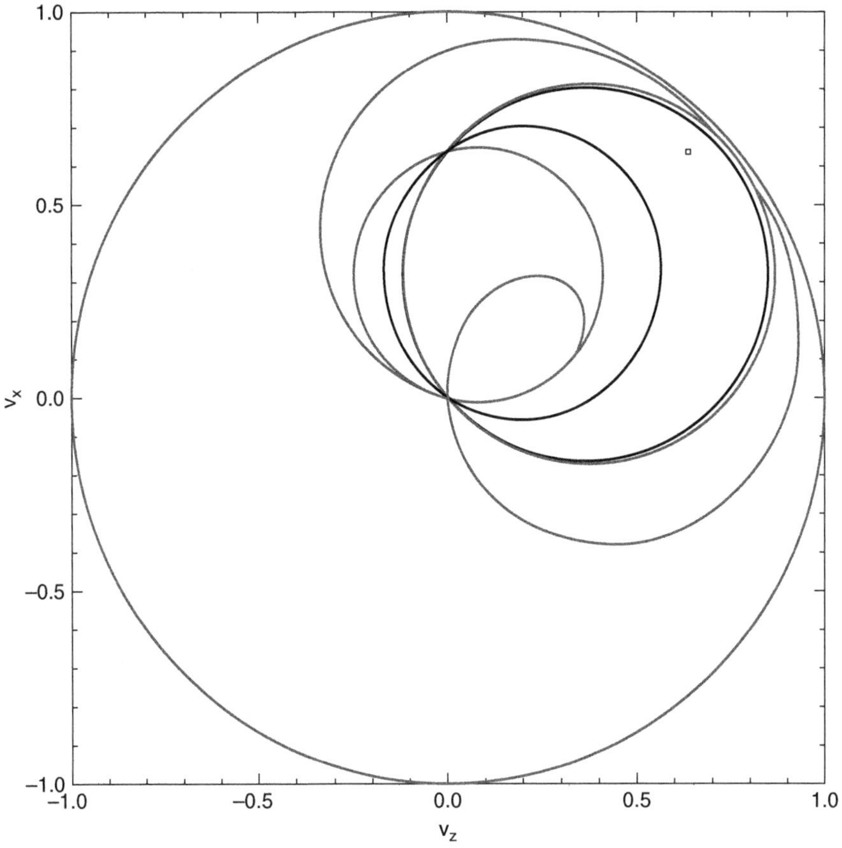

FIGURE 4.7. Phase diagram containing Alfvén, slow and fast magnetosonic signals, for plane waves as seen from a moving vantage point (from Keppens and Meliani, 2008).

The (relativistic) MHD description allows for seven linear waves, or characteristic speeds, namely an entropy wave, and pairs of slow magnetosonic, Alfvén, and fast magnetosonic speeds. All of these can be quantified in terms of phase and group speed expressions, and, even in the plasma rest frame, they display a clear anisotropic nature with slow, Alfvén, and fast speeds varying in magnitude as one changes direction with respect to the local uniform magnetic field three-vector **B**. Also, in contrast to the hydro case, phase and group speed diagrams differ significantly, even in the rest frame, although they are still related to each other by means of a Huygens construction (Keppens and Meliani, 2008). A phase diagram for a case with a horizontal (x-axis) magnetic field **B** is shown in Fig. 4.7, related to plane waves as emitted in a frame in relative motion of magnitude 0.9c along the diagonal. These phase speeds, here graphically drawn as curves in velocity space, encode again both relativistic Doppler and wave aberration aspects, which are at play for all wave families. Performing a Huygens construction on these diagrams leads to the group speed viewpoint, where e.g. the Alfvén wave signals become pure point perturbations traveling on and along the displaced magnetic field line, which was perturbed at the origin. It is to be stressed that even these linear wave aspects pose extreme demands on the capacity of numerical codes to properly distinguish the anisotropic propagation aspects, already at modest Lorentz factors (0.9c relates to $\Gamma \approx 2.29$ only). The phase speed quantifications are again an essential ingredient in most modern shock-capturing

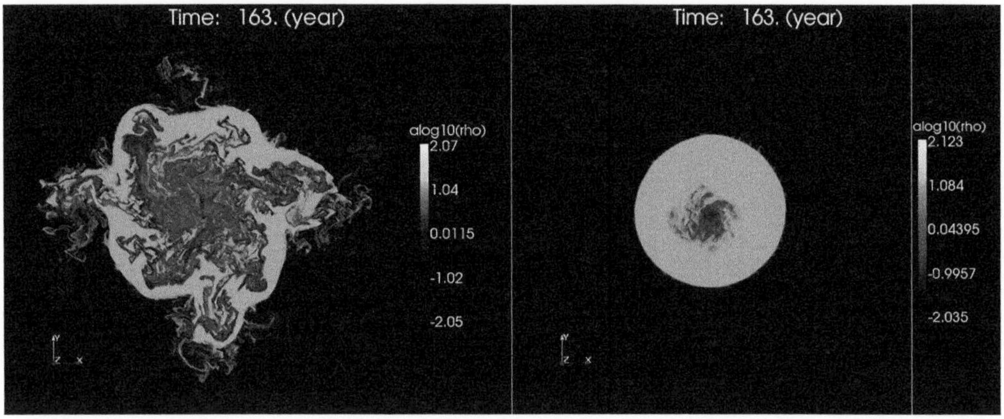

FIGURE 4.8. Cross sectional views of two-component jets. Left: a case liable to the relativistic Rayleigh–Taylor mode. Right: a case stable to this instability. From Meliani and Keppens (2009).

discretization schemes adopted, and they should ensure that, when discontinuities arise due to e.g. slow or fast wave family steepening, the jumps across the discontinuities obey the discrete equivalent of the Rankine–Hugoniot shock relations.

4.4.2 *Relativistic MHD Jet Dynamics from Simulations*

Relativistic MHD simulations targeting AGN jet physics gradually progressed to incorporate more realistic internal jet structures, magnetic acceleration, and jet launching scenarios. We highlight some representative dynamical studies in what follows. In Meliani and Keppens (2009), a two-component jet model was introduced, inspired by observations providing evidence for a radial stratification in AGN jets. Such a two-component structure is also expected from theoretical grounds, with a fast inner jet containing relativistically hot tenuous matter surrounded by a slower (Lorentz factor of a few) denser wind originating in the accretion disk. According to the current paradigm, each component is launched by a Blandford process (see Section 4.5). At distances far enough from the central regions, a special relativistic description suffices, and the two-component structure allows us to distribute the overall jet kinetic luminosity in different ratios over its inner and outer jet. The study showed that, when the inner jet inertia quantified by $\Gamma^2 \rho h + B^2$ exceeds the outer jet inertia, a pathway for a new relativistically enhanced Rayleigh–Taylor instability opens up. The effective gravity is due to centrifugal forces, as both jets rotate as a result of their launch mechanisms. Two-component jets that are initially liable to this instability, are seen to mix both components effectively.[3] This mixing causes a redistribution of angular momentum, and acts to slow down the overall jet significantly. In contrast, jets where the inner to outer jet energy content acts to have a larger effective inertia in the outer jet retain their two-component nature over many dynamically important timescales (see Fig. 4.8). In terms of the FR II to I AGN jet source dichotomy, this mechanism suggests that FR II jets with a two-component structure, as dictated by their launch conditions, can be slowed down to FR I type behavior when the relativistically enhanced Rayleigh–Taylor mode is operative. It should be noted

[3] Note that in a relativistic plasma, the inertia depends on the pressure via the specific enthalpy. As an additional effect, as the plasma surpasses relativistic temperatures, the adiabatic index of the mono-atomic gas transitions from the canonical value of 5/3 to the mass-less limit of 4/3, which in turn acts to increase the plasma inertia.

that relativistic flow is essential here, due to the $\Gamma^2 \rho h$ dependence of the effective inertia with the squared Lorentz factor acting to bring the light inner jet over the instability threshold due to its fast (Lorentz factors of order 10) progression. At the same time, the mechanism is important for both hydrodynamic as well as magnetohydrodynamic two-component jet setups.

The role of helical magnetic field topologies in kinetic energy dominated jets was studied for light, underdense cylindrical jets by Keppens *et al.* (2008). These jets also display a radial stratification, with central regions going up to a Lorentz factor of 22, but rather adopt axisymmetric conditions, preventing any disruptions due to non-axisymmetric mode development (such as the Rayleigh–Taylor mode discussed above). These simulations assumed that, at far enough distances, the jets are in a matter-dominated regime with magnetic fields at equipartition strengths. Still, the helical field responds to the appearance of internal jet beam shocks, which in turn result from vortices backflowing from the dynamically evolving contact surface between jet beam and ISM matter. Helicity changes are consistent with the diamond-shaped cross-shocks in the beam, and aid in jet re-acceleration, by pinching the jet flow.

That AGN jets almost necessarily end up in matter-dominated, cylindrically collimated, high Lorentz factor states (as used in Keppens *et al.*, 2008) has been argued by various analytic treatments, but has also been conclusively proven by axisymmetric relativistic MHD simulations. In Komissarov *et al.* (2007), a cleverly tailored numerical approach allowed us to follow magnetized jet acceleration over up to six orders of magnitude in the spatial scale. The jets were bounded by a prescribed-shape rigid boundary, which in essence sets the spatial distribution of density and pressure of a confining ambient medium. Within this funnel-shaped region, the solution strategy used shape-adapted coordinates and a grid-extension method designed to find steady solutions over increasingly larger regions. Cold, Poynting flux dominated jets were injected by rotating bottom boundary prescriptions appropriate for AGN jet launch conditions, where the cold matter starts at sub-Alfvénic speeds. The stationary solutions (see Fig. 4.9) demonstrated that the build-up (due to the rotation) of an azimuthal field component, and its associated magnetic pressure gradient, can act to accelerate the flow over large distances, beyond the classical fast magnetosonic surface. In these ideal relativistic MHD solutions, the jets developed a self-collimated core jet of cylindrical shape, and ended up in kinetic energy dominated states, with a final kinetic energy exceeding half the total energy flux. They augmented the analytic findings which usually adopt a restrictive self-similarity dependence, and verified the numerical solutions against known invariants along poloidal fieldlines, thereby quantifying their numerical accuracy. At the same time, they need to be revisited in full 3D settings to investigate stability against kink perturbations, and almost by construction leave out the time-dependent, shock-governed mixing aspects emphasized thus far.

A study targeting precisely these aspects has been performed by Porth (2013). Full 3D relativistic magnetized jets, as launched from a (boundary prescribed) Keplerian disk corona, demonstrated how self-collimation and acceleration (up to a modest Lorentz factor ≈ 2) could be demonstrated within the computational domain. The pitch profile as generated by the wound-up field lines realizes electrical force contributions counteracting the Lorentz force, in combination with letting the jet self-stabilize to kink displacements. In order to force such jets onto a more wiggly path, the medium through which they propagate should be clumpy (a high-density contrast of 100 to 1 was adopted for the clumps). In that case, the jet still stabilizes to kink deformations, but is forced to follow the path of least resistance. At the same time, the repeated interactions with the medium imprint a filamentary current layer structure throughout the jet, which

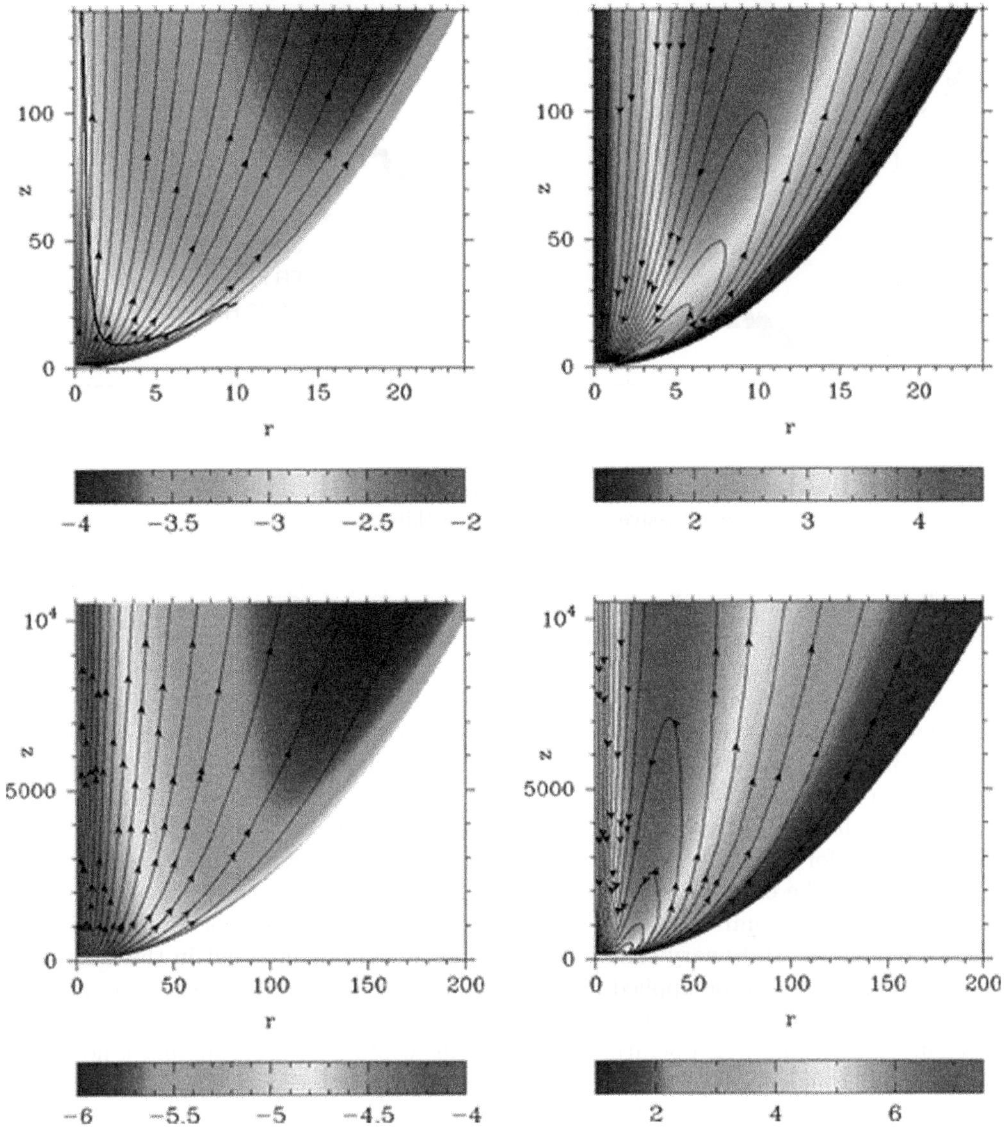

FIGURE 4.9. Close-in (top row) to further out (bottom row) views on the stationary relativistic MHD jet solutions from Komissarov *et al.* (2007). Right-hand panels show Lorentz factors and current lines. Left-hand panels show a laboratory jet density measure and poloidal field lines.

forms the basic ingredient for kinetic-scale mediated particle acceleration due to (chaotic) reconnection.

4.5 Jet Launching Aspects

4.5.1 *Scale-Invariance and the Fundamental Plane*

In the studies mentioned thus far, the special relativistic hydro or (ideal) MHD viewpoints on XRB or AGN jet physics did not address how these jets are ultimately connected to the underlying accretion disk–black hole configuration. At best, they prescribe a bottom

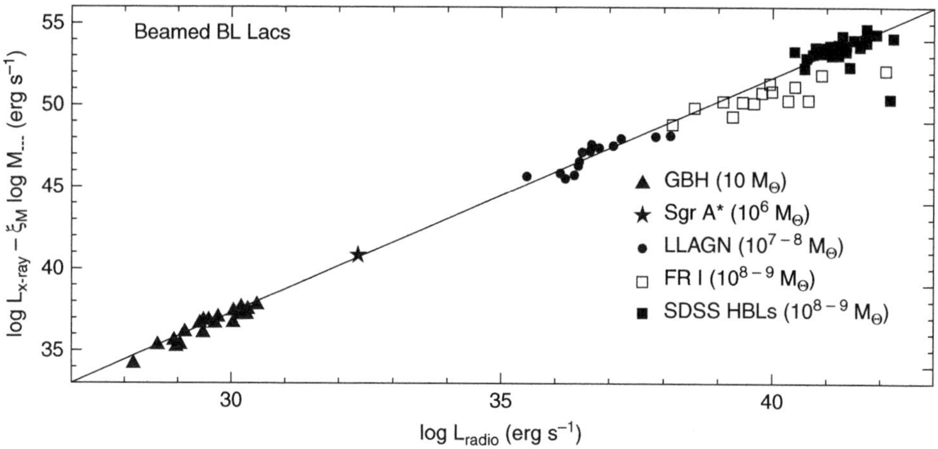

FIGURE 4.10. The fundamental plane for low-accretion rate black hole systems from stellar to supermassive masses, from Plotkin *et al.* (2012).

boundary magnetization and rotation profile inspired from self-similar analytical models for launching jets. As already mentioned for the two-component jet models, prevailing theories apply either to the magnetocentrifugal jet launch from a disk or to the electromagnetic energy extraction from the ergosphere of a rotating black hole. The former was introduced in a cold (zero temperature and pressure) self-similar Newtonian MHD solution by Blandford and Payne (1982), showing that a sufficiently bent magnetic field line anchored in a Keplerian rotating disk can act to magnetocentrifugally accelerate a jet-like outflow. The latter mechanism relates to a study by Blandford and Znajek (1977), where the equations for a force-free electromagnetic configuration (with only Poynting fluxes responsible for the energy transport) in a Kerr spacetime were shown to allow energy and angular momentum extraction from the rotating black hole. Both studies assume axially symmetric, stationary conditions and – as already pointed out in the Blandford and Payne paper – can be applied for any system with an accretion disk, from a stellar mass central object to a supermassive black hole.

That scale invariance is operating in all systems containing accreting black holes has been made explicit by the discovery of the so-called fundamental plane of black hole activity (Merloni *et al.*, 2003). This states that X-ray luminosity L_X, radio luminosity L_R, and black-hole mass correlate, across systems with central $\sim 10 M_\odot$ black holes as found in our Galaxy (galactic black holes or GBH), up to systems with $10^{10} M_\odot$ supermassive black holes (SMBH). The relation applies to sub-Eddington accreting black holes and, using modern regression analysis, has been updated (Plotkin *et al.*, 2012) to be expressed as

$$\log L_X = 1.45 \log L_R - 0.88 \log M_{\rm bh} - 6.07. \tag{4.4}$$

Figure 4.10 shows the relation for the 82 objects incorporated in the study, where GBHs, the black hole in the centre of our Milky Way (Sgr A*), low-luminosity AGN (LLAGN), FR I and various high-energy cutoff BL Lac objects from the Sloan Digital Sky Survey are plotted. The (slight) deviation for the FR I objects is explained by the fact that their X-ray jets are strongly affected by synchrotron cooling, an aspect neglected in the predicted scaling. The relation is a natural consequence when a scale-invariant jet description, depending on black-hole mass and accretion rate, is used to quantify the expected jet synchrotron emission (dominating L_R). The accretion flow itself, as well as optically thin synchrotron emission or inverse Compton emission from the jet, contribute to the X-ray luminosity.

4.5.2 *Newtonian MHD Models for Jet Launch*

That accretion disk–jet systems are ubiquitous is also evident from observations of protostar environments, especially the T-Tauri systems where central (proto)stars reach masses up to the solar value. Optical Hubble Space Telescope images provide direct views on the accretion disks, e.g. for HH30 where the disk is seen in absorption edge-on, with reflected light from the protostar on the further-out flaring disk. In emission lines, well-collimated jets can be detected over distances beyond 1000 AU, and knots within the jets move with supersonic speeds of a few hundred kilometers per second. Also for these systems, there is observational evidence for a disk–jet connection through proportionate jet and disk luminosities. In these Newtonian physics-governed disk–jet configurations, rotation measures at distances of 20–90 AU above the disk have resolved rotation reaching $\mathcal{O}(10)\,\mathrm{km\,s^{-1}}$ (Bacciotti *et al.*, 2002), and quantified its radial profile throughout the 30 AU wide jets (Dougados *et al.*, 2004). These are fully consistent with magneto-centrifugal launch theories, for which the Blandford and Payne prescription acts as prototype. This implies a magnetic origin for their launch and further collimation. In the more-variable accreting FU Ori system, Donati *et al.* (2005) reported the direct detection of a kG magnetic field in the inner disk region (down to 0.05 AU), where temperatures of order 10 000 K and number densities $n \approx 10^{23}\mathrm{m}^{-3}$ would translate in a plasma beta value of order unity, signaling equipartition field strengths. This all fits in the accepted paradigm that (equipartition) magnetic fields mitigate accretion rates, angular momentum transport and jet launch aspects, and form important ingredients for disk and jet equilibrium and (in)stability considerations.

Using numerical simulations in 2.5D (axisymmetric) MHD, Casse & Keppens (2002) could relax many of the restrictive assumptions (cold plasma, self-similarity, etc.) from the analytic approaches. By simulating how a geometrically thin, flaring disk in near-Keplerian rotation self-consistently deforms an initially vertical magnetic field distribution threading disk and surroundings, the authors showed that about 15% of the accreting material entering the domain at the outer radial boundary becomes deviated into a (hollow) jet flow. The magnetized accretion–ejection structure (MAES) that develops is a quasi-stationary state that can be scaled to a Young Stellar Object (YSO), accreting compact object (XRB systems), or an AGN, as it only sees the central object through its gravitational influence on the disk material. Most of this material accretes on the central regions, treated as an inner transparent boundary in the computational domain. The disk itself is in a resistive MHD regime, where the resistivity is crucial for allowing accreting matter to slip across the ultimately stationary magnetic field configuration. The magnetic torque inside the disk acts to slow down the rotational motion, to let gravity win from centrifugal forces and match the accretion rate. The regions above the disk and through the self-consistently formed jet are in an ideal MHD regime, where stationarity can be verified for the axisymmetric conditions by checking the expected invariants along poloidal field/streamlines. Analyzing the flow conditions throughout the jet, Alfvén and fast critical surfaces can be located, and the jet flow reaches fast magnetosonic speeds well before leaving the top boundary. The jet ejection mechanism is as expected from analytic theory, and shows that the pressure gradient at the disk surface aids to lift a fraction of the material out of the disk, precisely at the location where the magnetic torque changes sign: while it decelerated azimuthal motion in the disk, it accelerates it

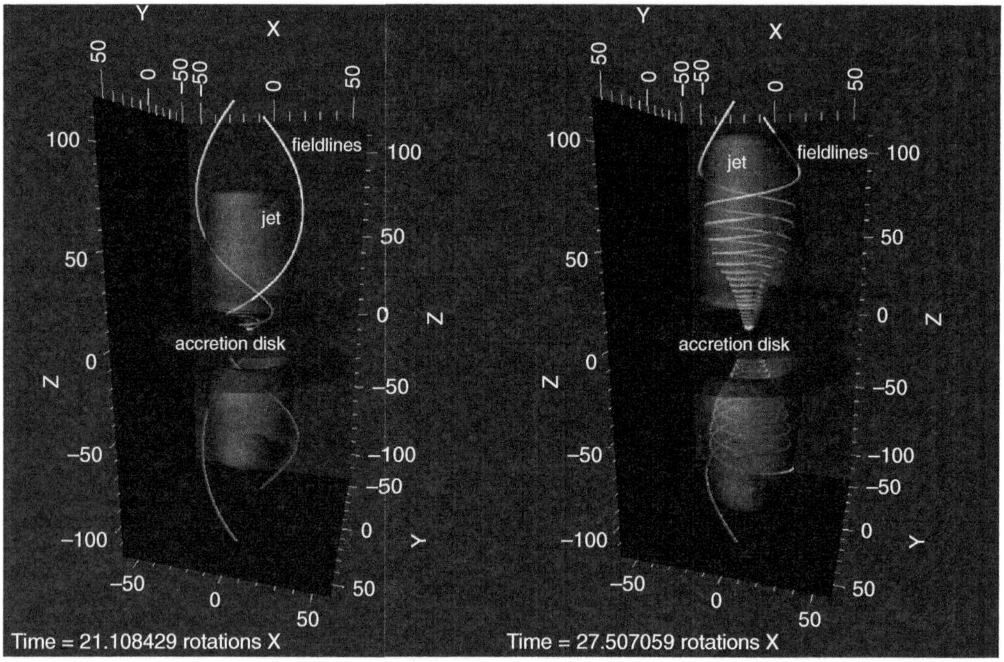

FIGURE 4.11. Snapshots from an axisymmetric MHD study where an accretion disk (gray isosurface) threaded by initially vertical magnetic field (white field lines) launches a self-collimated jet (gray density isosurface). Following Casse & Keppens (2002, 2004).

once above the disk. In combination, the jet then magneto-centrifugally accelerates to superfast magnetosonic velocities, thereby carrying off angular momentum into the jet. Selected views on the jet build-up are shown in Fig. 4.11, with density isosurfaces for disk versus jet material, and two jet-associated fieldlines. It is to be noted that the steady jet formation is preceded by a torsional Alfvén wave signal as soon as the simulation starts, as the vertical field inside the disk gets wound up by the near-Keplerian rotation. In the radial force balance through the jet, the magnetic hoop force is a dominant factor. The studies showed that, in addition to the sufficiently bent magnetic field configuration at the disk surface (known from the cold self-similar solution by Blandford and Payne, 1982), equipartition field strengths are needed throughout the jet base. By further improving the model in its energetics treatment (Casse & Keppens, 2004), it was possible to demonstrate that the jet luminosity rivals the energy liberated by the accretion process, with a hot jet emerging as the disk material is heated by both compression and Ohmic heating. As radiative losses were neglected, a radiatively inefficient MAES resulted, representative of systems with underluminous disks supporting bright and collimated jets, like M87.

The pioneering aspect of these simulations is the inclusion of the disk dynamics in the whole disk–jet system, even though the disk is treated using a parameterized spatio-temporally varying resistivity profile at anomalously high values (an α-type prescription, inspired from analytic viscous disk scalings by Shakura and Sunyaev, 1973). A follow-up study by Zanni *et al.* (2007) used adaptively refined (Flash code) computations, and highlighted the sometimes ambiguous role played by this resistivity prescription: reducing the resistivity values does not always allow for a steady jet launch outcome. Thereby, the influence of the numerical discretization, through numerical resistivity–viscosity effects, is sometimes hard to quantify, but is reduced when achieving sufficient resolution (e.g. through mesh-adaptivity). Another generalization (Meliani *et al.*, 2006a) combined the disk–jet setup with the idea of a two-component outflow, where an inner wind region related to the central object is driven by a hot stellar corona. At the same time, the α prescription was adopted for both viscosity and resistivity parameters, so viscous torques could aid in radially outward transport of angular momentum. Still, most angular momentum from the thin disk ended up being removed by Poynting flux. In the wind region (driven by a hot mass source along the polar axis) thermal and Lorentz forces control the inner regions, with the wind zone collimated in turn by the disk-driven jet. Recently, the role of (up/down) asymmetries in the accretion flow have been investigated by Sheikhnezami *et al.* (2012) and Fendt and Sheikhnezami (2013). Again, a key role in the dynamics, dampening/amplification of asymmetries, is played by the (ad hoc) parameterization of resistivity. However, once the resistivity is allowed dynamically to evolve with the flow, mass-outflow rates that differ by 10–30% between the bipolar jets could be observed for a range of symmetry-breaking perturbations applied to the disk.

Although these studies clarified many aspects of the disk–jet connection, many questions related to the ultimate jet stability (especially to kink-deformations) were left unanswered, by virtue of their assumption of axisymmetry. The same is true for details of mixing between jet/ISM matter, more time-dependent, shock-governed aspects in jet propagation, or the turbulent flows believed to underly the anomalously high visco-resistive disk behavior.

4.5.3 *General Relativistic Models*

While outflows from the outer regions of accretion disks surrounding black holes are well described by Newtonian models, the central few gravitational radii require a general relativistic perspective. Broadly speaking, close to the black hole, gravity will dominate causing accretion. In a Bondi type model (a spherically symmetric inflow turning supersonic close-in – see Bondi (1952) – as opposed to the Parker (1958) transonic solar wind solution), the radius where the inflow turns supersonic sets the accretion radius and defines the "plunging region." Rotation and angular momentum conservation can halt inflows, and induce deviations from spherical symmetry to form an outer torus–inner disk structure. The inner disk turbulence can again be prescribed phenomenologically through an α prescription. However, the strong radiation fields in the inner regions can repel the mass supply, and (temporarily) halt accretion. Electromagnetic influences can also dominate, and provide the Blandford–Znajek route to energy extraction when the central black hole rotates. For rotating (Kerr) black holes, both black-hole mass and spin-related length scale $a = J/M_{\rm bh}c$ for black-hole angular momentum J enter the $3+1$ space-time formulation, and define the region in between the spherical black-hole horizon and the ellipsoidally shaped border of the ergosphere, where corotation is enforced by frame-dragging. This allows the feeding of a Poynting dominated jet. A cartoon view of the possible near black-hole structure (Fig. 4.12, from chapter 4 of Boettcher *et al.* (2012)) shows the essential features, where (left) the disk is terminated at finite distance, while (right) a (thin disk, light gray) accretion flow is set up in the plunging region, causing magnetic flux accumulation onto the black hole.

4.5.4 *Stationary GRMHD Models*

General relativistic MHD (GRMHD) models for AGN jets have progressed to advanced full 3D, time-dependent numerical modeling, although mixed analytic–numerical approaches have also highlighted important aspects, especially for stationary

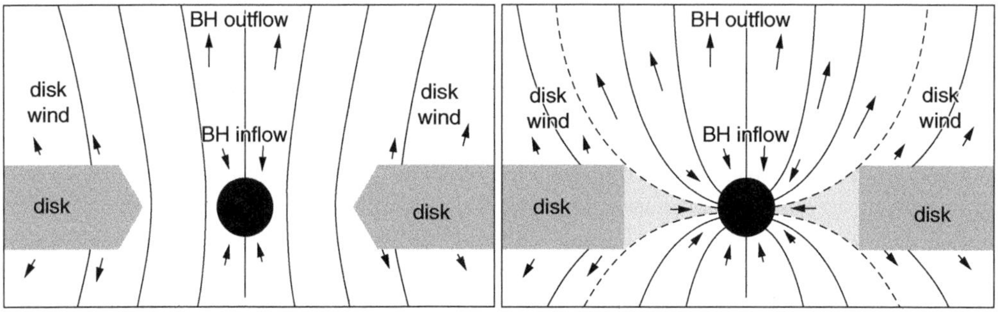

FIGURE 4.12. Cartoon view of the central black hole regions, without (left) and with (right) accretion in the plunging region. From Boettcher *et al.* (2012).

configurations. In the latter category, Meliani *et al.* (2006b) analyzed the collimation and acceleration properties of analytic, axisymmetric, stationary GRMHD solutions up to their asymptotic regime far away from a non-rotating, Schwarzschild black hole. While this excluded the Blanford–Znajek mechanism, it showed that a hot corona surrounding a black hole can drive a low Poynting flux spine (near-axis) jet. These jets can collimate to cylindrical geometry at large distances, by the usual magnetic pinching effect. They generalize meridional self-similar Newtonian models for jetted outflows, and the relativistic effects influence the launch region thermodynamics to increase the thermal driving of the flow. Another noteworthy effort is the early work by Fendt (1997), where a force-free, axisymmetric electromagnetic solution in Kerr geometry was computed by a finite-element based approach, giving a pure Poynting flux jet obeying the balance between electrical force and the Lorentz force throughout. The resulting stationary states represent global solutions, pass smoothly through the critical outer light-surface, and go beyond the low-spin parameter (a) solutions from the analysis of Blandford and Znajek (1977).

4.5.5 *Time-Dependent Simulations*

GRMHD numerical studies of 3D global time-dependent accretion flows, in fixed Schwarzschild to Kerr metrics, can currently be done in parametric fashion, whereby the spin value ($a = 0$ represents the Schwarzschild case), the initial magnetic field topology, and more numerical parameters are systematically varied (McKinney *et al.*, 2012). For a reference model taken from that study, Fig. 4.13 shows two cross-sectional snapshots in the top panel, as well as time history traces of several quantities, namely (i) the accretion rate on the black hole, $\dot{M}c^2(t)$, separated for the disk, the central jet, and the disk wind zones; (ii) a measure of the magnetic flux evolution near the black hole; and (iii) $\eta(t)$ to quantify an energy extraction efficiency. The sustained high values for the latter two quantities demonstrate that disk accretion can give rise to a near force-free magnetosphere in which the Blandford–Znajek mechanism operates efficiently. This can also be seen in the top left cross-sectional view in Fig. 4.13 containing the poles, where the thick magnetic field lines, connecting to the central black hole, are only lightly mass-loaded, so nearly all energy is carried there by Poynting fluxes. This model started from a weakly magnetized, geometrically thick torus surrounding a fast spinning black hole, and had originally only a poloidal field in alternating polarity. This non-equilibrium was rapidly distorted due to the development of magnetorotational and other magnetohydrodynamic instabilities, leading to a turbulent accretion flow. The advection of field towards

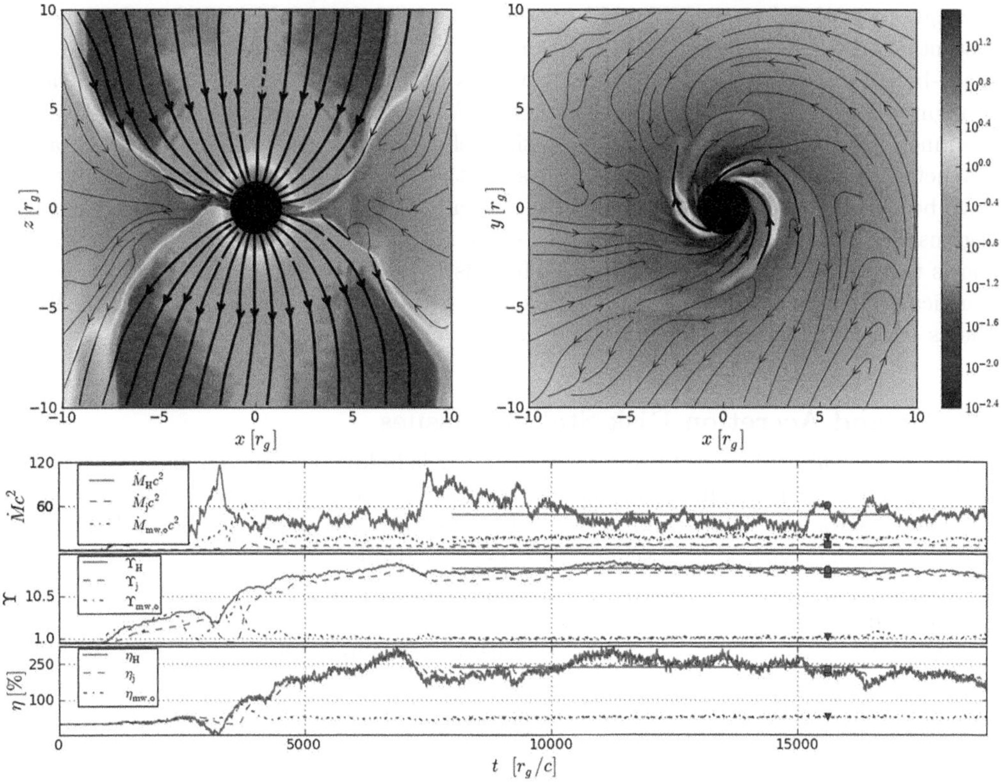

FIGURE 4.13. Snapshots (top row) from a 3D GRMHD run about a Kerr black hole, with time histories of accretion rate, black-hole flux evolution, and energy extraction efficiencies at bottom. From McKinney *et al.* (2012); see text for a discussion.

the central regions then leads to accumulated polar magnetic flux throughout the inner regions, forming a cylindrical region several Schwarzschild radii wide where the Poynting dominated jet resides. This forms a magnetic barrier, through which non-axisymmetric more-variable accretion onto the black hole takes place near the equator (see the equatorial cross-section at the top right of Fig. 4.13). These form the narrowing funnel-shaped accretion flows onto the black hole, and involve Rayleigh–Taylor/interchange dynamics permeating the accumulated field lines near the polar regions. They form geometrically thin inflows, aptly called "magnetically chocked accretion flows." From the parametric studies varying the field topologies, this type of configuration is typical for the higher rotation parameters, although they are also influenced by the initial disk thickness. Toroidal field configurations at high spin lead to more transient relativistic jets.

The case of spinning supermassive black-hole environments also allows us to question how the black-hole spin rate may be affected over time; this will differ for scenarios where the disk material is spiraling in the same direction or in the opposite direction (prograde to retrograde). One can also ask in which ways the disk angular momentum, jet direction, and black-hole spin axis are related. Using GRMHD simulations, similar to those discussed, McKinney *et al.* (2013) provided evidence for a magneto-spin alignment mechanism, by first starting a simulation about an untilted black hole with all three rotational axes then aligned in the quasi-stationary magnetically saturated state reached throughout a 40 Schwarzschild radii region. Subsequently, the black-hole spin axis was instantaneously tilted, and the simulation continued. The larger tilts caused the near black-hole magnetosphere regions to follow this abrupt change in the metric, as frame-dragging effects aided to react to this change on Alfvénic timescales. A magnetic torque on the inner disk regions results, and these ultimately get forced to reorient with the rotating magnetosphere. The new quasi-stationary state then shows an inner jet and disk part that have been forced to follow the black-hole spin adjustment, while the outer jet and disk regions still contain history of the original quasi-steady state. This numerical experiment shows that black-hole mass and spin evolutions (as yet only marginally explored) may drastically complicate the resulting 3D angular momentum, energy, and hence feedback effects of the AGN core regions on their galactic environments.

4.6 Jet and Accretion Disk Stability Issues

An aspect that challenges future research efforts is linking up the various studies focusing on different spatial regions or scales of the accretion disk–black hole–(two component) jet launch paradigm. This is particularly true for the role of physics beyond the ideal MHD viewpoint, with kinetic-scale processes responsible for accelerating particle populations to generate the high-energy radiation, which can already be investigated with idealized setups, highlighting details of reconnection processes, collisionless shocks and magnetic field generation via e.g. Weibel instability routes (Weibel, 1959). Even when sticking to a pure MHD approach, the aspect of relativistic jet acceleration and 3D stability especially requires scale-encompassing simulations from launch to termination. Topics of active research are (a) the details of how Poynting jets transform to matter-dominated AGN jets, with reconnection-aided acceleration processes along the way, and (b) why they manage to survive from violent disruptions through pressure gradient (interchange-type), shear-flow (Kelvin–Helmholtz) or current-driven (kink) instabilities on their megaparsec pathways. In the latter category, full 3D simulations in Newtonian (Keppens and Tóth, 1999; Baty and Keppens, 2002; Moll, 2009; Salvesen *et al.*, 2014) or relativistic (Mizuno *et al.*, 2007; Mizuno *et al.*, 2009; Mignone *et al.*, 2010; O'Neill *et al.*, 2012) MHD have highlighted that (i) local jet conditions can indeed cause rapid Kelvin–Helmholtz disruptions (Keppens and Tóth, 1999) and give insight into energy transfer through MHD turbulence (Salvesen *et al.*, 2014); (ii) nonlinear interactions of KH and current-driven modes may aid jet stability for sufficiently helical field topologies (Baty and Keppens, 2002); (iii) the toroidal field can decay along the jet path by kink-mode development and dissipation (Moll, 2009); (iv) relativistic jets with a spine-sheath structure may be stabilized against kink modes when a surrounding wind zone is present (Mizuno *et al.*, 2007); and (v) that also in relativistic settings, the details of the magnetic pitch profile dictates the kink instability linear to nonlinear development (Mizuno *et al.*, 2009; O'Neill *et al.*, 2012), although kink-unstable jets can still maintain a relativistic spine along their length (Mignone *et al.*, 2010). The fact that linear MHD instability development and interaction provides many routes to nonlinear turbulent regimes, allowing scale-free exchange between internal, kinetic, and magnetic energy reservoirs, means that to do justice to turbulent cascade processes one should exploit high resolution even in simple local box setups. In fact, the fair variety of linear MHD waves and instabilities in magnetized astrophysical jets, and even more so in magnetized accretion disks, calls for comprehensive quantifications of the MHD spectrum of all linear eigenmodes accessible to a specific stationary equilibrium configuration, preferably prior to full-scale nonlinear simulations. This allows magnetoseismic diagnosis of a gravitating, flowing (and rotating) plasma, a topic to which we turn next for accretion disk setups.

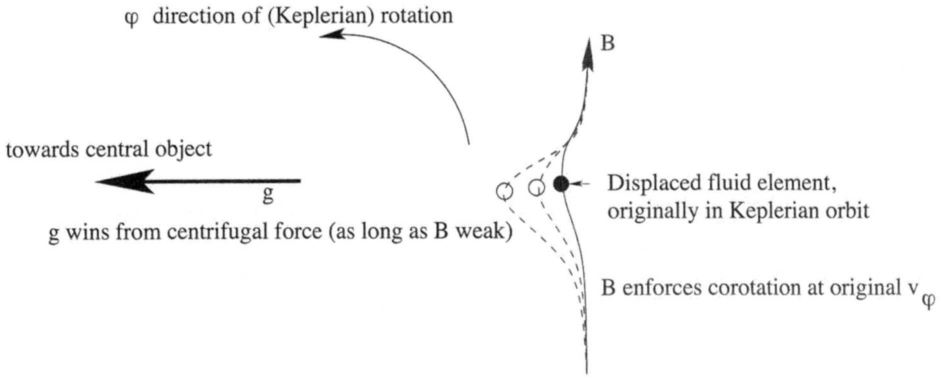

FIGURE 4.14. Cartoon view of the magnetorotational instability, with the central object at far left.

4.6.1 *Magnetoseismology Concepts*

Seismic probing of equilibrium configurations is of course well-known from geophysics, but has also been successfully used to determine the internal structure of the Sun with amazing accuracy. In the latter helioseismic variant, a solar model provides the run of temperature, pressure, and density with depth (assuming a static, spherically symmetric, unmagnetized equilibrium), which allows us to compute all linear pressure (p-) and gravity (g-) modes, describing plane waves that traverse the sphere as influenced by refraction throughout the Sun and reflection at its surface. The eigenoscillations, which obey the dispersion relation relating (spatial) eigenmode numbers with frequency, can be computed and contrasted with the observed frequency–spherical harmonic degree relations. Any mismatch between observed and computed frequencies is used to improve our knowledge of the background equilibrium stratification, and for the Sun has resulted in a better than few permille deviation between observations and theory in the end result (although incomplete knowledge of solar abundances has raised renewed discussions on it; see Chaplin & Basu, 2008). This is quite impressive, and exploits only an equilibrium structure where inward gravity is balanced by a pressure gradient in a 1D radial fashion. This technique is heavily used for all types of (variable) stars nowadays, yielding asteroseismic views on stellar internal structure and evolution. In principle, one can do the same for accretion disk configurations, where a Keplerian disk has centrifugal forces balanced by gravity. The introduction of (sheared) differential rotation does require the important switch from diagnosing static to stationary equilibrium configurations. Also, it has become evident that, in order to identify the modes responsible for the turbulent transport of angular momentum in disks (acting as anomalous visco-resistive factors), magnetic field effects need to be taken along, so that one really needs to perform linear eigenmode studies for stationary, gravitating, magnetized plasma equilibria.

In the case of accretion disks, linear MHD instabilities are the most likely route to magneto-turbulent accretion states. Differential rotation in a pure hydrodynamic Keplerian disk is stable according to the Rayleigh criterion, which predicts instability when the specific angular momentum Rv_φ decreases for increasing radius R. In a weakly magnetized setup, a linear MHD, weak-field shearing or magnetorotational instability (MRI), described by Velikhov (1959) and rediscovered by Balbus and Hawley (1991), can induce turbulent transitions and lead to effective accretion in the saturated, chaotic endstate. A crude description of its workings is given in Fig. 4.14, assuming a weak vertical field on which a fluid element becomes displaced from a Keplerian orbit towards the central object. Ideal MHD reasoning under axisymmetric stationary conditions means that the fluid element moves together with the field line, the magnetic field enforces corotation at the original azimuthal speed, such that gravity wins from centrifugal forces, and a run-away inward displacement results. Similar reasoning indicates that outwardly displaced fluid elements also undergo amplified displacement. The questions one should ask for proper accretion disk seismology studies are whether this MRI mode is the only instability relevant for accretion disk setups, whether non-axisymmetry provides alternative routes to instability, and to what extent does it survive when going to the (observed for T-Tauri stars and needed for magnetocentrifugal jet launch) equipartition field strengths. More fundamental questions to answer are its embedding in the full structure of MHD eigenmodes, which is well known for static, magnetized configurations as in fusion tokamak scenarios. There, the field of MHD spectroscopy (Goedbloed and Poedts, 2004; Goedbloed *et al.*, 2010) has allowed for detailed quantifications of the three-fold structure of slow, Alfvén, and fast modes in highly magnetized, non-gravitating, geometrically complex plasmas.

4.6.2 *Frieman–Rotenberg Formalism and Spectral Theory*

In fact, the theory to describe all linear waves and instabilities in ideal MHD, given an exact stationary, gravitating, magnetized plasma equilibrium, in any dimensionality (1D, 2D, 3D) has been known since 1960, and is governed by the equation of Frieman and Rotenberg (1960) given by

$$\rho\frac{\partial^2\boldsymbol{\xi}}{\partial t^2} + 2\rho\mathbf{v}\cdot\nabla\frac{\partial\boldsymbol{\xi}}{\partial t} + \underbrace{\nabla\Pi - \mathbf{B}\cdot\nabla\mathbf{Q} - \mathbf{Q}\cdot\nabla\mathbf{B} + \nabla\cdot(\rho\boldsymbol{\xi})\mathbf{g}}_{\text{Force operator } -\mathbf{F}(\boldsymbol{\xi})}$$
$$- \nabla\cdot[\rho\boldsymbol{\xi}(\mathbf{v}\cdot\nabla)\mathbf{v} - \rho\mathbf{v}\mathbf{v}\cdot\nabla\boldsymbol{\xi}] = 0\,. \quad (4.5)$$

In this equation, the equilibrium is allowed to have stationary flows $\mathbf{v}(\mathbf{x})$, a magnetic field variation through $\mathbf{B}(\mathbf{x})$, and density $\rho(\mathbf{x})$ and pressure $p(\mathbf{x})$. Their spatial $\mathbf{x}$-variation must obey the exact force balance between pressure gradients, Lorentz forces, inertial (e.g. centrifugal) effects, and gravity, here taken as given by an external gravitational acceleration $\mathbf{g}(\mathbf{x})$. The equation governs the spatio-temporal variation of a Lagrangian fluid element displacement $\boldsymbol{\xi}$, and this quantity also appears linearly in the Eulerian perturbation of total pressure Π and in the Eulerian perturbation of magnetic field $\mathbf{Q}$. When one is interested in eigenfrequencies, the temporal variation can be fixed to $\boldsymbol{\xi}(t) \propto \exp(-i\omega t)$, and eigenfrequencies ω are found for all solutions obeying the above PDE and suitable boundary conditions.

In the case of a static equilibrium where $\mathbf{v} = \mathbf{0}$, the eigenvalue determination is governed by the self-adjoint[4] operator $\mathbf{F}(\boldsymbol{\xi})$, and one needs to solve $-\rho\omega^2\boldsymbol{\xi} = \mathbf{F}(\boldsymbol{\xi})$. Owing to the self-adjointness of $\mathbf{F}(\boldsymbol{\xi})$, only real eigenvalues ω^2 can occur, so that stable waves ($\omega^2 > 0$) or instabilities ($\omega^2 < 0$) result. Note that pure adiabatic, hydrodynamic, stellar eigenoscillations for a static solar or stellar model indeed distinguish between stable, pressure-driven p-modes and g-modes that can either be stable or unstable, with their stability governed by a Schwarzschild criterion. In a homogeneous magnetized plasma, the same analysis yields three stable plane-wave-type families. In a finite-sized configuration with quantized mode numbers due to boundary conditions, these become a fast

[4] In the Hilbert space of solutions, the static force operator obeys $\int \boldsymbol{\eta}^* \cdot \mathbf{F}(\boldsymbol{\xi})\, dV = \int \boldsymbol{\xi} \cdot \mathbf{F}(\boldsymbol{\eta}^*)\, dV$.

(p-) mode sequence accumulating to infinite frequency squared, an infinitely degenerate Alfvén wave frequency, and a slow mode sequence accumulating to a slow clusterpoint frequency. These three accumulation points form the essential spectrum, and their degeneracy is lifted when inhomogeneity is introduced: both Alfvén and slow clusterpoints turn into continuous ranges of frequencies accessible to improper (infinitely localized) modes. Their ranges can, for simple 1D equilibria, easily be quantified from (i) the known run of equilibrium quantities; and (ii) fixed wavenumbers in the ignored directions. In static setups, one can use the self-adjointness of $\mathbf{F}$ to identify where Sturmian or anti-Sturmian eigenfrequency ranges may be situated, and these cluster sequences encompass the p-(fast) and g-mode families for the stratified unmagnetized case. Gravity can turn the Sturmian slow sequence to instability (Rayleigh–Taylor) but can also introduce discrete Alfvén mode sequences as additional g-modes. Their relation to the essential spectrum can be illustrated in the analytically tractable case of a slab of exponentially stratified magnetized atmosphere, where sound and Alfvén speed are constant with height (collapsing the essential spectrum to point values). There, one can show (Goedbloed *et al.*, 2010) that, for varying angle of the horizontal wavevector with the magnetic field, the bouyancy-mitigated Parker instability (Parker, 1958; long wavelength, wavevector aligned with the field) continuously changes from quasi-Parker instabilities up to an infinite sequence of unstable quasi-interchange modes (nearly perpendicular wavevector).

The full (mathematical) power of spectral theory governing physical eigenmode determination comes into play when using the Frieman–Rotenberg equation for moving $\mathbf{v} \neq \mathbf{0}$ equilibria, as applicable to jets and accretion disks. In that case, a generalized operator $\mathbf{G} = \mathbf{F}(\boldsymbol{\xi}) + \nabla \cdot [\rho\boldsymbol{\xi}(\mathbf{v}\cdot\nabla)\mathbf{v} - \rho\mathbf{v}\mathbf{v}\cdot\nabla\boldsymbol{\xi}]$, together with a Doppler–Coriolis operator $2\rho\mathbf{v}\cdot\nabla\frac{\partial\boldsymbol{\xi}}{\partial t} \equiv -2\omega U\boldsymbol{\xi}$, appear in the quadratic eigenvalue problem

$$-\rho\omega^2\boldsymbol{\xi} = \mathbf{G}(\boldsymbol{\xi}) - 2\omega U\boldsymbol{\xi}\,. \tag{4.6}$$

Both operators are self-adjoint, but allow the possibility of intrinsically complex eigenvalues $\omega \equiv \sigma + i\,\nu$, with overstable modes, e.g. driven by shear flow (Kelvin–Helmholtz). Unlike the case of static plasmas, where the transition to instability in the complex ω plane can happen only through the marginal frequency $\omega = 0$, the eigenmodes of stationary plasmas may enter the complex eigenfrequency plane from different locations along the real oscillation frequency σ-axis. While this may seemingly complicate the instability search, one can show that eigenvalue eigenvector pairs $(\omega, \boldsymbol{\xi})$, which obey equation (4.6), necessarily also obey the quadratic equation

$$\omega^2 - \omega\bar{V} - \bar{W} = 0\,, \tag{4.7}$$

which contains normalized versions of $V \equiv \int \boldsymbol{\xi}^* \cdot U\boldsymbol{\xi}\,dV$ and $W = -\int \boldsymbol{\xi}^* \cdot \mathbf{G}(\boldsymbol{\xi})\,dV$. This allows us to set up an algorithm to locate eigenvalues in the complex plane as follows. Let us assume a 1D variation for the sake of argument, with a left and right boundary condition given by $\mathbf{n}\cdot\boldsymbol{\xi} = 0$ for boundary normal $\mathbf{n}$. Pick an arbitrary $\omega \equiv \sigma + i\,\nu$, and solve the governing complex PDE system (numerically) for $\boldsymbol{\xi}(\mathbf{x})$ by starting at one boundary and integrating from the prescribed boundary behavior throughout the domain. The found function variation can be used to quantify $\bar{V}$ and $\bar{W}$. Eigenvalue–eigenfunction pairs will distinguish themselves from others by also obeying the other boundary condition, and such solutions will actually make $\bar{W}$ real. In that sense, curves given by $\mathrm{Im}(\bar{W}) = 0$ in the complex frequency plane contain the eigenvalues, and these curves are so-called solution paths. Another set of curves in the complex frequency plane can be identified by those frequencies that make the ratio of the radial Eulerian displacement to the total pressure perturbation Π, the so-called alternator, vanish. This alternator

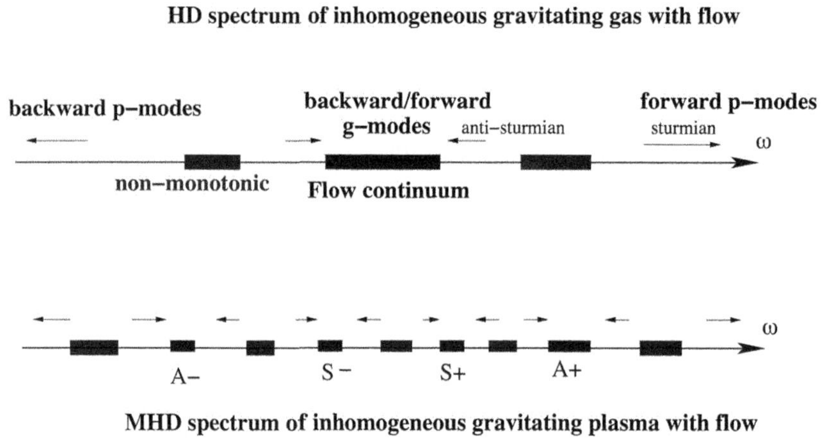

FIGURE 4.15. Hydro to magnetohydrodynamic eigenmode spectra organize about their essential spectrum, see text for a discussion. Adapted from Goedbloed *et al.* (2004a,b, 2010).

ratio is related to an impedance matching criterion, and a zero value guarantees fullfilment of the boundary condition $\mathbf{n} \cdot \boldsymbol{\xi} = 0$. In that sense, a spectral web of intersecting curves in the complex plane, namely solution paths and vanishing alternator paths, appears (Goedbloed, 2009a,b): only at their intersections can eigenfrequencies be found. Naturally, the basic essential spectrum of fast, Alfvén, and slow subspectra is still present to serve as organizing structure for labeling the various waves and instabilities, but in the presence of background flow, one must consider locally Doppler-shifted Alfvén and slow frequency ranges, which come into a backward and forward shifted variety. In that sense, the basic spectral structure is as illustrated in Fig. 4.15, contrasting a hydrodynamic spectrum of an inhomogeneous gravitating gas with flow, to an MHD case, in each case just emphasizing the stable wave organization along the real eigenoscillation axis (Goedbloed *et al.*, 2004a,b, 2010). The continuous ranges of backward and forward Doppler-shifted Alfvén and slow improper eigenfrequency ranges, together with the accumulation points at $\omega = \pm\infty$, degenerate in hydro to the flow continuum (Case, 1960) and accumulation points at infinite frequency for the p-modes (Goedbloed *et al.*, 2004a,b). Applications of this theory, demonstrating flow-driven modifications to Rayleigh–Taylor and Kelvin–Helmholtz instabilities for planar, sheared magnetic configurations have been shown by Goedbloed (2009a,b).

4.6.3 *Linear MHD Instabilities for Accretion Disks*

The theoretical considerations help to categorize the various ways to instability for an accretion disk (or astrophysical jet) equilibrium. For 1D disk configurations, with the MHD force balance in the disk equatorial plane about central mass M_* obeying

$$\left(p + \frac{B_\varphi^2 + B_Z^2}{2}\right)' = \rho\left(\frac{v_\varphi^2}{R} - \frac{GM_*}{R^2}\right) - \frac{B_\varphi^2}{R}, \tag{4.8}$$

one can indeed compute the eigenvalues throughout the complex plane, with an example taken from Keppens *et al.* (2002) and Goedbloed *et al.* (2010) in Fig. 4.16. This shows the eigenfrequency plane, together with the overlapping Alfvén and slow (in forward and backward variety indicated by $\Omega^{\pm}_{A,S}$) continua for axisymmetric modes with wavevector $\mathbf{k}$ for which the Doppler shift vanishes $\mathbf{k} \cdot \mathbf{v} = 0$. The equilibrium has a very weak, helical

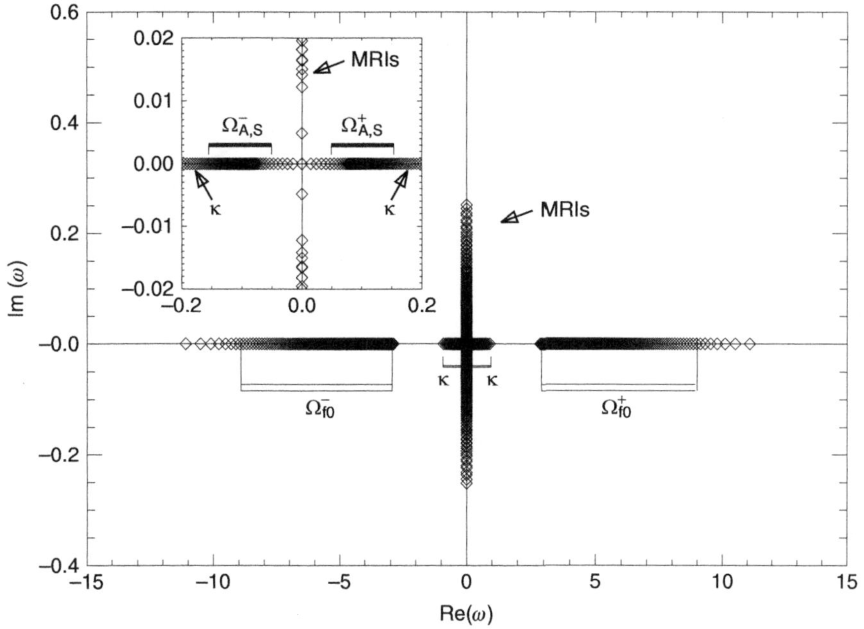

FIGURE 4.16. Eigenfrequency distribution for axisymmetric modes of a weakly magnetized accretion disk. From Keppens *et al.* (2002) and Goedbloed *et al.* (2010).

magnetic field, and linear instabilities (nearly, but not quite) align with the imaginary axis. Those are the MRI modes, associated with the slow subspectrum. A dense range of discrete stable wave modes linked to radial variation of the epicyclic frequency κ, is also present. When one draws the spectral web for the chosen equilibrium and wavenumber selection, the curve intersections perfectly correspond to the MRI eigenmodes. The accretion disk equilibrium studied here is parameterized by its power law dependencies on radius, and mimics the conditions through a thin, flaring disk. One can parametrically explore the location of all unstable modes throughout the complex frequency plane, for varying magnitude k of the wavevector, and obtain quantitative runs of the maximal growth rate out of the MRI sequence versus wavenumber, as shown in Fig. 4.17 (with slightly different parameter values than in Fig. 4.16).

The possibility to chart the complete MHD spectrum, as organized about the essential spectrum, also shows that lower plasma beta (β ratio of thermal to magnetic pressure) conditions allow for intricate non-axisymmetric overstabilities, which are keenly aware of the precise separation between forward and backward, partially to completely, overlapping Alfvén and slow frequency ranges. This is illustrated in Fig. 4.18, showing both the radial variation of the slow–Alfvén continua with radial distance (top panel), and the computed distribution of eigenmodes in the complex frequency plane (bottom panel), for non-axisymmetric modes. The equilibrium is at $\beta = 10$, but otherwise is similar to the previous figures. By varying systematically either mode numbers and/or equilibrium parameters, it is evident that these modes are a result of interacting cluster sequences of forward and backward slow/Alfvén modes (due to the modest β values it is difficult to clearly distinguish between slow and Alfvén modes, other than by their main polarization), and to call all of them MRI instabilities would be similar to categorizing all Belgian beers as of the "Stella"-variety.

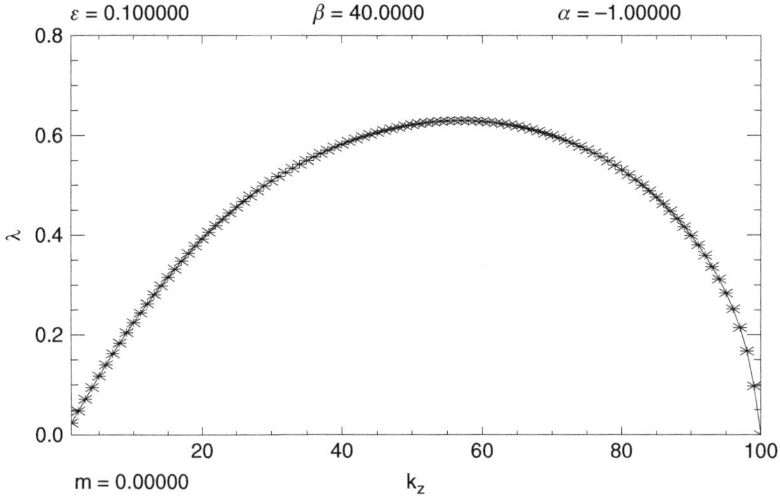

FIGURE 4.17. Maximal growth rate versus wavevector for axisymmetric MRI modes, for a weakly magnetized accretion disk similar to the case in Fig. 4.16.

4.6.4 *Equipartition Fields and Accretion Tori*

Based on the Frieman–Rotenberg formalism, one can compute full MHD spectra for realistic disk equilibria as shown above. Alternatively, various analytic results can be derived from it, e.g. for 1D disk equilibria as done by Blokland *et al.* (2007a,b). In that case, the governing eigenmode (with wavevector $\mathbf{k}$) equation becomes a second order ordinary differential equation for the radial component ξ_R of the Lagrangian perturbation $\boldsymbol{\xi}$, and

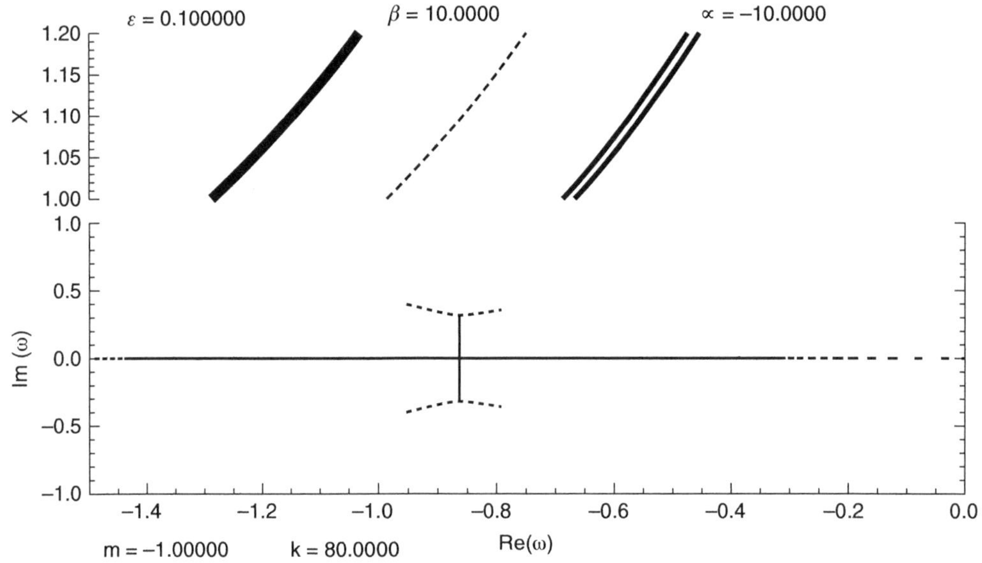

FIGURE 4.18. For a disk equilibrium at plasma $\beta = 10$, non-axisymmetric modes at given mode numbers show intricate interacting cluster sequences at the spectral web locations (bottom), with a clear relation to the four-fold essential spectrum of backward and forward propagating Alfvén/slow continua (top).

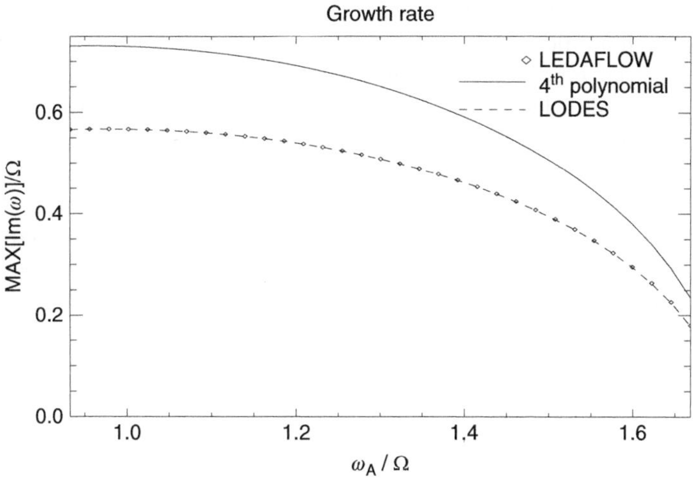

FIGURE 4.19. MRI growth rates for which a local dispersion relation neglecting toroidal field influences (solid line) overestimates the actual growth rates (shown to agree between a more complete local dispersion relation, dashed line, and a numerical result, diamonds). From Blokland *et al.* (2007a,b). Reproduced with permission © ESO.

it is singular at the local Doppler shifted Alfvén and slow frequency ranges. By means of a WKB analysis, a local dispersion relation can be found, which is a sixth-order polynomial in the Doppler-shifted eigenfrequency $\tilde{\omega} = \omega - \mathbf{k} \cdot \mathbf{v}$. The sixth-order matches with the forward and backward, slow, Alfvén, and fast possibilities. Under certain simplifying assumptions, such as when only axisymmetric modes are studied for a purely vertical magnetic field $\mathbf{B}$, this local dispersion relation allows us to deduce stability criteria for general field strength, formulated as

$$\frac{F^2}{\rho} \geq \frac{-k^2}{q^2 + k^2} \left[r(\Omega^2)' + \frac{\rho'}{\rho} V_g - \frac{\rho V_g^2}{\gamma p} \right] . \tag{4.9}$$

This formula contains the axial wavenumber k, its "radial" wavenumber q, and equilibrium variations of density ρ, pressure p, axial magnetic field (in the factor F), and centrifugal effects (factors Ω and V_g). These criteria can then be used for equipartition field strengths as well. The full sixth-order dispersion relation also quantifies effects due to dynamically important toroidal magnetic field components, and it was shown that axisymmetric MRI modes can persist to $\beta \approx 1$ conditions. Figure 4.19 shows a quantification of the growth rate of the most unstable MRI mode versus wavenumber (turned into a normalized Alfvén frequency). The lower curve shows the perfect agreement between the local, sixth-order dispersion relation and a full numerical solution of the governing linearized MHD equations, while the top curve quantifies the growth rate from the reduced dispersion relation without including toroidal field effects. The reduced relation overestimates the growth rate, as a strong toroidal field component is taken along.

Complementary results obtained using the local dispersion relation have quantified the contribution to the growth rate from convective to magneto-rotational influences, which both appear in the stability criterion from Eq. (4.9). Again looking at axisymmetric modes, van der Swaluw *et al.* (2005) concluded that $\beta \approx 1$ disks with dominant azimuthal field components B_φ necessarily have a dominant MRI contribution to the growth rate. A related study by Pessah and Psaltis (2005) also derived approximate local dispersion relations where compressibility, strong fields, and magnetic tension effects are included, although these authors did not start from the more elegant Frieman–Rotenberg formalism. They pointed out how curvature effects for $\beta \approx 0.01$ disks with dominant B_φ act to stabilize MHD overstabilities, especially for longer vertical wavelengths. Also, coupling of Alfvén and slow modes, for a broad range of field strengths, allows for additional instability routes. Future magnetoseismological computations are needed to show their close connection with the essential spectrum, as emphasized in the previous section.

All results for accretion disk seismology discussed so far adopt a 1D equilibrium configuration obeying Eq. (4.8), while the Frieman–Rotenberg formalism is valid for perturbations about full 3D MHD equilibria. Stationary (Newtonian) equilibria must obey the nonlinear force balance

$$-\nabla p + \mathbf{J} \times \mathbf{B} + \rho \mathbf{g} = \rho \mathbf{v} \cdot \nabla \mathbf{v} , \tag{4.10}$$

and it is nontrivial to compute magnetic configurations with general helical field topologies for accretion tori where both poloidal and toroidal flows are present.[5] For axisymmetric (2.5D) cases, one exploits the fact that the above equation can be rewritten to a second order PDE in terms of a flux surface label, augmented with an algebraic Bernoulli equation. Modern numerical approaches allow us to compute accurate accretion tori equilibria with significant poloidal and toroidal fields and flows. Example solutions showing density and poloidal Alfvén Mach number distribution in the poloidal cross-section (assumed symmetric about the midplane, only showing the top half) are given in Fig. 4.20. These are taken from Goedbloed *et al.* (2004a,b), and are computed with the FINESSE code (Belien *et al.*, 2002). The isoparametric, high-order finite element discretization adopts a given outer shape, and allows us to change the grid during the computation to ultimately align with the nested flux surfaces. It is also possible to transform the solution to a non-orthogonal coordinate representation in which the magnetic field lines, which trace out helical paths on the nested, torus-like flux surfaces, appear straight. This is critically important to allow a subsequent MHD spectral code, PHOENIX (Blokland *et al.*, 2007a,b), to map out all MHD eigenmodes allowed by the linearized MHD equations (i.e., the Frieman–Rotenberg formalism, augmented with trivial Eulerian entropy modes which appear at the local Doppler shift). The fact that the equilibrium is now intrinsically 2D (variation through the poloidal $[R, Z]$-plane) makes the quantification of the Doppler-shifted Alfvén and slow continuous parts of the spectrum nontrivial also, since the poloidal variation of the equilibrium can couple linear modes with different mode numbers in their poloidal angle variation, causing avoided crossings. These continuum modes are "improper" MHD eigenmodes localized on flux surfaces, and depending on the prevailing freedom in constructing equilibria, one can e.g. let the density be purely dependent on the flux surface label, making it constant on flux

[5] As a critical remark to the many efforts on global (GR)MHD simulations for accretion disks about black holes, it has become a fashion to start from exact hydrodynamic accretion tori (with stationary, axisymmetric rotational flows; see e.g. Hawley, 2000; Fishbone and Moncrief, 1976), and augment them with weak, purely poloidal field loops to initiate the time-dependent simulations. In fact, this configuration is never an exact MHD equilibrium, and must become violently unstable to kink deformations as soon as the magnetic field strength is raised. Moreover, such tori are known to be liable to non-axisymmetric, hydrodynamic instabilities as discovered by Papaloizou and Pringle (1984). Most numerical studies, e.g. those by McKinney *et al.* (2012), therefore always start at high plasma β values, of a few tens or higher. Actually, an analytic equilibrium solution for a Kerr metric, for an axisymmetric, magnetized torus with purely toroidal field and flow has been constructed by Komissarov (2006), for barotropic cases with constant angular momentum.

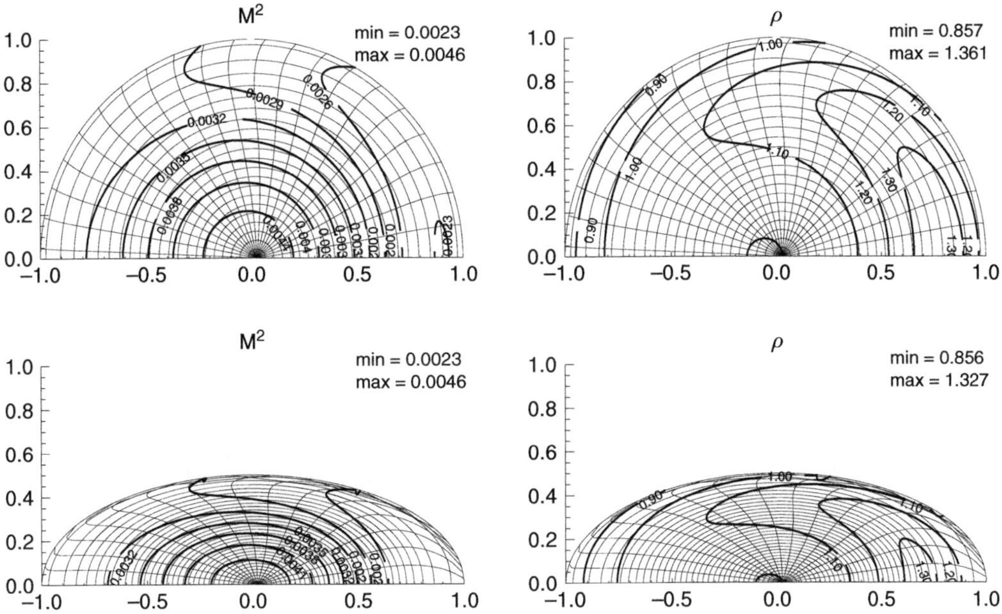

FIGURE 4.20. Variation of poloidal Alfvén Mach number and density throughout the (upper half of) the poloidal cross-section of magnetized accretion tori. The tori are axisymmetric with respect to the central object, assumed at far left. From Goedbloed *et al.* (2004a,b).

surfaces. In that case, Blokland *et al.* (2007a,b) demonstrated that convective continuum instabilities may arise (i.e. continuum modes becoming overstable) as governed by the precise variation of the Brunt–Väisäläa frequency projected on the flux surface. An example spectrum, showing only the continuum modes as located throughout the complex eigenfrequency plane is shown in Fig. 4.21 (left panel), where the CCI modes reach Alfvénic growth rates. Note that gaps appear in oscillation frequency ranges between the stable continuum ranges, and these are the preferred frequency ranges for discrete global mode sequences accumulating to the gap edges.

While Blokland *et al.* (2007a,b) analyzed thin to thick disks of unit plasma β with only toroidal rotation included, the most advanced MHD spectroscopic study for up to strongly magnetized disks with toroidal and poloidal flows, and otherwise general field topologies, has been provided by Goedbloed *et al.* (2004a,b). Augmented with analytic results on equilibria and linear continuum modes, a new class of instabilities was identified, the so-called trans-slow Alfvén continuum (TSAC) modes. They are the result of intricate coupling schemes between two Alfvén continuum branches, and four slow continuum branches at specific neighboring poloidal mode numbers. Figure 4.21 (right panel) again plots these continuum modes throughout the eigenfrequency plane, here computed for a thick accretion disk about a massive object. Just like the CCI modes, these TSAC modes are completely different from the traditional MRI modes: they come about when poloidal equilibrium flows exceed the local slow magnetosonic speed.

4.6.5 *Outlook for Accretion Disk Studies*

Our extensive discussion on linear Newtonian MHD modes clearly showed that what ultimately drives accretion flows and turbulence within accretion disk or tori configurations is as yet incompletely understood. Parametric surveys of the full MHD spectrum of waves

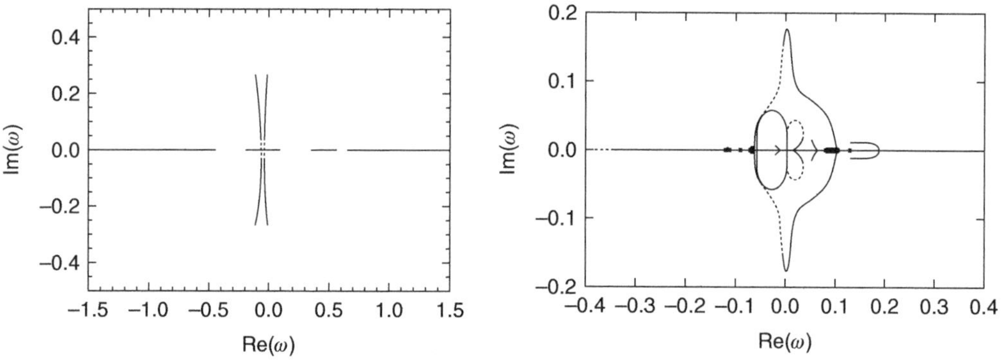

FIGURE 4.21. *Left:* The continuum (slow and Alfvén) eigenfrequencies for a thick accretion torus, with unstable modes due to the convective continuum instability (CCI). From Blokland *et al.* (2007a,b). *Right:* The continuum eigenmodes for a magnetized accretion torus, with also trans-slow Alfvén continuum (TSAC) instabilities. From Goedbloed *et al.* (2004a,b). Reproduced with permission © ESO.

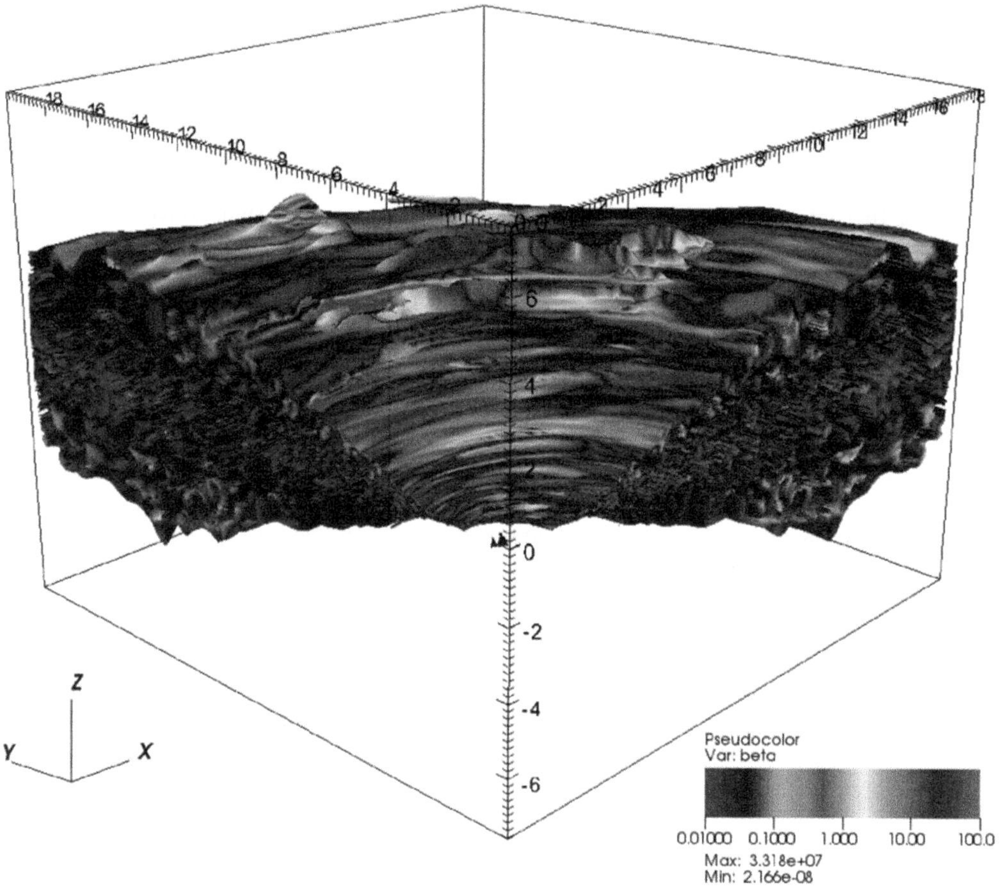

FIGURE 4.22. Distribution of plasma beta in a turbulent, thin accretion disk section. From global simulations in Beckwith *et al.* (2011).

and instabilities for 1D to more-dimensional configurations is called for, especially because the astrophysical literature almost invariably refers to MRI related activity, irrespective of the field topology or strength. We highlighted several linear MHD instability routes that rely on the intricate spectral structure in the ideal MHD equations. We have only partial notions of how nonideal effects (resistivity, viscosity) and nonadiabatic conditions complicate the mode structure (with e.g. the entropy modes yielding a pathway to thermal instability; Field, 1965), and how relativistic effects modify or enrich the spectrum (with e.g. new linear modes like the relativistically enhanced Rayleigh–Taylor instability for two-component setups; Meliani and Keppens, 2009). At the same time, fully global (GR)MHD simulations of accretion disks are now routinely undertaken, and Fig. 4.22 shows an example distribution of the plasma beta conditions in a geometrically thin disk section as simulated by Beckwith *et al.* (2011), where the achieved saturated state is characterized by efficient angular momentum transport radially outwards, and where the disk is surrounded by a magnetized corona with suprathermal fields (low β). The fact that such significant variations in plasma conditions arise in different disk regions, means that linear instabilities of all kinds are likely involved, and their detailed interactions and nonlinear behavior all contribute to the resulting dynamics. Their close connection with dynamo-generated fields, turbulent energy transfer, and with disk wind/jet studies, is only beginning to emerge in modern studies.

4.7 Conclusions

Motivated by observationally established relations between AGN activity and feedback, with a central role for kinetic feedback through large-scale AGN jets, we surveyed theoretical and computational findings related to magnetized, relativistic jet descriptions and their linkage to the accretion flows onto supermassive black holes. We exploited analogies with stellar mass black hole environments and protostar systems. Since the spatial and temporal scales involved are impossible to treat self-consistently from near black-hole regions to the far intergalactic medium, our current understanding gathers many results obtained under simplifying assumptions or adopting purely local conditions. Future challenges remain in accretion disk seismic studies, in distinguishing the role played by different MHD instabilities in disk and jet nonlinear evolutions, in establishing full 3D (two-component) accretion–ejection configurations covering launch to termination in their asymptotic flow regime, and showing how near black-hole conditions (Poynting dominated) connect to bulk plasma flows far out. Energy transfer mechanisms should thereby involve more details of particle acceleration processes and radiative effects.

Acknowledgments

These results were obtained in the framework of the projects GOA/2009/009 (KU Leuven) and G.0238.12 (FWO-Vlaanderen). We acknowledge funding from EC Seventh Framework Programme (FP7/2007-2013) under grant agreement SWIFF (project no. 263340, www.swiff.eu) along with the KU Leuven fellowship F+/11/027. This research benefits from the Interuniversity Attraction Poles Programme initiated by the Belgian Science Policy Office (IAP P7/08 CHARM). Part of the simulations used infrastructure of the VSC, the Flemish Supercomputer Center, funded by the Hercules foundation and the Flemish government, department EWI. RK acknowledges discussions within the ISSI team on "Flow driven instabilities of the sun–earth system," and the kind hospitality of the IAC winter school organization.

REFERENCES

Bacciotti, F., Ray, T. P., Mundt, R., Eislöfel, J., and Solf, J.: 2002, *ApJ* **576**, 222–231

Balbus, S. A., and Hawley, J. F.: 1991, *ApJ* **376**, 214–222

Baty, H., and Keppens, R.: 2002, *ApJ* **580**, 800–814

Beckwith, K., Armitage, P. J., and Simon, J. B.: 2011, *MNRAS* **416**, 361–382

Beliën, A. J. C., Botchev, M. A., Goedbloed, J. P., van der Holst, B. and Keppens, R.: 2002, *JCP* **182**, 91–117

Blandford, R. D., and Payne, D. G.: 1982, *MNRAS* **199**, 883–903

Blandford, R. D., and Znajek, R. L.: 1977, *MNRAS* **179**, 433–456

Blokland, J. W. S., Keppens, R., and Goedbloed, J. P.: 2007a, *A&A* **467**, 21–35

Blokland, J. W. S., van del Holst, B., Keppens, R., and Goedbloed, J. P.: 2007b, *JCP* **226**, 509–533

Boettcher, M., Harris, D. E., and Krawczynski, H.: 2012, in M. Boettcher, D. E. Harris, and H. Krawczynski (Ed.), *Relativistic Jets From Active Galactic Nuclei*, Wiley-VCH, ISBN 978-3-527-41037-8

Bondi, H.: 1952, *MNRAS* **112**, 195–204

Case, K. M.: 1960, *Phys. Fluids* **3**, 43–148

Casse, F., and Keppens, R.: 2002, *ApJ* **581**, 988–1001

Casse, F., and Keppens, R.: 2004, *ApJ* **601**, 90–103

Chaplin, W. J., and Basu, S.: 2008, *SPh* **251**, 53–75

Donati, J. F., Paletou, F., Bouvier, J., and Ferreira, J.: 2005, *Nat.* **438**, 466–469

Dougados, C., Cabrit, S., Ferreira, J., *et al.*: 2004, *Astrophys. Space Sci.* **293**, 45–52

Fabian, A. C.: 1999, *MNRAS* **308**, L39–L43

Fabian, A. C.: 2012, *ARA&A* **50**, 455

Fabian, A. C., Sanders, J. S., Allen, S. W., *et al.*: 2003, *MNRAS* **344**, L43–L47

Fanaroff, B. L., and Riley, J. M.: 1974, *MNRAS* **167**, 31–35

Fendt, C.: 1997, *A&A* **319**, 1025–1035

Fendt, C., and Sheikhnezami, S.: 2013, *ApJ* **774**, 12

Ferrarese, L., and Merritt, D.: 2000, *ApJ* **539**, 9–12

Field, G. B.: 1965, *ApJ* **142**, 531–567

Fishbone, L. G., and Moncrief, V.: 1976, *ApJ* **207**, 962–976

Frieman, E., and Rotenberg, M.: 1960, *Rev. Modern Phys.* **32**, 898–902

Goedbloed, J. P.: 2009a, *Phys. of Plasmas* **16**, 13

Goedbloed, J. P.: 2009b, *Phys. of Plasmas* **16**, 14

Goedbloed, J. P., and Poedts, S.: 2004, in *Principles of MHD. With Application to Laboratory and Astrophysical Plasmas*, Cambridge University Press

Goedbloed, J. P., Beliën, A. J. C., van der Holst, B., and Keppens, R.: 2004a, *Phys. of Plasmas* **11**, 28–54

Goedbloed, J. P., Beliën, A. J. C., van der Holst, B., and Keppens, R.: 2004b, *Phys. of Plasmas* **11**, 4332–4340

Goedbloed, J. P., Keppens, R., and Poedts, S.: 2010, in *Advanced MHD. With application to Laboratory and Astronomical Plasmas*, Cambridge University Press

Gopal-Krishna, and Wiita, P. J.: 2002, *NewAR* **46**, 357–360

Graham, A. W., Onken, C. A., Athanassoula, E., and Combes, F.: 2011, *MNRAS* **412**, 2211–2228

Hawley, J. F.: 2000, *ApJ* **528**, 468–479

Keppens, R., and Meliani, Z.: 2008, *Phys. of Plasmas* **15**, 102103

Keppens, R., and Tóth, G.: 1999, *Phys. of Plasmas* **6**, 1461–1469

Keppens, R., Casse, F., and Goedbloed, J. P.: 2002, *ApJ* **569**, L121–L126

Keppens, R., Meliani, Z., van der Holst, B., and Casse, F.: 2008, *A&A* **486**, 663–678

Keppens, R., Meliani, Z., van Marle, A. J., *et al.*: 2012, *JCP* **231**, 718

Komissarov, S. S.: 2006, *MNRAS* **368**, 993–1000

Komissarov, S. S., Barkov, M. V., Vlahakis, N., and Königl, A.: 2007, *MNRAS* **380**, 51

Krause, M., Alexander, P., Riley, J., and Hopton, D.: 2012, *MNRAS* **427**, 3196

Marscher, A. P.: 2006, *Relativistic Jets: The Common Physics of AGN, Microquasars, and Gamma-Ray Bursts*, AIPC, 856, 1

McKinney, J. C., Tchekhovskoy, A., and Blandford, R. D.: 2012, *MNRAS* **423**, 3083

McKinney, J. C., Tchekhovskoy, A., and Blandford, R. D.: 2013, *Science* **339**, 49

Meliani, Z. and Keppens, R.: 2009, *ApJ* **705**, 1594

Meliani, Z., Casse, F., and Sauty, C.: 2006a, *A&A* **460**, 1

Meliani, Z., Sauty, C., Vlahakis, N., Tsinganos, K., and Trussoni, E.: 2006b, *A&A* **447**, 797

Meliani, Z., Keppens, R., and Giacomazzo, B.: 2008, *A&A* **491**, 321

Mendygral, P. J., Jones, T. W., and Dolag, K.: 2012, *ApJ* **750**, 166

Merloni, A., Heinz, S., and di Matteo, T.: 2003, *MNRAS* **345**, 1057

Mignone, A., Rossi, P., Bodo, G., Ferrari, A., and Massaglia, S.: 2010, *MNRAS* **402**, 7

Mirabel, I. F. and Rodríguez, L. F.: 1998, *Nat.* **392**, 673

Mizuno, Y., Hardee, P., and Nishikawa, K.-I.: 2007, *ApJ* **662**, 835

Mizuno, Y., Lyubarsky, Y., Nishikawa, K.-I., and Hardee, P. E.: 2009, *ApJ* **700**, 684

Moll, R.: 2009, *A&A* **507**, 1203

Monceau-Baroux, R., Keppens, R., and Meliani, Z.: 2012, *A&A* **545**, A62

Monceau-Baroux, R., Porth, O., Meliani, Z., and Keppens, R.: 2014, *A&A* **561**, A30

O'Neill, S. M., Beckwith, K., and Begelman, M. C.: 2012, *MNRAS* **422**, 1436

Papaloizou, J. C. B. and Pringle, J. E.: 1984, *MNRAS* **208**, 721

Parker, E. N.: 1958, *ApJ* **128**, 664

Pessah, M. E. and Psaltis, D.: 2005, *ApJ* **628**, 879

Plotkin, R. M., Markoff, S., Kelly, B. C., Körding, E., and Anderson, S. F.: 2012, *MNRAS* **419**, 267

Porth, O.: 2013, *MNRAS* **429**, 2482

Reynolds, C. S., Heinz, S., and Begelman, M. C.: 2002, *MNRAS* **332**, 271

Salvesen, G., Beckwith, K., Simon, J. B., O'Neill, S. M., and Begelman, M. C.: 2014, *MNRAS* **438**, 1355

Shakura, N. I. and Sunyaev, R. A.: 1973, *A&A* **24**, 337

Sheikhnezami, S., Fendt, C., Porth, O., Vaidya, B., and Ghanbari, J.: 2012, *ApJ* **757**, 65

Silk, J. and Rees, M. J.: 1998, *A&A* **331**, L1

Sutter, P. M., Yang, H.-Y.K., Ricker, P. M., Foreman, G., and Pugmire, D.: 2012, *MNRAS* **419**, 2293

Tonry, J. and Davis, M.: 1979, *AJ* **84**, 1511

van der Swaluw, E., Blokland, J. W. S., and Keppens, R.: 2005, *A&A* **444**, 347

Velikhov, E. P., *Soviet Phys.-JETP Lett.* **36**, 995

Weibel, E. S.: 1959, *Phys. Rev. Lett.* **2**, 83

Wurster, J. and Thacker, R. J.: 2013, *MNRAS* **431**, 2513

Zanni, C., Ferrari, A., Rosner, R., Bodo, G., and Massaglia, S.: 2007, *A&A* **469**, 811

5. Magnetic Fields in Galaxies

RAINER BECK

5.1 Introduction

Magnetism is a fundamental force with special properties.

- Most baryonic matter is ionized. Magnetic fields are easy to generate.
- Magnetic monopoles do not exist or are extremely rare. Magnetic fields are hard to destroy.
- Magnetic fields need illumination. Magnetic fields are difficult to observe.

The scarceness of data leaves many questions on cosmic magnetic fields.

- When and how were the first fields generated?
- Did significant fields exist before galaxies formed?
- How and how fast were the fields amplified?
- How did fields affect the evolution of stars, planets, galaxies and galaxy clusters?
- How strongly is intergalactic space magnetized?

Magnetic fields are often ignored in astrophysics, in particular in models of galaxies and the interstellar medium (ISM), although they are a major agent in the ISM and in galaxy halos, and important for the structure and evolution of galaxies.

- Magnetic fields contribute significantly to the total pressure which balances the gas disk of galaxies against gravitation (Fletcher & Shukurov, 2001).
- Magnetic turbulence distributes energy from supernova explosions within the ISM (Subramanian, 1998).
- Magnetic reconnection is a possible heating source for the ISM and halo gas (Birk et al., 1998).
- Magnetic fields increase angular momentum transport and hence the gas inflow rate in barred galaxies (Beck et al., 2005; Kim & Stone, 2012).
- Magnetic fields affect the dynamics of the turbulent ISM (de Avillez & Breitschwerdt, 2005) and the gas flows in spiral arms (Gómez & Cox, 2002).
- Magnetic fields make the gaseous spiral arms more patchy and drive gas outflows into the halo (Pakmor & Springel, 2013).
- The shock strength in spiral density waves is decreased and structure formation is reduced in the presence of strong fields (Dobbs & Price, 2008; Fletcher et al., 2011).
- Magnetic fields stabilize gas clouds and reduce the star-formation efficiency to the observed low values (Vázquez-Semadeni et al., 2005; Price & Bate, 2008).
- Magnetic fields are essential for the onset of star formation as they enable the removal of angular momentum from the protostellar cloud via ambipolar diffusion (Heitsch et al., 2004).
- Strong fields may shift the initial stellar mass spectrum towards the more massive stars (Mestel, 1990).
- Magnetic fields control the density and distribution of cosmic rays in the ISM (Strong et al., 2000).
- Cosmic rays accelerated in star-forming regions can provide the pressure to drive galactic outflows and buoyant loops of magnetic fields via the Parker instability (Hanasz et al., 2002).
- Outflows from galaxies can significantly magnetize the intergalactic medium (Kronberg et al., 1999).

- Magnetic fields in the halos of galaxy clusters can affect thermal conduction and hence the dynamics and evolution of the intracluster medium (Balbus & Reynolds, 2008; Parrish & Quataert, 2008).

These results are based on modeling. Some of them were found to be consistent with observations, if available, others still need testing.

5.2 Measuring Magnetic Fields in Galaxies

Magnetic fields need illumination to be detectable. *Polarized emission* at optical, infrared, submillimeter, and radio wavelengths holds the clue to measuring magnetic fields in galaxies.

5.2.1 *Basic Magnetic Field Components*

The total magnetic field can be separated into a regular component and a turbulent component. A regular field has a well-defined direction within the telescope beam, while a turbulent field frequently reverses its direction within the telescope beam. In other words, the coherence scale of regular fields is much larger than the turbulent scale. Turbulent fields can be isotropic turbulent (i.e. the same dispersion in all three spatial dimensions) or anisotropic turbulent (i.e. different dispersions). Anisotropic turbulent fields can be generated from isotropic turbulent fields by compressing or shearing gas flows. Observations with limited spatial resolution cannot distinguish turbulent fields generated by turbulent gas flows from fields tangled by small-scale gas motions; observations with higher spatial resolution can distinguish them, based on their different power spectra.

Polarized synchrotron emission emerges from the components of ordered fields in the sky plane. Anisotropic turbulent, anisotropic tangled, and regular fields cannot be distinguished by polarization observations because polarization angles are ambiguous by 180°. Faraday rotation and the longitudinal Zeeman effect are sensitive to the field direction and hence trace regular fields.

Table 5.1 summarizes the different field components and the methods of their observation.

5.2.2 *Optical and Near-Infrared Polarization*

Elongated dust grains can be oriented with their major axis perpendicular to the field lines by paramagnetic alignment (the *Davis–Greenstein effect*) or, more efficiently, by radiative torque alignment (Hoang & Lazarian, 2008). When particles on the line of sight to a star are oriented with their major axis perpendicular to the line of sight (and the field is oriented in the same plane), the different levels of extinction along the major and the minor axis leads to polarization of the starlight, with the E-vector pointing parallel to the magnetic field. This is the basis for measuring magnetic fields with optical and near-infrared polarization, by observing individual stars or diffuse starlight. Extinction is most efficient for grains of sizes similar to the wavelength. These small particles are only aligned in the medium between molecular clouds, not in the dense clouds themselves (Cho & Lazarian, 2005).

The detailed physics of the alignment is complicated and not fully understood. The degree of polarization p generated in a volume element along the line of sight of extent L is proportional to $(B_{\mathrm{tot},\perp}^2\, L)$, but also depends on the magnetic properties, density, and temperature of the grains (Ellis & Axon, 1978).

TABLE 5.1. Magnetic field components and their observational signatures.

Field component	Notation	Geometry	Observational signature
Total field	$B_{\mathrm{tot}}^2 = B_{\mathrm{turb}}^2 + B_{\mathrm{reg}}^2$	3D	Total synchrotron intensity, corrected for inclination
Total field perpendicular to the line of sight	$B_{\mathrm{tot},\perp}^2 = B_{\mathrm{turb},\perp}^2 + B_{\mathrm{reg},\perp}^2$	2D	Total synchrotron intensity
Turbulent or tangled field	$B_{\mathrm{turb}}^2 = B_{\mathrm{iso}}^2 + B_{\mathrm{aniso}}^2$	3D	Total synchrotron emission, partly polarized, corrected for inclination
Isotropic turbulent or tangled field perpendicular to the line of sight	$B_{\mathrm{iso},\perp}$	2D	Unpolarized synchrotron intensity, beam depolarization, Faraday depolarization
Isotropic turbulent or tangled field along the line of sight	$B_{\mathrm{iso},\parallel}$	1D	Faraday depolarization
Ordered field perpendicular to the line of sight	$B_{\mathrm{ord},\perp}^2 = B_{\mathrm{aniso},\perp}^2 + B_{\mathrm{reg},\perp}^2$	2D	Intensity and vectors of radio, optical, IR, submillimeter polarization
Anisotropic turbulent or tangled field perpendicular to the line of sight	$B_{\mathrm{aniso},\perp}$	2D	Intensity and vectors of radio, optical, IR, submillimeter polarization, Faraday depolarization
Regular field perpendicular to the line of sight	$B_{\mathrm{reg},\perp}$	2D	Intensity and vectors of radio, optical, IR, submillimeter polarization, Goldreich–Kylafis effect
Regular field along the line of sight	$B_{\mathrm{reg},\parallel}$	1D	Faraday rotation and depolarization, longitudinal Zeeman effect

Starlight polarization yields the orientation of large-scale magnetic fields in the Milky Way (Fig. 5.5), and in the spiral galaxies M 51 (Scarrott et al., 1987) and NGC 6946 (Fig. 5.1). The major shortcoming when applying this method to extended sources like gas clouds or galaxies is that light can also be polarized by scattering, a contamination that is unrelated to magnetic fields and difficult to subtract.

5.2.3 *Far-Infrared and Submillimeter Polarization*

Linearly polarized emission from elongated dust grains at far-infrared and submillimeter wavelengths is not affected by polarized scattered light. The B-vector is parallel to the magnetic field. The field structure can be mapped in gas clouds of the Milky Way, e.g. in the massive star formation site W51 (Tang et al., 2009), and in galaxies, e.g. in the halo of the starburst galaxy M 82 (Greaves et al., 2000).

5.2.4 *Chandrasekhar–Fermi Method*

In the interstellar medium there is competition between the magnetic force (total field strength B) and the kinetic force by turbulent gas motions (turbulent velocity v_{turb} and density ρ). In a strong field the field lines are straight and the dispersion of polarization angles σ is small. According to Chandrasekhar & Fermi (1953), the total field strength

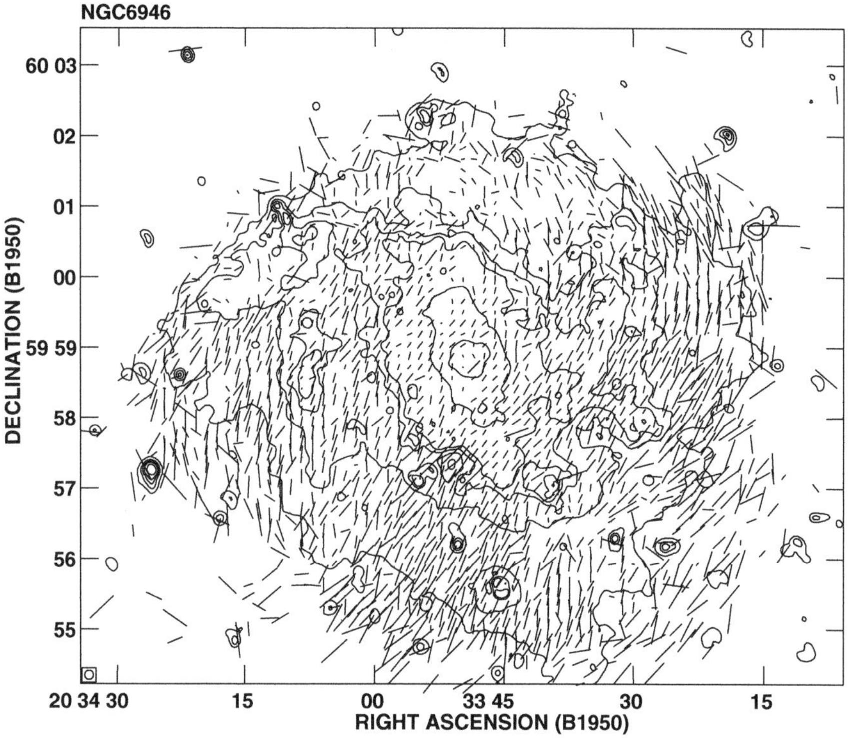

FIGURE 5.1. Optical emission (contours) and polarization E-vectors of NGC 6946, observed with the polarimeter of the Landessternwarte Heidelberg at the MPIA Calar Alto 1 m telescope. The polarization vectors are partly along the spiral arms (compare with the radio polarization B-vectors in Fig. 5.11). Polarization due to light scattering is obvious in the southern part (from Fendt et al., 1998). Reproduced with permission © ESO.

$B_{\text{tot},\perp}$ in the sky plane can be estimated by σ:

$$B_{\text{tot},\perp} \simeq \left(\frac{4}{3}\pi\rho\right)^{1/2} v_{\text{turb}}/\sigma\,. \tag{5.1}$$

Application to starlight polarization in the Milky Way yielded field strengths of about $7\,\mu$G, in agreement with other methods (Section 5.4.1). The Chandrasekhar–Fermi method was improved by correcting for the observational errors in polarization angle and signal integration (Hildebrand et al., 2009; Houde et al., 2009, 2011). It can also be applied to radio polarization data (Houde et al., 2013).

5.2.5 *Longitudinal Zeeman Effect*

The Zeeman effect is the most direct method of remote sensing of magnetic fields. It has been used in optical astronomy since the first detection of magnetic fields in sunspots of the Sun. The radio detection was first made in the 21 cm line of neutral hydrogen (HI). In the presence of a regular magnetic field $B_{\text{reg},\parallel}$ along the line of sight, the spectral line is split into two components (longitudinal Zeeman effect). The two components are circularly polarized of opposite sign. The frequency shift is 2.8 MHz/Gauss for the HI line and higher for molecular lines like OH, CN, or H_2O (Heiles & Crutcher, 2005). Zeeman splitting of the HI lines was used to measure the field strength in gas clouds of the Milky

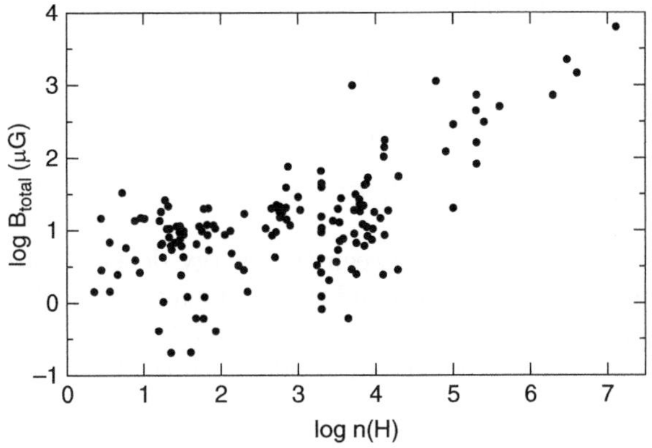

FIGURE 5.2. Zeeman measurements of the magnetic fields $B_{\rm tot}$ in gas clouds plotted against the hydrogen volume density n_H (in $\rm cm^{-3}$). To derive the total field, each measured line-of-sight component was multiplied by a factor of two which is the average correction factor for a large sample (from Crutcher et al., 2010). Reproduced with permission © AAS.

Way (Fig. 5.2; Crutcher et al., 2010) and Zeeman splitting of the OH line for starburst galaxies (Robishaw et al., 2008).

The interpretation of the Zeeman effect is hampered by its weakness, requiring high line intensities and careful correction of instrumental polarization. Most results have been obtained for the HI line that traces the diffuse (warm) gas. Suitable molecular lines tracing denser gas (e.g. OH and CN) are much weaker.

5.2.6 *Transversal Zeeman and Goldreich–Kylafis Effects*

Magnetic fields perpendicular to the line of sight generate two shifted lines in addition to the main unshifted line, all linearly polarized (transversal Zeeman effect). These lines cannot be resolved for observations in the Milky Way and external galaxies, and no net polarization is observed under symmetric conditions. Detection of linearly polarized lines becomes possible for unequal populations of the different sublevels, a gradient in optical depth or velocity, or an anisotropic velocity field (Goldreich & Kylafis, 1981). Depending on the line ratios, the orientation of linear polarization can be parallel or perpendicular to the magnetic field orientation. The effect was detected in molecular clouds, star-forming regions, outflows of young stellar systems and supernova remnants of the Milky Way. It was also observed in the ISM of the galaxy M 33 (Li & Henning, 2011), where it is consistent with the field orientations along the spiral arms as measured by polarized radio continuum emission.

The major obstacle in the interpretation of the Goldreich–Kylafis effect is its $\pm 90°$ ambiguity that needs to be solved with help of additional dust or radio polarization data.

5.2.7 *Radio Synchrotron Emission and Equipartition Field Strength*

The intensity of *total synchrotron emission* (Section 5.5.3; examples in Figs. 5.14, 5.16, and 5.18) is a measure of the number density of cosmic-ray electrons (CRE) in the relevant energy range and of the strength of the total magnetic field component in the sky plane. Synchrotron emission at a frequency ν is emitted by a range of cosmic-ray electrons (CRE) with average energy E (in GeV) in a total magnetic field with a component perpendicular to the line of sight of strength $B_{\mathrm{tot},\perp}$ (in μG):

$$\nu \simeq 16\,\mathrm{MHz}\; E^2\; B_{\mathrm{tot},\perp}\,. \tag{5.2}$$

The lifetime of CRE due to synchrotron losses is:

$$t_{\mathrm{syn}} = 8.35\times 10^9\,\mathrm{yr}\; B_{\mathrm{tot},\perp}^{-2}\, E^{-1} = 1.06\times 10^9\,\mathrm{yr}\; B_{\mathrm{tot},\perp}^{-1.5}\,\nu^{-0.5}\,. \tag{5.3}$$

The assumption of equipartition between the energy densities of the cosmic rays (by integrating over their energy spectrum) and the total magnetic field allows us to calculate the total field strength from the synchrotron intensity (Beck & Krause, 2005; Arbutina et al., 2012). The synchrotron intensity I_{syn} scales with the total magnetic field strength as $I_{\mathrm{syn}} \propto B_{\mathrm{tot},\perp}^{3+\alpha}$. Vice versa, $B_{\mathrm{tot},\perp}$ scales as:

$$B_{\mathrm{tot},\perp} \propto \left(I_{\mathrm{syn}}\,(K_0+1)\,/\,L\right)^{\,1/(3+\alpha)}\,, \tag{5.4}$$

where $B_{\mathrm{tot},\perp}$ is the strength of the total field perpendicular to the line of sight, α is the synchrotron spectral index ($I_{\mathrm{syn}} \propto \nu^{-\alpha}$), and L is the effective pathlength through the source. A value of K_0 is the ratio of number densities of CR protons and electrons. A value of $K_0 \simeq 100$ is a reasonable assumption in the star forming regions of the disk (Bell, 1978). (For an electron–positron plasma, $K_0 = 0$.) The input parameters, especially the pathlength L and the ratio K_0, are uncertain. Changing one of these values by a factor b changes the field strength by a factor of $b^{\pm 1/(3+\alpha)}$.

Issues with the equipartition estimate are as follows.

(1) Equation 5.4 can be applied only for steep radio spectra with $\alpha > 0.5$. For flatter spectra, the integration over the energy spectrum of the cosmic rays has to be restricted to a limited energy interval.

(2) Radio continuum emission also includes a thermal contribution that needs to be subtracted. At low radio frequencies thermal emission is negligible, but thermal absorption may reduce the synchrotron intensity in star-forming regions.

(3) In dense gas, e.g. in starburst regions, secondary positrons and electrons may be responsible for most of the radio emission via pion decay. Notably, the ratio of protons to secondary electrons is also about 100 for GHz–emitting electrons (Lacki & Beck, 2013).

(4) Owing to the highly nonlinear dependence of I_{syn} on $B_{\mathrm{tot},\perp}$, the average equipartition value $B_{\mathrm{tot},\perp}$ derived from synchrotron intensity is biased towards high field strengths and hence is an overestimate if B_{tot} varies along the line of sight or across the telescope beam (Beck et al., 2003). For $\alpha = 1$ and constant density of CRE, the overestimation factor g of the total field is:

$$g = (\langle B_{\mathrm{tot},\perp}^2\rangle)^{1/2}\,/\langle B_{\mathrm{tot},\perp}\rangle = (1+a^2)^{1/2}\,, \tag{5.5}$$

where $\langle\ \rangle$ indicates the averages along the line of sight and the beam. $a = (\langle \delta B_{\mathrm{tot},\perp}^2\rangle)^{1/2}\,/\langle B_{\mathrm{tot},\perp}\rangle$ is the amplitude of the field fluctuations relative to the mean field. For the equipartition case (see Appendix A in Stepanov et al., 2014):

$$g = (\langle B_{\mathrm{tot},\perp}^4\rangle)^{1/4}\,/\langle B_{\mathrm{tot},\perp}\rangle = \left(1+\frac{8}{3}a^2+\frac{8}{9}a^4\right)^{1/4}\,. \tag{5.6}$$

For strong fluctuations of $a = 1$, the overestimation factor is $g = 1.41$ for the non-equipartition and 1.46 for the equipartition case.

(5) Synchrotron radiation in the outer disk and halo of a galaxy is emitted by CR electrons that propagated far away from the places of origin. The ratio K increases because energy losses of aging CRE are more severe than those of CR protons. Using the standard value K_0 underestimates the total magnetic field by a factor of $(K/K_0)^{1/(3+\alpha)}$ in these regions (Beck & Krause, 2005).

(6) The general concept of energy equipartition has often been questioned. For example, a correlation analysis of radio continuum maps from the Milky Way and the nearby galaxy M 33 by Stepanov et al. (2014) showed that equipartition cannot hold for scales smaller than about 1 kpc, which is understandable in view of the typical propagation length of CRE (see Section 5.5.2). Arguments for equipartition on larger scales come from the joint analysis of radio continuum and γ-ray data, allowing an independent determination of magnetic field strengths, e.g. in the Milky Way (Strong et al., 2000), the Large Magellanic Cloud (LMC; Mao et al., 2012) and in M 82 (Yoast-Hull et al., 2013).

5.2.8 *Polarized Radio Synchrotron Emission*

Linearly polarized synchrotron emission (examples in Figs. 5.11, 5.12, and 5.13) emerges from ordered fields in the sky plane. As polarization angles are ambiguous by 180°, they cannot distinguish *regular fields*, defined to have a constant direction within the telescope beam, from *anisotropic turbulent or tangled fields*, which reverse their direction within the telescope beam. Unpolarized synchrotron emission indicates *isotropic turbulent or tangled fields* that have random directions in 3D and have been amplified by turbulent gas flows (see Table 5.1).

The intrinsic degree of linear polarization p_0 of synchrotron emission is:

$$p_0 = (1+\alpha)/(5/3+\alpha), \tag{5.7}$$

where α is the spectral index of the synchrotron emission. In spiral galaxies, typical values are $\alpha = 0.8$–1.0 and $p_0 = 73$–75%.

The observed degree of polarization is smaller because of the contribution of unpolarized thermal emission, which may dominate in star-forming regions, by wavelength-dependent Faraday depolarization (Section 5.2.11), and by wavelength-independent "beam depolarization" due to variations of the field orientation within the beam.

Magnetic fields in galaxies preserve their direction only over a coherence scale that depends on field tangling and/or on the properties of turbulence. In the case of isotropic turbulent fields with a constant coherence length d, the degree of synchrotron polarization can be described as:

$$p = p_0 \, N^{-1/2}, \tag{5.8}$$

where $N \simeq D^2 L f/d^3$ is the number of cells (assumed to be spherical) with diameter d and filling factor f within the volume defined by the telescope beam with linear size D at the galaxy's distance and pathlength L; p_0 is the polarization in a single cell. Typical degrees of polarization observed in nearby galaxies with spatial resolutions of about 500 pc at high frequencies (where Faraday depolarization is small) are 1–5% in spiral arms (where isotropic turbulent fields dominate) (e.g. in NGC 6946, Beck, 2007; and in M 33, Tabatabaei et al., 2008). With $L \simeq 1000$ pc and $f \simeq 0.5$, we get $d \simeq 50$ pc, consistent with estimates from Faraday depolarization at long wavelengths (Eq. (5.20)).

If the medium is pervaded by an isotropic turbulent field $B_{\rm iso}$ plus an ordered field $B_{\rm ord}$ (regular and/or anisotropic turbulent) that has a constant orientation in the volume observed by the telescope beam, and if the density of CRE is constant, the degree of polarization is reduced by wavelength-independent geometrical depolarization (Burn, 1966):

$$p = p_0 \,/(1 + q^2)\,, \tag{5.9}$$

where $q = B_{\mathrm{iso},\perp} \,/\, B_{\mathrm{ord},\perp}$ ($B_{\mathrm{iso},\perp} = \sqrt{2/3}\, B_{\mathrm{iso}}$), while for the case of equipartition between the energy densities of magnetic field and cosmic rays (Sokoloff et al., 1998):

$$p = p_0 \,(1 + 3.5\, q^2) \,/\, (1 + 4.5\, q^2 + 2.5\, q^4)\,, \tag{5.10}$$

giving less geometrical depolarization (larger p) than for the non-equipartition case.

5.2.9 *Faraday Rotation and RM Synthesis*

The polarization plane is rotated in a magnetized thermal plasma by *Faraday rotation.* As the rotation angle is sensitive to the sign of the field direction, only regular fields give rise to Faraday rotation, while the Faraday rotation contributions of turbulent fields largely cancel along the line of sight. Measurements of the Faraday rotation from multi-wavelength observations (example in Fig. 5.15) yield the strength and direction of the average regular field component along the line of sight. If Faraday rotation is small (in galaxies typically at wavelengths shorter than a few centimeters), the B-vector of polarized emission gives the intrinsic field orientation in the sky plane, so that the magnetic pattern can be mapped directly (Section 5.5.3).

The Faraday *rotation measure RM* is defined as the slope of the observed variation of Faraday rotation angle $\Delta\chi$ with the square of wavelength λ:

$$\Delta\chi = \Delta\lambda^2\, RM\,, \tag{5.11}$$

where RM is measured in rad m^{-2}. RM is constant if $\Delta\chi$ is a linear function of λ^2, e.g. if one or more Faraday-rotating regions are located in front of the emitting region (a *Faraday screen*). In other cases, RM needs to be replaced by the *Faraday depth (FD)* (Burn, 1966):

$$\Delta\chi = \Delta\lambda^2\, FD\,, \quad \text{with } FD = 0.812 \int B_{\mathrm{reg},\parallel}\, n_{\mathrm{e}}\, dl\,. \tag{5.12}$$

FD is measured in rad m^{-2}, the line-of-sight magnetic field $B_{\mathrm{reg},\parallel}$ in μG, the thermal electron density n_{e} in cm^{-3}, and the line of sight l in pc.

A nonlinear variation of $\Delta\chi$ with λ^2 and hence a variation of FD with λ^2 occurs in case of:

(1) emission and rotation in the same region where the distribution of electrons or regular magnetic field strength is not symmetric, or field reversals occur, or the field is helical (Sokoloff et al., 1998);
(2) Faraday depolarization (Section 5.2.11);
(3) several distinct emitting and rotating regions with different values of FD are located within the beam or along the line of sight, or the source has internal structure in FD, and the resolution in Faraday space is sufficiently high to resolve them (see below).

In these cases, multi-channel spectro-polarimetric radio data are needed, which are Fourier-transformed by the *RM synthesis* algorithm (Brentjens & de Bruyn, 2005) to obtain a *Faraday spectrum* (previously called *Faraday dispersion function*). This shows the intensity of polarized emission and its polarization angle as a function of FD. If the medium has a relatively simple structure, Faraday spectra at each position of a source can reveal the 3D structure of the magnetized interstellar medium (*Faraday tomography*). Regular fields, field reversals and turbulent fields can be recognized from Faraday spectra (Bell et al., 2011; Frick et al., 2011; Beck et al., 2012b). The shape of the Faraday spectrum averaged over many lines of sight can be used to extract the global properties of the galactic magnetic field (Ideguchi et al., 2017). Helical fields can also imprint characteristic features in the Faraday spectrum (Brandenburg & Stepanov, 2014; Horellou & Fletcher, 2014).

The distribution of the frequency channels across the total band and the channel width of the observation defines the *Rotation Measure Spread Function* (RMSF; Brentjens & de Bruyn, 2005), which represents the "telescope beam" in Faraday space. The half-power width of the RMSF, the resolution in Faraday space, decreases with increasing λ^2 span of the observation. The highest and lowest wavelengths determine the resolution, while the sampling of the wavelength range determines the sidelobe level of the RMSF. With a high resolution in Faraday space and a low sidelobe level, different FD components and internal FD structure of the sources become visible in the Faraday spectrum. The Faraday spectrum can be deconvolved with the RMSF ("dirty beam"), which is similar to cleaning of radio synthesis data in intensity space (Heald, 2009).

A *grid of RM measurements* of polarized background sources is another powerful tool to measure magnetic field patterns in galaxies (Stepanov et al., 2008). About 10 RM values from sources behind a galaxy's disk are sufficient to recognize a simple large-scale field pattern, if the Galactic foreground contribution is constant and the background sources have no intrinsic RMs.

5.2.10 *Faraday Rotation and Dispersion of Pulsar Signals*

The arrival times of pulsar signals are delayed in a cloud of ionized gas, which is proportional to the *dispersion DM*:

$$DM = \int n_e \, dl \,, \tag{5.13}$$

so that:

$$RM/DM = 0.812 \langle B_{\mathrm{reg},\|} \, n_{\mathrm{e}} \rangle / \langle n_{\mathrm{e}} \rangle \,, \tag{5.14}$$

which allows us to compute the field strength $B_{\mathrm{reg},\|}$ (in μG) if the fluctuations in field strength and in electron density are *uncorrelated.* The above estimate suffers from various problems.

(1) If the field fluctuations of relative amplitude $a = (\langle \delta B^2_{\mathrm{reg},\|} \rangle)^{1/2} / \langle B_{\mathrm{reg},\|} \rangle$ are *correlated* with n_{e} with a correlation coefficient q, the observed RM is modified (Beck et al., 2003):

$$RM = RM_0 \left(1 + \frac{2}{3} q \, (a^2/(1+a^2) \right) , \tag{5.15}$$

where $RM_0 = 0.812 \langle B_{\mathrm{reg},\|} \rangle \langle n_{\mathrm{e}} \rangle L$ (where L is the pathlength). For a perfect correlation ($q = 1$), as expected for compression by planar shock fronts, and strong fluctuations of $a \simeq 1$, the standard estimate is too large by 33%. *The larger the correlated fluctuations, the larger is the observed RM.*

(2) Observations of external galaxies indicate that the regular magnetic field B_{reg} and n_{e} are *anticorrelated* on kpc scales (Beck, 2007). B_{reg} is strongest in the interarm regions and weaker in the spiral arms where n_{e} is largest (see Fig. 5.11). If the fluctuations are perfectly anticorrelated on small scales ($q = -1$), as expected for local pressure equilibrium, and again assuming $a \simeq 1$, the standard estimate is too small by 33%. *The larger the anticorrelated fluctuations, the smaller is the observed RM.*

(3) If there are *N field reversals* along the line of sight, the average regular field between the reversals is larger than the standard estimate by a factor of $\simeq (N+1)$. The pulsar RM data have been corrected for large-scale reversals detected in the Milky Way, but there may be more reversals on smaller scales.

5.2.11 *Faraday Depolarization*

The Faraday rotation of $\Delta\chi$ within one individual channel of bandwidth $\Delta\nu$ at wavelength λ_0 (frequency ν_0) leads to *bandwidth depolarization*:

$$p = p_0 \,|\sin(\Delta\chi)/(\Delta\chi)|\,, \tag{5.16}$$

where $\Delta\chi = 2\,\lambda_0^2\,RM\,\Delta\nu\,/\nu_0$. To avoid bandwidth depolarization, narrow channels are needed at low frequencies. The maximum detectable rotation measure is $RM_{\rm max} \simeq \sqrt{3}/\Delta\nu$.

In a single region containing cosmic-ray electrons, thermal electrons, and purely *regular magnetic fields*, wavelength-dependent Faraday depolarization occurs because the polarization planes of waves from the far side of the emitting layer are more rotated than those from the near side. This effect is called *differential Faraday rotation.* For one single layer with a symmetric distribution of thermal electron density and field strength along the line of sight the degree of polarization is reduced by (Burn, 1966; Sokoloff et al., 1998):

$$p = p_0 \,|\sin(2\,RM\,\lambda^2)/(2\,RM\,\lambda^2)|\,, \tag{5.17}$$

where RM is the observed rotation measure, which is half of the Faraday depth FD through the whole layer. Note that p varies periodically with wavelength. With $|RM| = 100\,{\rm rad\,m^{-2}}$, typical for normal galaxies, $p = 0$ occurs at wavelengths of $\lambda = 12.5\,N^{1/2}$ cm, where $N = 1, 2, \ldots$ At each zero point the polarization angle jumps by 90°. Observing at a fixed wavelength hits zero points at certain values of the intrinsic RM, giving rise to depolarization canals along the level lines of RM (Shukurov & Berkhuijsen, 2003). At wavelengths just below that of the first zero point in p, only the central layer of the emitting region is observed, because the emission from the far side and that from the near side cancel (their rotation angles differ by 90°). Beyond the first zero point, only a small layer on the near side of the disk remains visible.

Applying RM synthesis (Section 5.2.9) to polarization data of an emitting and Faraday-rotating region reveals an extended component in the Faraday spectrum. However, regions broader than $FD_{\rm max} \simeq \pi/\lambda_{\rm min}^2$ cannot be recovered, where $\lambda_{\rm min}$ is the minimum wavelength of the observations (Brentjens & de Bruyn, 2005); only two "horns" remain at the edges of the FD structure (Beck et al., 2012b). This problem is similar to the missing short baselines in synthesis imaging.

For multiple emitting + rotating layers, the Faraday spectrum contains several extended components, and Eq. (5.17) is no longer applicable.

Faraday rotation in *helical fields* has a completely different behavior and may lead to an increase of the degree of polarization with increasing wavelength in certain wavelength ranges (Brandenburg & Stepanov, 2014; Horellou & Fletcher, 2014).

Turbulent fields also cause wavelength-dependent depolarization, called *Faraday dispersion* (Sokoloff et al., 1998; Arshakian & Beck, 2011). For an emitting and Faraday-rotating region (internal dispersion):

$$p = p_0\,(1 - \exp(-S))/S\,, \tag{5.18}$$

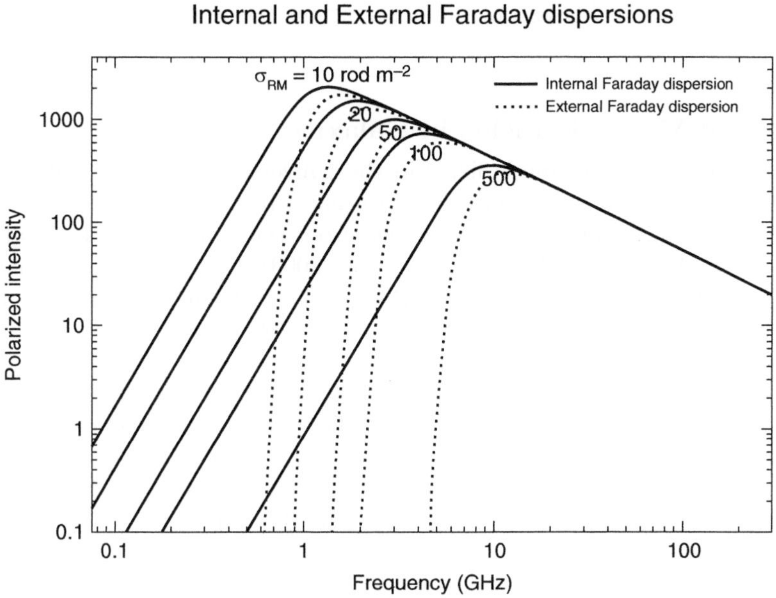

FIGURE 5.3. Spectrum of polarized intensity (in arbitrary units) for a synchrotron source with spectral index $\alpha = 0.9$ (visible as a straight line at high frequencies) and depolarization by internal (solid line) and external (dashed line) Faraday dispersion at different levels of $\sigma_{\rm RM}$. To the left of the maxima, the medium is "Faraday thick."

where $S = 2\sigma_{\rm RM}^2\,\lambda^4$. $\sigma_{\rm RM}$ is the dispersion in rotation measure and depends on the turbulent field strength along the line of sight, the turbulence scale, the thermal electron density, and the pathlength through the medium.

External dispersion occurs in a turbulent Faraday-rotating (but not emitting) medium in the foreground if the diameter of the telescope beam at the distance of the screen is larger than the turbulence scale:

$$p = p_0 \exp(-S)\,. \tag{5.19}$$

The main effect of Faraday dispersion is that the interstellar medium becomes "Faraday thick" for polarized radio emission beyond a wavelength, depending on $\sigma_{\rm RM}$, and only a front layer remains visible in polarized intensity. Typical values for galaxy disks are $\sigma_{\rm RM} = 20\text{–}100\,{\rm rad\,m^{-2}}$, while star-forming regions can have dispersions of $\sigma_{\rm RM} \simeq 1000\,{\rm rad\,m^{-2}}$ (Arshakian & Beck, 2011). The maximum in the spectrum of polarized intensity is located between 1.5 GHz and 15 GHz (Fig. 5.3).

In a random-walk approach, we may write $\sigma_{\rm RM} = 0.812\,B_{\rm turb,\|}\,n_{\rm e}\,d\,N^{1/2}$ (Beck, 2007), where $B_{\rm turb,\|} = \sqrt{1/3}\;B_{\rm turb}$ and $n_{\rm e}$ are the turbulent field strength and electron density within a cell of size d, and $N = L\,f/d$ is the number of cells along the line of sight L with a volume filling factor f. The average electron density along the line of sight is $\langle n_{\rm e}\rangle = n_{\rm e}/f$, so that we get:

$$\sigma_{\rm RM} = 0.812\,B_{\rm turb,\|}\,\langle n_{\rm e}\rangle\,\sqrt{L\,d/f}\,. \tag{5.20}$$

The average value for galaxy disks of $\sigma_{\rm RM} \simeq 50\,{\rm rad\,m^{-2}}$ is consistent with typical ISM values of $\langle n_{\rm e}\rangle = 0.03\,{\rm cm^{-3}}$, $B_{\rm turb} = 10\,\mu{\rm G}$, $L = 1000\,{\rm pc}$, $f = 0.5$ and $d = 50\,{\rm pc}$.

The dispersion $\sigma_{\rm RM}$ leads to a "Faraday forest" of N components in the Faraday spectrum. If N is not large, the components are possibly resolvable with very high Faraday resolution, hence a wide λ^2 span of the observations (Beck et al., 2012b; Bernet et al., 2012).

5.3 Origin of Magnetic Fields in Galaxies

The origin of the first magnetic fields in the Universe is a mystery (Widrow, 2002). *Seed fields* may be "primordial," generated during a phase transition in the early Universe (Caprini et al., 2009), or may originate from the time of cosmological structure formation by the Weibel instability (Lazar et al., 2009), from injection by the first stars or jets generated by the first black holes (Rees, 2005) or from the Biermann mechanism in the first supernova remnants (Hanayama et al., 2005), or from plasma fluctuations (Schlickeiser, 2012; Schlickeiser & Felten, 2013). The non-detection of GeV γ-ray emission with the FERMI satellite from blazars, which were observed at TeV energies with the HESS observatory, may indicate that the secondary particles are deflected by intergalactic fields of least 10^{-16} G strength and a high volume filling factor (Dolag et al., 2011). However, this interpretation is not generally accepted because fluctuations of the IGM plasma may also disperse the γ-ray emission (Broderick et al., 2012).

Numerical models of evolving galaxies show fast field amplification of a weak seed field with the help of differential rotation (Wang & Abel, 2009; Kotarba et al., 2009), possibly supported by the magneto-rotational instability (MRI) (Pakmor & Springel, 2013). In these models the magnetic field is *ordered*, forming spiral arm segments, but it has frequent reversals in azimuthal and radial directions and hence is not regular. In the model by Pakmor & Springel (2013) the magnetic pressure exceeds the thermal pressure after about 1 Gyr of evolution and suppresses star formation.

Another source of field amplification is turbulence in the gas driven by supernova explosions, called the *small-scale dynamo*. In protogalaxies this mechanism can amplify weak seed fields to several μG strength (the energy level of turbulence) within less than 10^8 yr (Schleicher et al., 2010; Beck et al., 2012a). The resulting field is *turbulent*.

The most promising mechanism to sustain magnetic fields and generate *regular fields* in the interstellar medium of galaxies is the $\alpha - \Omega$ dynamo (Beck et al., 1996). It is based on differential rotation (Ω), turbulence, and helical gas flows (α*-effect*), driven by supernova explosions (Ferrière, 1996; Gressel et al., 2008). The mean-field approximation of the $\alpha - \Omega$ dynamo was supported by high-resolution MHD modeling (Gent et al., 2013).

Dynamo-type fields are described by modes with different azimuthal symmetries in the disk plane and two different vertical symmetries (even or odd parity) perpendicular to the disk plane. Several modes can be excited in the same object. In flat rotating objects like galaxy disks, the strongest mode S0 consists of a toroidal field of *axisymmetric spiral* shape within the disk, without sign reversals across the equatorial plane, and a weaker poloidal field of even-symmetry structure. Azimuthal dynamo modes can be identified observationally from the pattern of polarization angles and of RMs of the diffuse polarized emission of galaxy disks (see Section 5.5.4).

Mean-field solutions of the $\alpha - \Omega$ dynamo in galaxy disks predict that turbulent fields are ordered and become large-scale regular fields within a few 10^9 yr (Arshakian et al., 2009), but field reversals from the early phases may survive until today (Moss et al., 2012). Global numerical models of galaxies (Gissinger et al., 2009; Hanasz et al., 2009; Kulpa-Dybeł et al., 2011; Siejkowski et al., 2014) confirmed the basic results of the mean-field solutions.

The $\alpha - \Omega$ dynamo generates large-scale helicity with a non-zero mean in each hemisphere. As total helicity is a conserved quantity, the dynamo is quenched by the small-scale fields with opposite helicity unless these are removed from the system (Shukurov

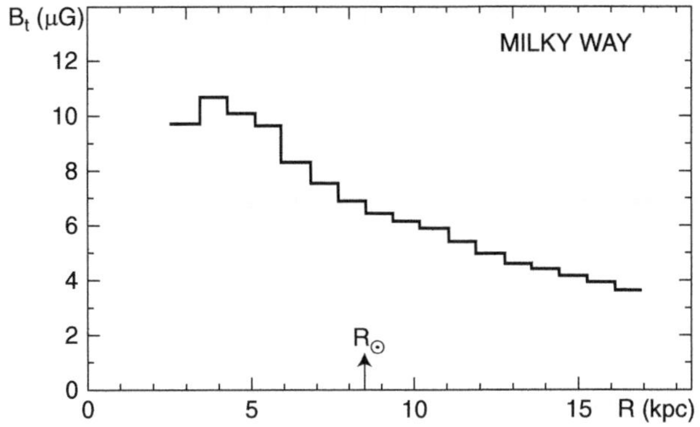

FIGURE 5.4. Strength of the total magnetic field in the Galaxy, averaged from the deconvolved surface brightness of the synchrotron emission at 408 MHz (Beuermann et al., 1985), assuming energy equipartition between magnetic field and cosmic-ray energy densities. The Sun is assumed to be located at $R = 8.5\,\mathrm{kpc}$ (from Berkhuijsen, private communication). The figure has been reproduced with permission © ESO.

et al., 2006). Outflows are essential for effective $\alpha - \Omega$ dynamo action. On the other hand, strong outflows can suppress dynamo action, leaving a range of outflow velocities for optimal dynamo action (Rodrigues et al., 2015). As outflows are stronger above spiral arms, the $\alpha - \Omega$ dynamo is expected to be more efficient between the spiral arms (Chamandy et al., 2015).

5.4 Magnetic Fields in the Milky Way

5.4.1 *Field Strengths in the Milky Way*

Modeling surveys of the total synchrotron and γ-ray emission from the Milky Way yield field strengths near the Sun of about $5\,\mu\mathrm{G}$ of the isotropic turbulent field, about $2\,\mu\mathrm{G}$ of the anisotropic turbulent field and about $2\,\mu\mathrm{G}$ of the regular field (Orlando & Strong, 2013), adding up to a total field strength of about $6\,\mu\mathrm{G}$. This is in excellent agreement with the Voyager data (see below), the Zeeman splitting data of low-density gas clouds (Fig. 5.2) and pulsar RM data (Sections 5.4.2 and 5.4.3). In the inner Galaxy the total field strength is about $10\,\mu\mathrm{G}$ (Fig. 5.4). In the synchrotron filaments near the Galactic Center the total field strength is about $100\,\mu\mathrm{G}$ (Crocker et al., 2010).

Measurements from the Voyager 2 spacecraft in the heliosheath indicate that the surrounding interstellar magnetic field is 4–5 μG strong and oriented at an angle of about 30° from the Galactic plane (Opher et al., 2009), probably because the ISM field twists close to the heliosphere. Voyager 1 crossed into interstellar space in 2012 and measured a smooth increase in field strength to $5.62 \pm 0.01\,\mu\mathrm{G}$ (Burlaga et al., 2013).

5.4.2 *Large-Scale Magnetic Fields in the Milky Way*

Optical polarization data of about 5500 selected stars in the Milky Way yielded the orientation of the large-scale magnetic field near the Sun (Fosalba et al., 2002) (Fig. 5.5), which is mostly parallel to the Galactic plane and oriented along the local spiral arm. A homogeneous region of alignment, with high polarization values, was seen towards the anticenter (Galactic longitude $l \simeq 140°$). Well-aligned magnetic field vectors are also seen along the North Polar Spur that extends into the upper disk from $l \simeq 30°$.

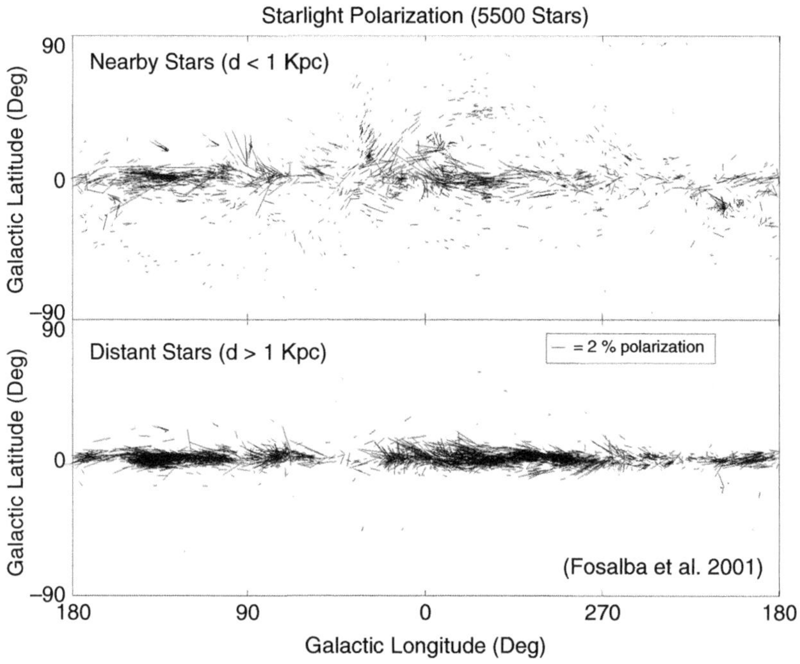

FIGURE 5.5. Optical starlight polarization for two distance intervals (from Fosalba et al., 2002). Reproduced with permission © AAS.

Starlight polarization for stars within 40 pc of the Sun indicates that the local ISM field points towards $l \simeq 50°$, $b \simeq 25°$ (Frisch et al., 2012). The local field is probably parallel to a fragment of the expanding shell of the Loop I superbubble.

The all-sky maps of polarized synchrotron emission at 1.4 GHz from the Milky Way from DRAO and Villa Elisa and at 22.8 GHz from WMAP, and the Effelsberg RM survey of polarized extragalactic sources, were used to model the regular Galactic field (Sun et al., 2008; Sun & Reich, 2010). One large-scale *field reversal* is required at about 1–2 kpc from the Sun towards the Milky Way's center. The refined model by Jansson & Farrar (2012a) (Fig. 5.7) is based on the WMAP survey and more than 40 000 extragalactic Faraday rotation measures (from the NVSS catalog by Taylor et al. (2009) and other sources, Fig. 5.6) and also needs a field reversal.

Pulsars are ideal objects to deduce the Galactic magnetic field because their RMs provide field directions at many distances from the Sun (Noutsos, 2012). Since most pulsars are concentrated along the Galactic plane, they sample the field in the disk. Combination of RM and dispersion measure (DM) data of pulsars (Eq. (5.14)), and assuming uncorrelated fluctuations, gives an average strength of the local regular field of $2.1 \pm 0.3\,\mu$G and about $4\,\mu$G at 3 kpc Galactocentric radius (Han et al., 2006).

RM data from extragalactic radio sources (Fig. 5.6) and pulsars were used to model the Galactic magnetic field (Nota & Katgert, 2010; Van Eck et al., 2011, see Fig. 5.8). As the distances of most pulsars are not well known, only gross field patterns can be reliably detected. The local magnetic field in the Perseus arm is clockwise. A large-scale magnetic field reversal is present between the Scutum-Crux–Sagittarius arm and the Carina–Orion arm (Fig. 5.8). This reversal could be part of a bisymmetric spiral (BSS) field structure. However, the data are inconsistent with any simple field configuration (Men et al., 2008).

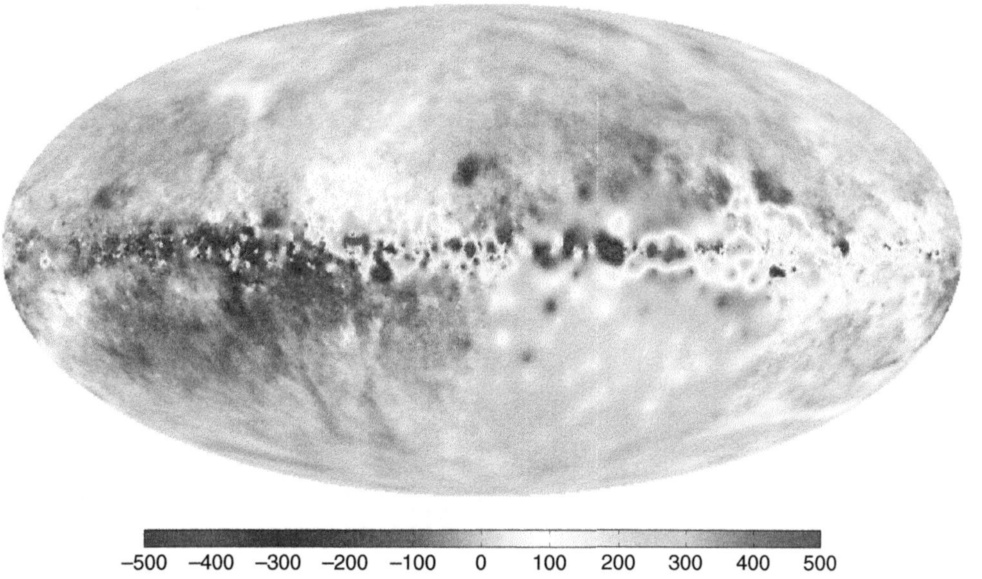

FIGURE 5.6. All-sky map of rotation measures in the Milky Way, reconstructed from the RM data of about 40 000 polarized extragalactic sources from the VLA NVSS survey (Taylor et al., 2009) and other catalogues. This is a black-and-white version of the original image from Oppermann et al. (2012). Light gray: positive RM, dark gray: negative RM. Reproduced with permission © AAS.

A satisfying explanation for the reversal in the Milky Way is still lacking. Firstly, it may be restricted to a thin layer near to the plane and therefore hardly visible in the average RM data of external galaxies along the line of sight. Secondly, the reversal in the Milky Way may be of limited azimuthal extent and would be difficult to observe in external galaxies with present-day telescopes. Thirdly, the reversal in the Milky Way may be part of a disturbed field structure, e.g. due to interaction with the Magellanic clouds, or a relic from the seed field (Moss et al., 2012). In conclusion, the overall structure of the regular field in the disk of the Milky Way is not yet known (Noutsos, 2009). A larger sample of pulsar and extragalactic RM data can be expected from the LOFAR and SKA radio telescopes (Section 5.6).

5.4.3 *Small-Scale Magnetic Fields in the Milky Way*

Present-day polarization observations of external galaxies are restricted to angular resolutions of $10''$–$15''$ because at higher resolutions polarized intensities are too low. Before much more sensitive telescopes like the SKA become operational, the small-scale structure of interstellar magnetic fields has to be investigated by polarization observations in the Milky Way.

From the dispersion of pulsar RMs, the Galactic magnetic field was found to have a significant turbulent component with a mean strength of 5–6 μG (Rand & Kulkarni, 1989; Ohno & Shibata, 1993; Han et al., 2004). Magnetic turbulence occurs over a large spectrum of scales, with the largest scale determined from pulsar RMs of $l_{\rm turb} \simeq 55\,{\rm pc}$ (Rand & Kulkarni, 1989) or $l_{\rm turb} \simeq 10 - 100\,{\rm pc}$ (Ohno & Shibata, 1993). These values are consistent with the size of turbulent cells of $d \simeq 50\,{\rm pc}$ derived from depolarization data in external galaxies (Sections 5.2.8 and 5.2.11). In five regions in the southern Galactic plane, Haverkorn et al. (2006) determined the structure function of RM fluctuations that

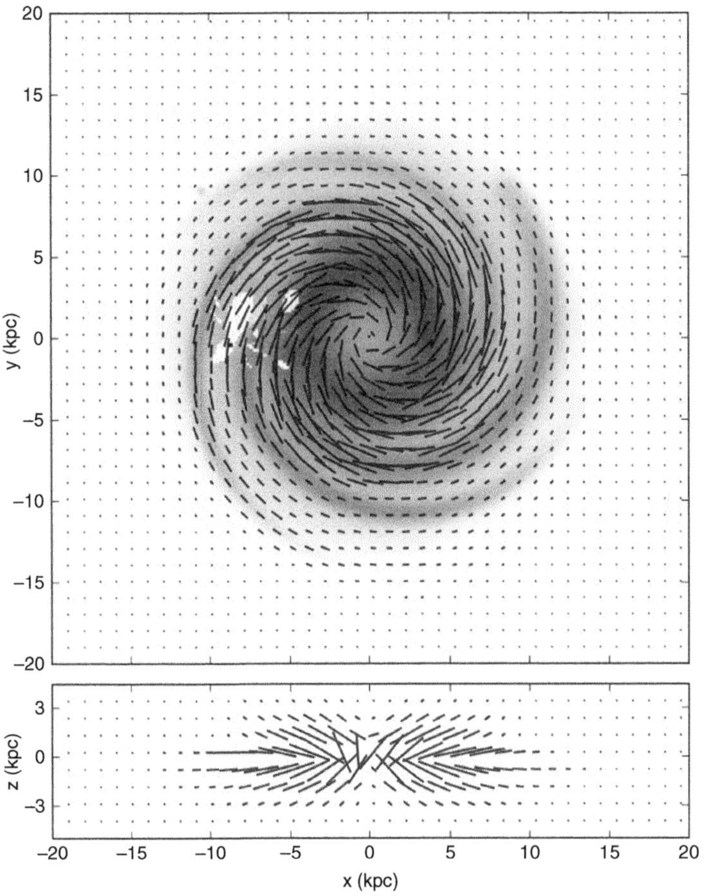

FIGURE 5.7. Model of the magnetic field in the Milky Way (face-on and edge-on views), derived from the polarized synchrotron emission observed at 22.8 GHz with WMAP and more than 40 000 extragalactic Faraday rotation measures (from Jansson & Farrar, 2012a). Reproduced with permission © AAS.

reveals an outer scale of about 100 pc in the interarm regions, while the spiral arm regions do not show correlated fluctuations.

5.4.4 *Magnetic Fields in the Halo of the Milky Way*

The detection with the AUGER observatory of ultrahigh-energy cosmic rays (UHECRs, energies beyond $3\,10^{19}$ eV) reaching the Earth and the possibly anisotropic distribution of their arrival directions (Abreu et al., 2010) calls for a proper model of particle propagation. As UHECR particles are deflected by large-scale regular fields and scattered by turbulent fields, the structure and the extent of the fields in the disk and especially in the halo of the Milky Way are necessary parameters for a reliable propagation model. However, our present knowledge does not allow safe conclusions.

The vertical full equivalent thickness of the thick radio synchrotron disk of the Milky Way at 408 MHz is about 3 kpc near the Sun (Beuermann et al., 1985; scaled to a distance to the Galactic Center of 8.5 kpc), corresponding to an exponential scale height of about

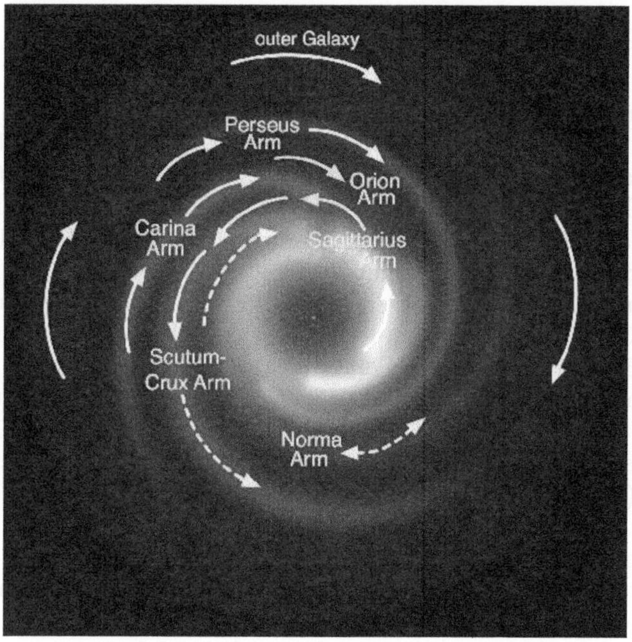

FIGURE 5.8. Model of the magnetic field in the Milky Way, derived from Faraday rotation measures of pulsars and extragalactic sources. Generally accepted results are indicated by gray vectors, while white vectors are not fully confirmed (from Jo-Anne Brown, private communication).

1.5 kpc. This indicates a scale height of the total magnetic field of at least 6 kpc, while the best-fit Gaussian scale height of the random field in the model by Jansson & Farrar (2012b) is about 3 kpc. The γ-ray data indicate that the "vertical boundary" of the cosmic-ray halo may be 4 kpc or more (Ackermann et al., 2012). Confining them needs a magnetic field of about twice this extent, similar to the size of halos around external galaxies (Section 5.5.7).

The signs of RM of extragalactic sources and of pulsars at high Galactic longitudes ($l = 90° - 270°$) are the same above and below the plane (Fig. 5.6): The local magnetic field is symmetric, while the RM signs towards the inner Galaxy ($l = 270° - 90°$) are *opposite* above and below the plane. This could be assigned to local features (Wolleben et al., 2010) or to an odd-symmetry halo field (Sun & Reich, 2010). However, the local regular Galactic field, according to RM data from extragalactic sources, has no significant vertical component towards the northern Galactic pole and only a weak vertical component of $B_z \simeq 0.3\,\mu$G towards the south (Mao et al., 2010). This is neither consistent with an odd-symmetry halo field nor with an even-symmetry halo field as found in several external galaxies (Section 5.5.8). On the other hand, modeling the diffuse polarized emission and RM gave evidence for an X-shaped vertical field component (Fig. 5.7, bottom), similar to that in external galaxies (Section 5.5.7). The vertical filaments in the central region could be part of a poloidal field component (Ferrière, 2009). Similar filaments in external galaxies would be hard to observe due to insufficient resolution.

In summary, the Milky Way's disk and halo fields are more complicated than predicted by dynamo models. The view on external spiral galaxies can help.

5.5 Magnetic Fields in External Galaxies

5.5.1 *Field Strengths*

The average equipartition strength (Eq. (5.4)) of *total fields* (corrected for inclination) for a sample of 74 spiral galaxies is $B_{\text{tot}} = 9 \pm 2\,\mu\text{G}$ (Niklas, unpublished PhD thesis, 1995). The average strength of 21 bright galaxies observed between 2000 and 2010 is $B_{\text{tot}} = 17 \pm 3\,\mu\text{G}$ (Fletcher, 2010). Dwarf galaxies host fields of similar strength as spirals if their star-formation rate per volume is similarly high (Chyży et al., 2011). Blue compact dwarf galaxies are radio bright with equipartition field strengths of 10–20 μG (Klein et al., 1991). Radio-faint galaxies like M 31 (Fig. 5.14) and M 33 have weaker total magnetic fields (about 6 μG), while gas-rich spiral galaxies with high star-formation rates, like M 51 (Fig. 5.12), M 83 (Fig. 5.13) and NGC 6946 (Fig. 5.11), have total field strengths of 20–30 μG in their spiral arms. The strongest total fields of 50–100 μG are found in starburst galaxies like M 82 (Adebahr et al., 2013), in the "Antennae" NGC 4038/9 (Chyży & Beck, 2004), in nuclear starburst regions like in NGC 253 (Heesen et al., 2011a), and in barred galaxies (Beck et al., 2005).

The strength of *ordered fields* as observed by polarized synchrotron emission varies strongly between galaxies, from 10–15 μG in M 51 and the magnetic arms of NGC 6946 (Section 5.5.3) to about 5 μG in the star-forming ring of M 31 (Fletcher et al., 2004). The average strength of the ordered fields of 21 bright galaxies observed since 2000 is $B_{\text{ord}} = 5 \pm 3\,\mu\text{G}$ (Fletcher, 2010). Irregular starburst galaxies show spots of ordered fields (Heesen et al., 2011b). No ordered fields have been detected so far in dwarf irregular galaxies (Chyży et al., 2011).

If energy losses of cosmic-ray electrons (CRE) are significant, especially in starburst regions or massive spiral arms, the equipartition values are lower limits and are probably underestimated in starburst galaxies by a factor of a few (Thompson et al., 2006). Field strengths of 0.5–18 mG were detected in starburst galaxies by the Zeeman effect in the OH megamaser emission line at 18 cm wavelength (Robishaw et al., 2008). These values refer to highly compressed gas clouds and are not typical of the diffuse interstellar medium.

The relative importance of various competing forces in the interstellar medium can be estimated by comparing the corresponding *energy densities*. The mean energy densities of the total (mostly turbulent) magnetic field and the cosmic rays are $\simeq 10^{-11}$ erg cm^{-3} in NGC 6946 (Fig. 5.9), M 63, M 83 and NGC 4736 (Basu & Roy, 2013) and $\simeq 10^{-12}$ erg cm^{-3} in M 33 (Tabatabaei et al., 2008). In all cases this is similar to the energy density of the turbulent gas motions across the whole star-forming disk and about 10 times larger than that of the ionized gas (but still 500–1000 times smaller than the energy density of the general rotation of the neutral gas). In conclusion, magnetic fields are dynamically important for the ISM. The total magnetic energy density may even dominate in the outer galaxy where the equipartition field strength is an underestimate due to energy losses of the CRE. The energy density of the regular magnetic field decreases even more slowly than that of the total field. Although the star-formation activity is low in the outer disk, the magneto-rotational instability (MRI) may serve as the source of turbulence required for dynamo action (Sellwood & Balbus, 1999).

In the case of energy equipartition, the radial scale length of the total field in the disk of mildly inclined galaxies, or the vertical scale height in the halo of edge-on galaxies, is at least $(3 + \alpha)$ times (where $\alpha \simeq 0.8$ is the typical synchrotron spectral index) larger than the synchrotron scale length of typically 3–4 kpc (Beck, 2007; Basu & Roy, 2013). The resulting value of 11–15 kpc is a lower limit because the CRE lose their energy with distance from the star-forming disk and the equipartition assumption yields too small values for the field strength.

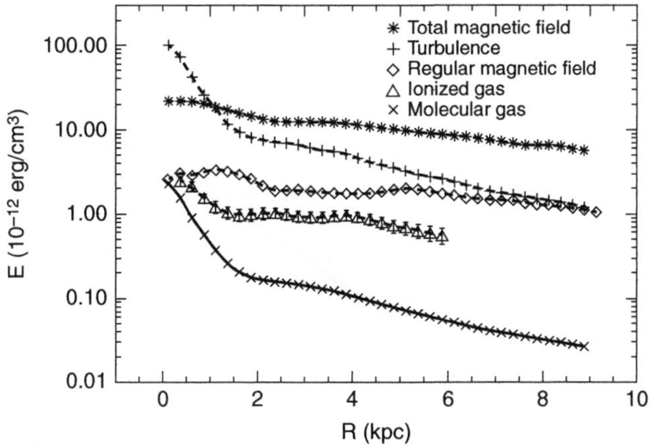

FIGURE 5.9. Radial variation of the energy densities in NGC 6946: total magnetic field E_B $(B_{tot}^2/8\pi)$, regular magnetic field $(B_{reg}^2/8\pi)$, turbulent motion of the neutral gas E_{turb} $(0.5\,\rho_n\,v_{turb}^2$, where $v_{turb} \approx 7$ km s$^{-1})$, thermal energy of the ionized gas E_{th} $(1.5\,n_e\,k\,T_e)$ and thermal energy of the molecular gas E_n $(0.5\,\rho_n\,k\,T_n)$, determined from observations of synchrotron and thermal radio continuum and the CO and HI line emissions (from Beck, 2007). Reproduced with permission © ESO.

The galactic fields may extend far out into intergalactic space. A large radial scale length would mean that magnetic fields affect the global rotation of the gas in the outer parts of spiral galaxies (Ruiz-Granados et al., 2010; Jałocha et al., 2012), possibly explaining some of the flattening of the rotation curves.

The measured extent of galactic magnetic fields is limited by the energy loss processes of CRE diffusing outwards from their locations of origin. At low frequencies synchrotron and inverse Compton losses are smaller, so that the radio disks should be larger than at high frequencies. First LOFAR results for the galaxy M 51 at 150 MHz reveal a steepening of the radial distribution of synchrotron emission, located at about 10 kpc radius, a few kpc beyond a sharp decrease in the star-formation rate (Mulcahy et al., 2014). Synchrotron loss in M 51 is high, so that the CRE cannot illuminate any extended magnetic field in the outer disk beyond about 15 kpc radius. LOFAR observations at even lower frequencies are under investigation.

5.5.2 *The Radio–Infrared Correlation*

The integrated luminosity of the total radio continuum emission at centimeter wavelengths (frequencies of a few GHz), which is mostly of non-thermal synchrotron origin, and the infrared (IR) luminosity of star-forming galaxies are tightly correlated. This correlation is one of the tightest correlations known in astronomy. The tightness needs multiple feedback mechanisms that are not yet understood (Lacki et al., 2010). The correlation extends over five orders of magnitude (Bell, 2003) and is valid in starburst galaxies to redshifts of at least 4 (Seymour et al., 2008). The validity of the correlation requires that total (mostly turbulent) magnetic fields and star formation are connected. The detection of strong radio emission in distant galaxies (which is at least partly of synchrotron origin) demonstrates that magnetic fields existed already in young galaxies with strengths of several 100 μG (Murphy, 2009). A breakdown of synchrotron emission and of the correlation is expected beyond a critical redshift when inverse Compton loss

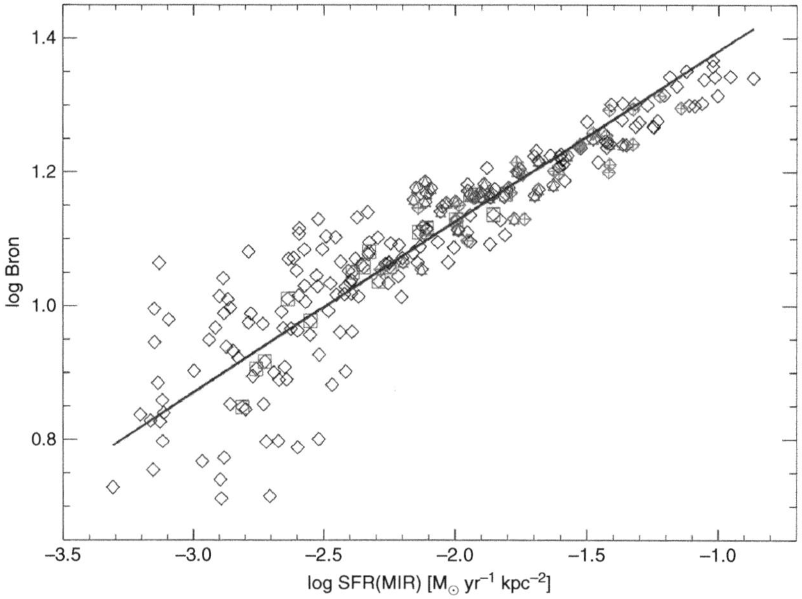

FIGURE 5.10. Correlation between the strength of the total equipartition field (dominated by the turbulent field) and star-formation rate per area (determined from the 24 μm infrared intensities) within the galaxy NGC 4254 (from Chyży, 2008). Reproduced with permission © ESO.

of the cosmic-ray electrons dominates synchrotron loss; this critical redshift will give us information about the field evolution in young galaxies (Schleicher & Beck, 2013).

The total radio and IR intensities *within* galaxies are also highly correlated. The exponent of the correlation in M 51 was found to be different in the central region, spiral arms, and interarm regions (Dumas et al., 2011; Basu et al., 2012), and it also differs between galaxies due to differences in diffusion length of the CRE (Berkhuijsen et al., 2013). The magnetic field and its structure play an important role to understand the correlation (Tabatabaei et al., 2013a,b; Heesen et al., 2014). The radio–infrared correlation can be presented as a correlation between turbulent field strength and star-formation rate (Fig. 5.10; Tabatabaei et al., 2013b; Heesen et al., 2014). Turbulent fields in spiral arms are probably generated by turbulent gas motions related to star formation activity (small-scale dynamo, Section 5.3). In contrast, the ordered field is either uncorrelated with the star-formation rate (Chyży, 2008) or anticorrelated in galaxies where the ordered field is strongest in interarm regions with low star formation (Fig. 5.11).

A cross-correlation analysis for M 33 based on wavelet transforms showed that the radio–FIR correlation holds at scales <1 kpc (Tabatabaei et al., 2007). The correlation in the Large Magellanic Cloud (LMC) breaks down below scales of about 50 pc (Hughes et al., 2006). The propagation of CRE away from their sources in star-forming regions is probably responsible for the breakdown scale, and the propagation length seems to depend on the field structure (Tabatabaei et al., 2013a).

5.5.3 *Structure of Ordered Magnetic Fields in Spiral Galaxies*

Ordered (regular and/or anisotropic) fields traced by polarized synchrotron emission are observed to form spiral patterns in almost every spiral galaxy (Beck & Wielebinski, 2013), even in galaxies with a star-forming ring (Chyży & Buta, 2008), in flocculent galaxies

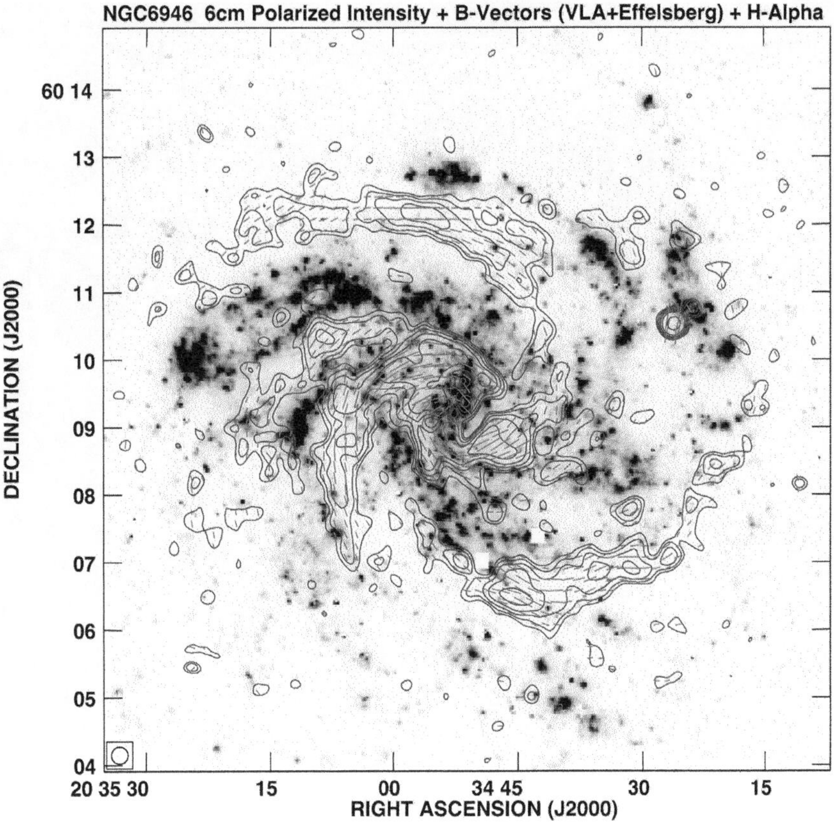

FIGURE 5.11. Polarized radio emission (contours) and B-vectors of NCC 6946, combined from observations at 6 cm wavelength with the VLA and Effelsberg telescopes and smoothed to 15″ resolution (from Beck, 2007), overlaid onto an Hα image from Anne Ferguson. Reproduced with permission © ESO.

without massive spiral arms (Soida et al., 2002), in the central regions of galaxies and in circum-nuclear gas rings of barred galaxies (Beck et al., 2005).

In galaxies with strong density waves like M 51 (Fig. 5.12) and M 83 (Fig. 5.13) enhanced ordered fields sometimes occur at the inner edges of the inner optical arms where the cold molecular gas is densest (Patrikeev et al., 2006). These are probably anisotropic turbulent fields generated by the compression of the density wave. An analysis of dispersions of the radio polarization angles at 6 cm in M 51 shows that the ratio of the correlation lengths parallel and perpendicular to the local ordered magnetic field is about 2 (Houde et al., 2013).

In galaxies without strong density waves the ordered fields do not coincide with the gaseous or the optical spiral arms. The typical degree of radio polarization at small wavelengths is only a few percent within the spiral arms; hence the field in the spiral arms must be mostly tangled or randomly oriented within the telescope beam on scales of a few tens of pc (Eq. (5.8)).

Ordered fields are strongest (10–15 μG) in the regions *between* the spiral arms and oriented parallel to the adjacent spiral arms, in some galaxies forming *magnetic arms*, like in IC 342 (Krause, 1993; Beck, 2015) and NGC 6946 (Fig. 5.11), with exceptionally high degrees of polarization (up to 50%). These are probably generated by the $\alpha - \Omega$ dynamo (Chamandy et al., 2015; Moss et al., 2015).

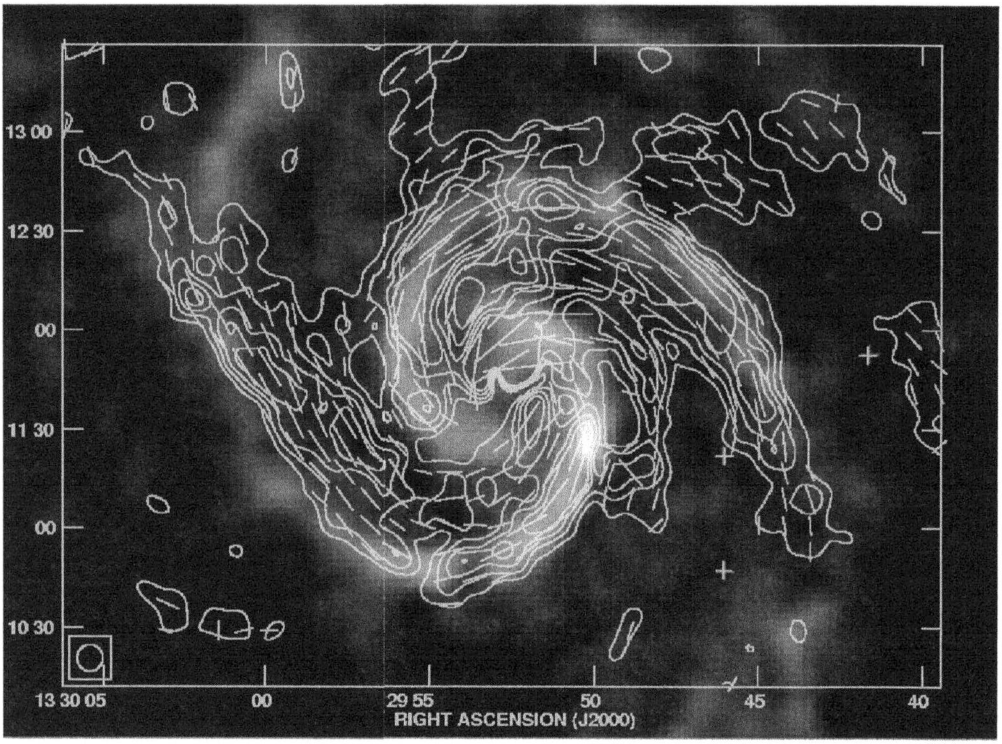

FIGURE 5.12. Polarized radio emission (contours) and B-vectors of M 51, combined from observations at 6 cm wavelength with the VLA and Effelsberg telescopes at 8″ resolution (from Fletcher et al., 2011). The background image shows the CO line emission of molecular gas from Helfer et al. (2003). Reproduced with permission © AAS.

The observed magnetic spiral patterns indicate a general decoupling between magnetic fields and the gas flow, as predicted by $\alpha - \Omega$ dynamo action. At present, no other model can explain the magnetic spiral patterns in the many types of galaxies.

At wavelengths of around 20 cm, most of the polarized emission from the far side of the disk and halo is Faraday-depolarized and the emission from the front side dominates. A striking asymmetry of the polarized emission occurs along the major axis of 12 spiral galaxies with inclinations of less than about 60°, observed with sufficiently high sensitivity. The emission is always much weaker around the kinematically receding side (positive radial velocities) of the major axis (Urbanik et al., 1997; Braun et al., 2010; Vollmer et al., 2013). This asymmetry is still visible at 11 cm wavelength, but disappears at smaller wavelengths where the polarized emission from the far side becomes observable. In strongly inclined galaxies, both sides of the major axis become Faraday-depolarized at around 20 cm. Modeling of a combination of disk and halo fields, as predicted by $\alpha - \Omega$ dynamo models (Section 5.2), can explain the asymmetry (Braun et al., 2010).

At even longer wavelengths, Faraday effects depolarize the synchrotron emission almost completely. With help of RM synthesis applied to 90 cm data from the Westerbork Synthesis Radio Telescope, an extremely low average degree of polarization of $0.21 \pm 0.05\%$ was measured in the star-forming "ring" of M 31 (Gießübel et al., 2013). No polarized emission could be detected from M 51 with the Low Frequency Array (LOFAR) at around 2 m wavelength – total Faraday depolarization occurs at this long wavelength (Mulcahy et al., 2014).

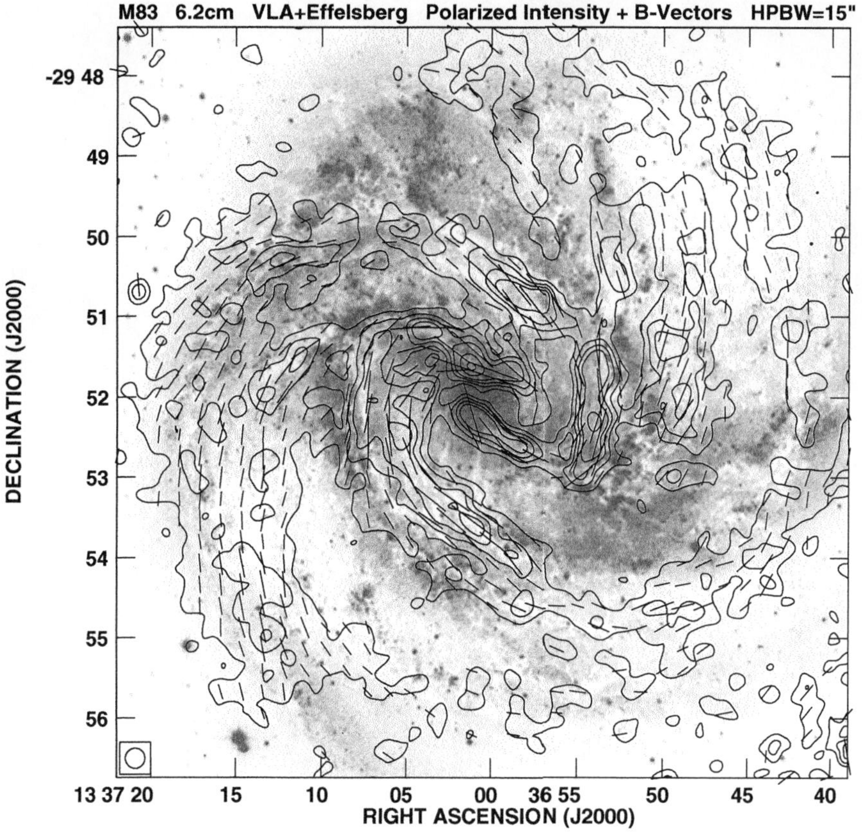

FIGURE 5.13. Polarized radio emission (contours) and B-vectors of M 83, combined from observations at 6 cm wavelength with the VLA and Effelsberg telescopes and smoothed to 15″ resolution (from Frick et al., 2016), overlaid onto an optical image from Dave Malin, Anglo Australian Observatory (copyright: MPIfR Bonn and AAO).

A list of radio polarization observations of galaxies can be found in Beck & Wielebinski (2013). Total intensity and polarization images are compiled in: http://www.mpifr-bonn.mpg.de/atlasmag

5.5.4 *Regular Fields in Galactic Disks*

Spiral fields are generated by compression, by shear in interarm regions, or by dynamo action (Section 5.3). Large-scale patterns (azimuthal modes) of Faraday rotation measures (RM) are signatures of regular fields generated by the $\alpha - \Omega$ dynamo and can be identified in RM derived from polarized emission of the galaxy disks at several frequencies (Krause, 1990) or in RM data of polarized background sources (Stepanov et al., 2008). If several modes are superimposed, a Fourier analysis of the RM variation with azimuthal angle is needed. The resolution of present-day observations is sufficient to identify two to three modes.

The disks of about a dozen nearby spiral galaxies reveal large-scale RM patterns. The Andromeda galaxy M 31 (Fig. 5.14) is the prototype of a dynamo-generated axisymmetric spiral disk field, with a striking sinusoidal RM variation along the star-forming ring (Fig. 5.15), which is a strong indication of an axisymmetric spiral field (Fletcher et al., 2004). Other candidates for a dominating axisymmetric disk field (mode $m = 0$) are the

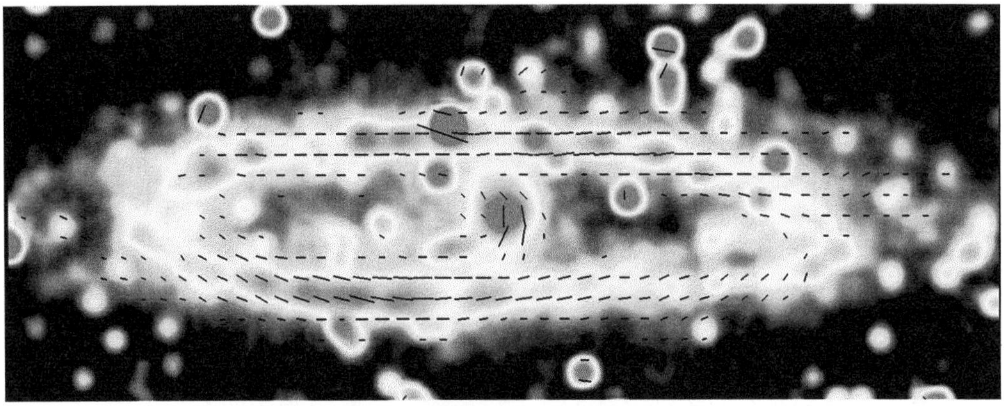

FIGURE 5.14. Total radio emission (image) and B-vectors (corrected for Faraday rotation in the foreground of the Milky Way) in the Andromeda galaxy (M 31), observed at 6 cm with the Effelsberg telescope at 3′ resolution (from Gießübel, 2012). Reproduced with permission © MPIfR Bonn.

nearby spirals IC 342 (Beck, 2015) and NGC 253 (Heesen et al., 2009b). The axisymmetric field in the irregular Large Magellanic Cloud (LMC) is almost azimuthal (small pitch angle) (Gaensler et al., 2005; Mao et al., 2012). Dominating bisymmetric spiral fields (mode $m = 1$) are rare, as predicted by dynamo models, but possibly exist in M 81 (Krause et al., 1989a; Sokoloff et al., 1992). Faraday rotation in NGC 6946 and in other similar galaxies with magnetic arms can be described by a superposition of two azimuthal modes ($m = 0$ and $m = 2$) with about equal amplitudes where the quadrisymmetric ($m = 2$) mode is phase shifted with respect to the density wave (Beck, 2007). For several other galaxies, three modes ($m = 0$, 1 and 2) are necessary to describe the data (Table 5.2).

The spiral pattern of magnetic fields cannot be solely the result of $\alpha - \Omega$ dynamo action. If the beautiful spiral pattern of M 51 seen in radio polarization (Fig. 5.12) were

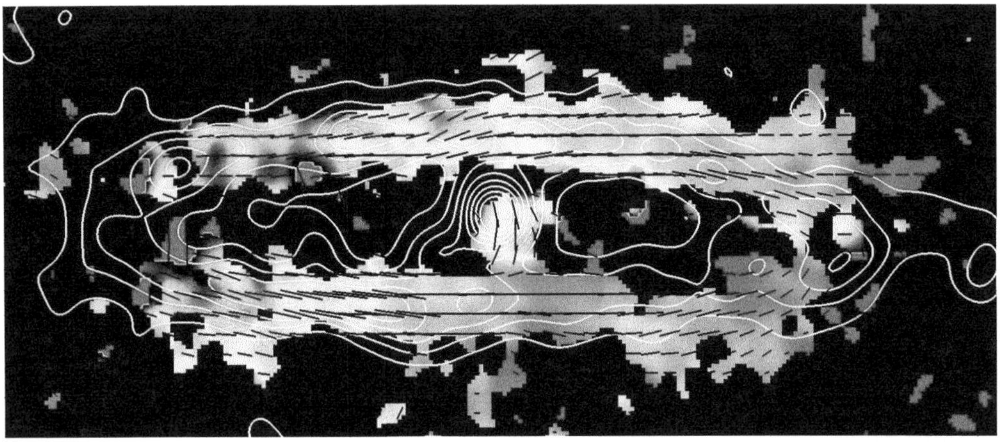

FIGURE 5.15. Faraday rotation measures between 6 cm and 11 cm (shading) and total radio emission at 6 cm (contours) in the Andromeda galaxy (M 31), observed with the Effelsberg telescope at 5′ resolution (from Berkhuijsen et al., 2003). Reproduced with permission © MPIfR Bonn.

TABLE 5.2. Decomposition of regular magnetic fields in galaxies into azimuthal modes of order m. Columns 2–4 give the approximate amplitudes relative to the dominating mode. A dash indicates modes that were not sought, whereas a zero indicates that no signature of this mode was found. Modes of order higher than 2 cannot be detected with present-day telescopes. This is an updated version of a table in Fletcher (2010).

Galaxy	$m = 0$	$m = 1$	$m = 2$	Reference
M 31	1	0	0	Fletcher et al. (2004)
M 33	1	1	0.5	Tabatabaei et al. (2008)
M 51 (disk)	1	0	0.5	Fletcher et al. (2011)
M 51 (halo)	0	1	0	Fletcher et al. (2011)
M 81	0.5	1	–	Sokoloff et al. (1992)
NGC 253	1	–	–	Heesen et al. (2009b)
NGC 1097	1	1	1	Beck et al. (2005)
NGC 1365	1	1	1	Beck et al. (2005)
NGC 4254	1	0.5	–	Chyży (2008)
NGC 4414	1	0.5	0.5	Soida et al. (2002)
NGC 6946	1?	–	1?	Ehle & Beck (1993); Rohde et al. (1999)
IC 342	1	–	–	Sokoloff et al. (1992); Beck (2015)
LMC	1	–	–	Gaensler et al. (2005); Mao et al. (2012)

only due to a regular field in the disk, its line-of sight component should generate a conspicuous large-scale pattern in RM, but this is not observed (Fletcher et al., 2011). This means that a large amount of the ordered field is *anisotropic turbulent* and probably generated by compression and shear of the non-axisymmetric gas flows in the density-wave potential. The anisotropic turbulent field is strongest at the positions of the prominent dust lanes on the inner edge of the inner gas spiral arms, due to compression of turbulent fields in the density-wave shock. A regular field (composed of modes $m = 0$ and $m = 2$) also exists in the disk of M 51, but is much weaker than the anisotropic turbulent field (Fletcher et al., 2011). Notably, RMs around 20 cm, tracing only the polarized emission from regions nearest to the observer, show a large-scale pattern that indicates a bisymmetric regular field in the halo of M 51 (Fletcher et al., 2011; Mao et al., 2015).

In the disks of many other galaxies no clear pattern of Faraday rotation was found. Either several high-order modes are superimposed and cannot be distinguished with the limited sensitivity and resolution of present-day telescopes, or the timescale for the generation of large-scale modes is longer than the galaxy's lifetime (Arshakian et al., 2009). Also field injection by strong star-formation activity may perturb the generation of the large-scale regular field (Moss et al., 2012).

By comparing the signs of the RM distribution and the velocity field on both sides of a galaxy's major axis, the inward and outward directions of the radial component of the axisymmetric spiral field can be distinguished (Krause & Beck, 1998). Dynamo models predict that both signs should have the same probability, which is confirmed by observations. The axisymmetric fields of M 31, IC 342, NGC 253, and the axisymmetric field component in NGC 6946 point inwards, while those of NGC 891, NGC 4254, and NGC 5775 (Krause, 2009), and the axisymmetric component of the disk field in M 51 (Fletcher et al., 2011) point outwards.

5.5.5 *Large-Scale Field Reversals in Galactic Disks*

Large-scale field reversals at certain radial distances from a galaxy's center, like that observed in the Milky Way (Fig. 5.8), have not been detected in external galaxies,

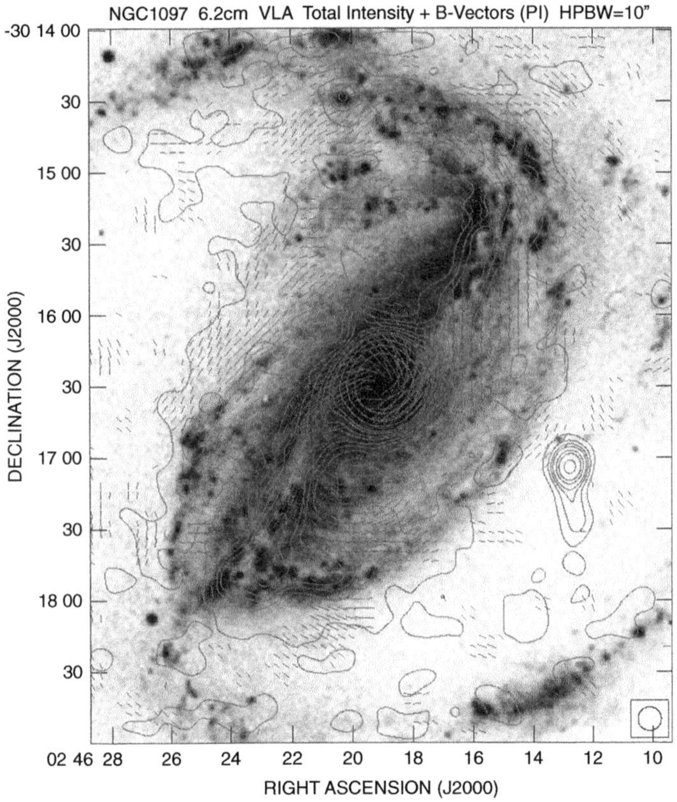

FIGURE 5.16. Total radio emission (contours) and B-vectors of the barred galaxy NGC 1097, observed at 6 cm wavelength with the VLA and smoothed to 10″ resolution (from Beck et al., 2005). The background optical image is from Halton Arp.

although high-resolution RM maps of Faraday rotation are available for many spiral galaxies. In the barred galaxy NGC 7479, where a jet serves as a bright polarized background source and high-resolution observations were possible with high signal-to-noise ratio, several reversals on the 1–2 kpc scale were detected in the foreground disk of the galaxy (Laine & Beck, 2008). NGC 6946 reveals a field reversal with an extent of 0.6 kpc, coinciding with a hole in neutral hydrogen (Heald, 2012).

5.5.6 *Barred Galaxies*

In galaxies with massive bars the magnetic field lines (Fig. 5.16) follow a pattern that resembles the flow of cold gas in a bar potential. As the gas rotates faster than the bar pattern of a galaxy, a shock occurs in the cold gas, which has a small sound speed, while the flow of warm, diffuse gas is only slightly compressed but sheared. The ordered field is also hardly compressed, probably coupled to the diffuse gas and strong enough to affect its flow (Beck et al., 2005). The ordered field is also strong in the upstream region (south of the center in Fig. 5.16), oriented almost perpendicular to the bar and aligned with the dust filaments seen on the optical image in the background. The polarization pattern in barred galaxies can be used as a tracer of shearing gas flows in the sky plane and complements spectroscopic measurements of radial velocities.

The central regions of barred galaxies are often sites of ongoing intense star formation and strong magnetic fields that can affect gas flows. NGC 1097 hosts a bright ring with

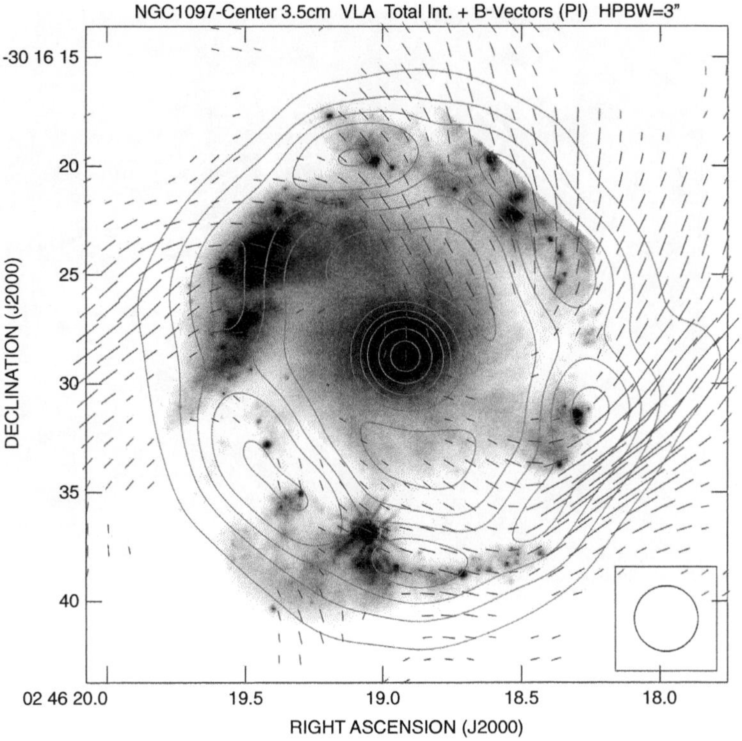

FIGURE 5.17. Total radio emission (contours) and B-vectors in the circumnuclear ring of the barred galaxy NGC 1097, observed at 3.5 cm wavelength with the VLA at 3″ resolution. The background optical image is from the HST (from Beck et al., 2005).

about 1.5 kpc diameter and an active nucleus in its center (Fig. 5.17). The ordered field in the ring has a spiral pattern and extends to the nucleus. The orientation of the innermost spiral field agrees with that of the spiral dust filaments visible in optical images. Magnetic stress in the circumnuclear ring due to the strong total magnetic field (about 50 μG) can drive gas inflow (Balbus & Hawley, 1998) at a rate of several solar masses per year, which is sufficient to fuel the activity of the nucleus (Beck et al., 2005). MHD modeling confirmed that magnetic fields can strongly enhance the mass gas inflow rate (Kim & Stone, 2012).

5.5.7 *Halos Around Edge-On Spiral Galaxies*

Nearby galaxies seen *edge-on* generally show a disk-parallel field near the disk plane (Dumke et al., 1995). As a result, polarized emission can also be detected from distant, unresolved galaxies if the inclination is larger than about 20° (Stil et al., 2009). This opens a new method to search for ordered fields in distant galaxies.

High-sensitivity radio polarization observations of several edge-on galaxies, like NGC 253 (Heesen et al., 2009b), NGC 891 (Krause, 2009), NGC 4631 (Fig. 5.18) and NGC 5775 (Tüllmann et al., 2000; Soida et al., 2011), revealed vertical field components in the halo forming an X-shaped pattern that may be related to dynamo action (Moss et al., 2010) or to outflows. Ferrière & Terral (2014) presented analytical models of divergence-free magnetic fields with X-patterns.

Edge-on spiral galaxies observed with high resolution have exponential scale heights of the radio halo of about 1.8 kpc (Krause, 2014). Because the average field strengths

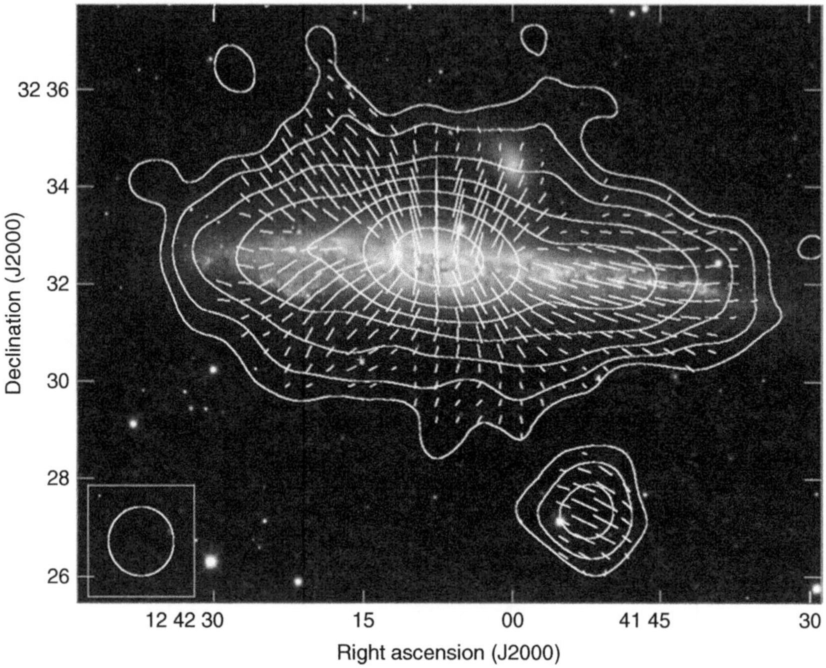

FIGURE 5.18. Total radio emission (contours) and B-vectors of the edge-on galaxy NGC 4631, observed at 3.6 cm with the Effelsberg telescope at 85″ resolution (from Mora & Krause, 2013). The background optical image is from the Digital Sky Survey. Reproduced with permission © ESO.

and hence lifetimes of CRE are different in these galaxies, an outflow speed increasing with the average field strength and hence with the star-formation rate could balance the smaller electron lifetime (Krause, 2009).

In the exceptionally large radio halos around the interacting galaxies M 82 (Adebahr et al., 2013; Reuter et al., 1994) and NGC 4631 (Golla & Hummel, 1994; Mora & Krause, 2013) the scale height varies considerably between different radii. A few magnetic spurs could be resolved, connected to star-forming regions. These observations support the idea of a strong galactic outflow (*galactic wind*) that is driven by regions of star formation in the disk.

The stronger magnetic field in the central regions leads to larger synchrotron loss, leading to the "dumbbell" shape of many radio halos, e.g. in NGC 253 (Heesen et al., 2009a) and NGC 4565 (Krause, 2009). From the radio scale heights at several frequencies and the corresponding lifetimes of cosmic-ray electrons (due to synchrotron, IC, and adiabatic losses) a wind speed of about 300 km s^{-1} was measured for the electrons in the halo of NGC 253 (Heesen et al., 2009a). NGC 253 and NGC 4388 host central starburst regions driving a nuclear outflow (Heesen et al., 2011a; Damas-Segovia et al., 2016).

5.5.8 *Large-Scale Fields in Halos of Spiral Galaxies*

While the azimuthal symmetry of the magnetic field is known for many galaxies, the vertical symmetry (even or odd) is much harder to determine. The field of odd symmetry reverses its sign above and below the galactic plane. In mildly inclined galaxies, the RMs of diffuse polarized emission from even- and odd-symmetry fields differ by a factor of about two, while the RMs of background sources are close to zero for even symmetry (because the RM contributions from the two halves cancel) but large for odd symmetry. Background RMs in the LMC field indicate an even-symmetry field (Mao et al., 2012). The symmetry type is best visible in strongly inclined galaxies, via the RM sign above and below the plane. Indications for even patterns were found in NGC 891 (Krause, 2009) and NGC 5775 (Soida et al., 2011).

Above the central starburst region of NGC 253, an outwards-directed helical field of about 20 μG strength, extending to at least 1 kpc height, could be identified in the gas outflow cone with help of high-resolution RM mapping (Heesen et al., 2011a). This field may help to confine the outflow.

5.5.9 *Interacting Galaxies*

Gravitational interaction between galaxies leads to asymmetric gas flows, compression, shear, enhanced turbulence, and outflows. Magnetic fields can become aligned along the compression front or perpendicular to the velocity gradients. Such gas flows make turbulent fields highly anisotropic.

The classical interacting galaxy pair is NGC 4038/39, the "Antennae" (Chyży & Beck, 2004). It shows bright, extended radio emission filling the whole system. In the interaction region between the galaxies, where star formation has not yet started, and at the northeastern edge of the system, the magnetic field is partly ordered, probably the result of compression and shearing motions along the tidal tail. Particularly strong, almost unpolarized emission comes from a region of violent star formation, hidden in dust. The average total magnetic field is stronger than in normal spirals, but the mean degree of polarization is unusually low, implying that the fields are tangled. The Antennae galaxies have been modeled by Kotarba et al. (2010) who could reproduce the main features observed in the radio images.

Interaction with a dense intergalactic medium also imprints unique signatures onto magnetic fields and thus the radio emission. The Virgo cluster is a location of especially strong interaction effects (Fig. 5.19), and almost all cluster galaxies observed so far show asymmetries of their polarized emission because the outer magnetic fields were compressed (Vollmer et al., 2007, 2013; Weżgowiec et al., 2007, 2012). Ordered fields are an excellent tracer of past interactions between galaxies or with the intergalactic medium (Damas-Segovia et al., 2016).

5.5.10 *Flocculent and Irregular Galaxies*

Flocculent galaxies have disks but no grand-design spiral structure. Nevertheless, magnetic spiral patterns are observed in all flocculent galaxies, indicating that the $\alpha - \Omega$ dynamo works independently of density waves. Ordered magnetic fields with strengths similar to those in grand-design spiral galaxies have been detected in the flocculent galaxies M 33 (Tabatabaei et al., 2008), NGC 3521 and NGC 5055 (Knapik et al., 2000) and in NGC 4414 (Soida et al., 2002), and also the mean degree of polarization (corrected for the differences in spatial resolution) is similar in grand-design and flocculent galaxies.

Radio continuum maps of irregular, slowly rotating galaxies may reveal strong total magnetic fields, e.g. in the Magellanic-type galaxy NGC 4449 (Fig. 5.20), with a partly ordered field of about 7 μG strength, a spiral pattern in the northeast and a radial pattern in the central region (Chyży et al., 2000). Faraday rotation shows that this ordered field is partly regular and the $\alpha - \Omega$ dynamo is operating (Siejkowski et al., 2014).

Dwarf irregular galaxies with almost chaotic rotation do not have any regular fields and only spots of faint polarized emission (Heesen et al., 2011b). The turbulent field

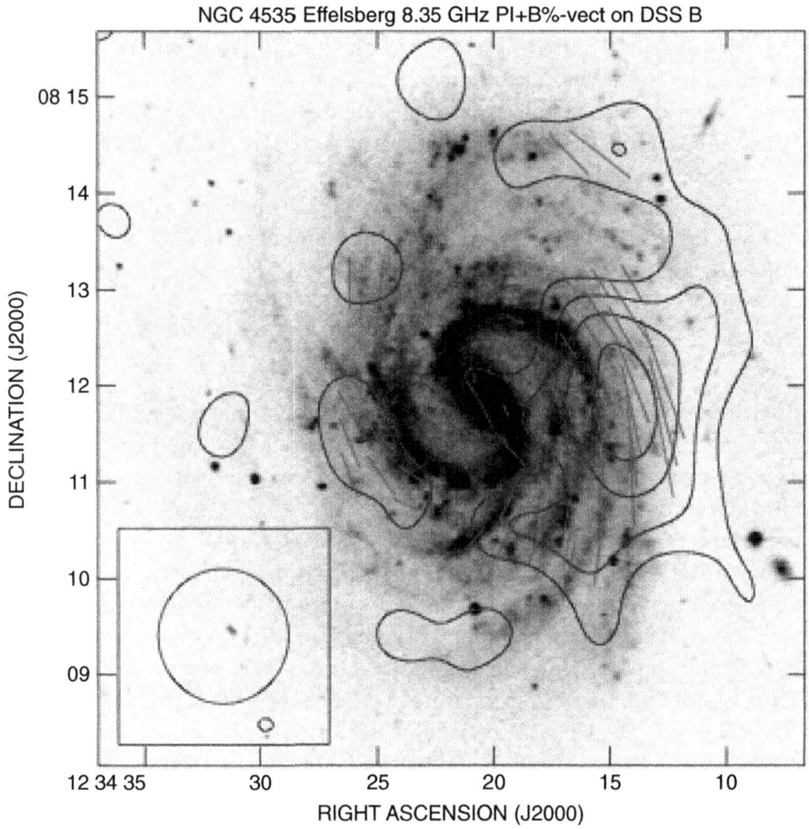

FIGURE 5.19. Polarized radio emission (contours) and B-vectors of the Virgo galaxy NGC 4535, observed at 3.6 cm with the Effelsberg telescope at 90″ resolution (from Weżgowiec et al., 2012). The background optical image is from the Digital Sky Survey. Reproduced with permission © ESO.

strengths are generally smaller than in spiral galaxies (Chyży et al., 2011), except for starburst dwarfs, e.g. NGC 1569 with 10–15 μG field strength (Kepley et al., 2010), where star formation activity is sufficiently high for the operation of the small-scale dynamo. Dwarf galaxies with weak star formation are expected to reveal a steeper radio–infrared correlation than spiral galaxies (Schleicher & Beck, 2016).

5.5.11 *Early-Type and Spheroidal Galaxies*

Spiral galaxies of type Sa and S0 and elliptical galaxies without an active nucleus have very little star formation and hence produce few cosmic rays that emit only weak synchrotron emission (Nyland et al., 2017). The only deep observation of an Sa galaxy, M 104 with a prominent dust ring, revealed weak, ordered magnetic fields (Krause et al., 2006). Large-scale regular magnetic fields may exist in differentially rotating galaxies without star formation because turbulence in the ISM can also be generated by the magneto-rotational instability (MRI; Sellwood & Balbus, 1999). Their detection may become possible via RM grids of background sources with future radio telescopes.

The search for synchrotron emission from nearby dwarf spheroidal galaxies without star formation is an attractive possibility to detect decaying weakly interactive massive particles (WIMPs), candidates of dark matter (Colafrancesco et al., 2007).

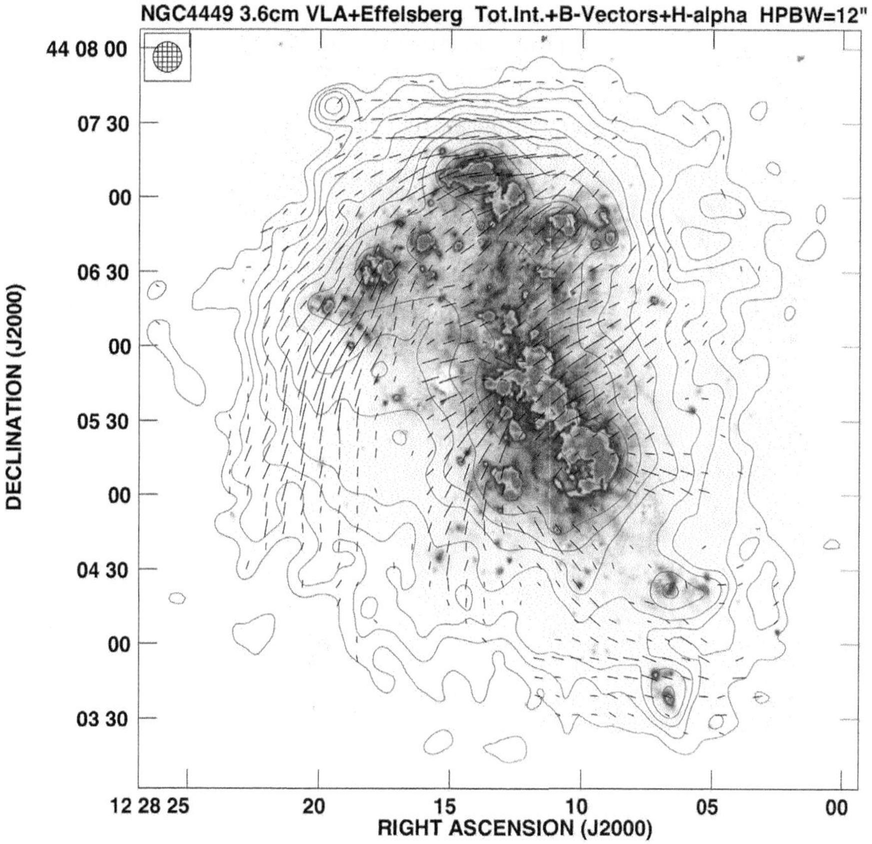

FIGURE 5.20. Total radio emission (contours) and B-vectors of the dwarf irregular galaxy NGC 4449, observed at 3.6 cm with the VLA at 12″ resolution (from Chyży et al., 2000). The background Hα image is from Dominik Bomans (Bochum University). Reproduced with permission © ESO.

5.5.12 *Summary*

Table 5.3 gives a concise description of the structures of ordered magnetic fields in various galaxy types as observed from radio polarization observations. "Turbulent" stands for "isotropic turbulent or tangled." "Regular field" refers to measurements of Faraday rotation.

5.6 Outlook

Future high-resolution, high-sensitivity observations at high frequencies with the *Jansky Very Large Array* (VLA) and the planned *Square Kilometre Array* (SKA) will directly map the detailed field structure and the interaction with the gas (Beck et al., 2015). Construction of the first phase of the SKA is planned to start in 2018. The low-frequency radio telescopes *Low Frequency Array* (LOFAR) (van Haarlem et al., 2013) and the *Murchison Widefield Array* (MWA) (Tingay et al., 2013) are suitable instruments to search for extended synchrotron radiation in outer galaxy disks and halos and the transition region to intergalactic space. Several galaxies have already been observed with LOFAR (Beck et al., 2013; Mulcahy et al., 2014). Low frequencies are also ideal to search for small Faraday rotation measures from weak interstellar and intergalactic fields.

TABLE 5.3. Typical field structures in nearby galaxies

Galaxy type	Magnetic field structure	Regular field
Sc galaxy with strong density wave	Spiral at inner arm edge and in interarm regions, turbulent in arms	Strong, fluctuating
Sb or Sc galaxy with weak or moderate density wave	Spiral in interarm regions, turbulent + ordered in arms	Moderate, large-scale
Barred Sc galaxy	Ordered + turbulent along bar, spiral outside bar	Moderate
Sa galaxy	Ordered + turbulent	Detected
S0 galaxy	Not detected	Not detected
Interacting spiral	Ordered, asymmetric	Weak
Flocculent Sc or Sd galaxy	Spiral + turbulent in disk	Weak
Irregular galaxy	Turbulent in star-forming regions + segments of ordered field	Weak
Starburst dwarf galaxy	Turbulent in star-forming regions	Not detected
Spheroidal dwarf galaxy	Not detected	Not detected
E galaxy (non-active nucleus)	Turbulent if star-formation occurs	Not detected

Total synchrotron emission, the signature of total magnetic fields, can be detected with the SKA out to very large redshifts for starburst galaxies, depending on luminosity and magnetic field strength (Fig. 5.21). However, for fields weaker than $3.25\,\mu\mathrm{G}\ (1+z)^2$, energy loss of CRE is dominated by the inverse Compton effect with CMB photons, so that their energy is converted mostly into X-rays but not into radio waves. On the other hand, for strong magnetic fields the energy range of the electrons emitting at 1.4 GHz shifts to low energies where ionization and bremsstrahlung losses become dominant. The maximum redshift until synchrotron emission is detected will constrain models of the evolution of magnetic fields in young galaxies (Schleicher & Beck, 2013).

If polarized emission from galaxies is too weak to be detected, the method of *RM grids* towards background QSOs can still be applied. Here, the distance limit is given

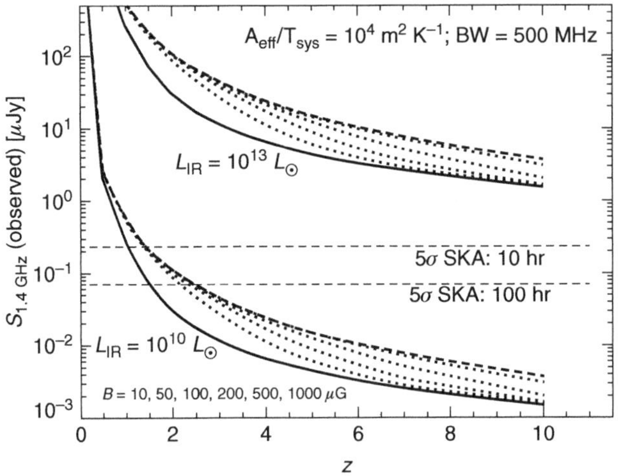

FIGURE 5.21. Total synchrotron emission at 1.4 GHz as a function of redshift z and magnetic field strength B, and the 5σ detection limits for 10 h and 100 h integration time with the full SKA (from Murphy, 2009). Reproduced with permission © AAS.

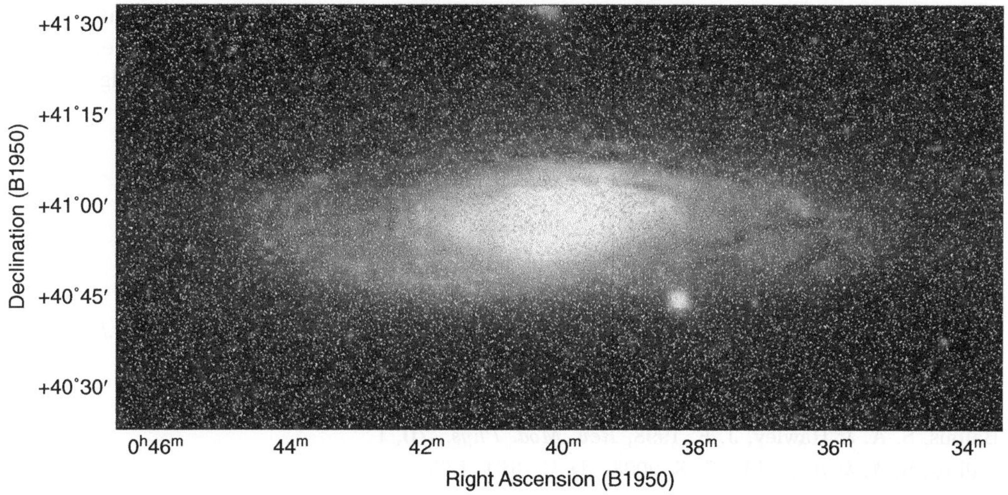

FIGURE 5.22. Simulation of RMs towards background sources (white points) in the region of M 31 observable with the SKA within 1 h. Black and white version of a color image that combines optical emission, diffuse radio continuum and diffuse polarized emission (from Gaensler, private communication).

by the polarized flux of the background QSO which can be much higher than that of the intervening galaxy. Significant regular fields of several μG strength were detected in distant galaxies (Bernet et al., 2008, 2013; Mao et al., 2017). Mean-field solutions of the $\alpha - \Omega$ dynamo theory predict RMs from evolving regular fields at $z \leq 3$ with increasing coherence scale until $z \leq 1$ (Arshakian et al., 2009). Note that the RM values are reduced by the redshift dilution factor of $(1+z)^{-2}$, so that detecting regular fields at high redshifts needs low-frequency observations.

Faraday rotation in the direction of polarized lobes of radio galaxies allows us to determine the strength and pattern of a regular field in an intervening galaxy (Kronberg et al., 1992). RM data indicate that halos of ionized gas extend to several 10 kpc distance from galaxies (Bernet et al., 2013). A reliable model for the field structure in and around galaxies needs RM values from a large number of polarized background sources, hence large sensitivity and high survey speed.

The SKA "Magnetism" Key Science Project plans to observe an all-sky survey at around 1 GHz with 1 h integration per field that will allow detection of about 10 million discrete extragalactic sources and measure their RMs, about 300 RMs per square degree (Johnston-Hollitt et al., 2015). More than 1000 RM values are expected in the area of M 31 (Fig. 5.22) and will allow the detailed reconstruction of the 3D field structure in this galaxy. Applying this method to other galaxies, simple patterns of regular fields can be recognized out to distances of about 100 Mpc (Stepanov et al., 2008).

Acknowledgements

The author would like to thank all colleagues who have pursued the studies of magnetic fields in the Milky Way and in galaxies over the past 40 years, especially Richard Wielebinski, Elly M. Berkhuijsen, Marita Krause, Patricia Reich, Wolfgang Reich, Sui Ann Mao and Aritra Basu at MPIfR, Andrew Fletcher and Anvar Shukurov at Newcastle University, Chris Chyży and Marek Urbanikat at Kraków University, Dmitry Sokoloff at Moscow University, Peter Frick and Rodion Stepanov at ICMM Perm, Ralf-Jürgen Dettmar at Bochum University, Detlef Elstner at AIP Potsdam and Katia Ferrière at Toulouse University. Support from DFG Research Unit FOR1254 is acknowledged.

REFERENCES

Abreu, P., Aglietta, M., Ahn, E. J., et al. 2010, *Astropart. Phys.*, **34**, 314

Ackermann, M., Ajello, M., Atwood, W. B., et al. 2012, *ApJ*, **750**, 3

Adebahr, B., Krause, M., Klein, U., et al. 2013, *A&A*, **555**, A23

Arbutina, B., Urošević, D., Andjelić, M. M., Pavlović, M. Z., & Vukotić, B. 2012, *ApJ*, **746**, 79

Arshakian, T. G. & Beck, R. 2011, *MNRAS*, **418**, 2336

Arshakian, T. G., Beck, R., Krause, M., & Sokoloff, D. 2009, *A&A*, **494**, 21

Balbus, S. A. & Hawley, J. F. 1998, *Rev. Mod. Phys.*, **70**, 1

Balbus, S. A. & Reynolds, C. S. 2008, *ApJL*, **681**, L65

Basu, A. & Roy, S. 2013, *MNRAS*, **433**, 1675

Basu, A., Roy, S., & Mitra, D. 2012, *ApJ*, **756**, 141

Beck, A. M., Lesch, H., Dolag, K., et al. 2012a, *MNRAS*, **422**, 2152

Beck, R. 2007, *A&A*, **470**, 539

Beck, R. 2015, *A&A*, **578**, A93

Beck, R., Anderson, J., Heald, G., et al. 2013, *Astronomische Nachrichten*, **334**, 548

Beck, R., Bomans, D., Colafrancesco, S., et al. 2015, *Advancing Astrophysics with the Square Kilometre Array (AASKA14)*, 94

Beck, R., Brandenburg, A., Moss, D., Shukurov, A., & Sokoloff, D. 1996, *ARA&A*, **34**, 155

Beck, R., Fletcher, A., Shukurov, A., et al. 2005, *A&A*, **444**, 739

Beck, R., Frick, P., Stepanov, R., & Sokoloff, D. 2012b, *A&A*, **543**, A113

Beck, R. & Gaensler, B. M. 2004, NewAR, 48, 1289

Beck, R. & Krause, M. 2005, *Astronomische Nachrichten*, **326**, 414

Beck, R., Shukurov, A., Sokoloff, D., & Wielebinski, R. 2003, *A&A*, **411**, 99

Beck, R. & Wielebinski, R. 2013, in *Planets, Stars and Stellar Systems. Volume 5: Galactic Structure and Stellar Populations*, ed. T. D. Oswalt & G. Gilmore (New York: Springer), 641

Bell, A. R. 1978, *MNRAS*, **182**, 443

Bell, E. F. 2003, *ApJ*, **586**, 794

Bell, M. R., Junklewitz, H., & Enßlin, T. A. 2011, *A&A*, **535**, A85

Berkhuijsen, E. M., Beck, R., & Hoernes, P. 2003, *A&A*, **398**, 937

Berkhuijsen, E. M., Beck, R., & Tabatabaei, F. S. 2013, *MNRAS*, **435**, 1598

Bernet, M. L., Miniati, F., Lilly, S. J., Kronberg, P. P., & Dessauges-Zavadsky, M. 2008, *Nat.*, **454**, 302

Bernet, M. L., Miniati, F., & Lilly, S. J. 2012, *ApJ*, **761**, 144

Bernet, M. L., Miniati, F., & Lilly, S. J. 2013, *ApJL*, **772**, L28

Beuermann, K., Kanbach, G., & Berkhuijsen, E. M. 1985, *A&A*, **153**, 17

Birk, G. T., Lesch, H., & Neukirch, T. 1998, *MNRAS*, **296**, 165

Brandenburg, A. & Stepanov, R. 2014, *ApJ*, **786**, 91

Braun, R., Heald, G., & Beck, R. 2010, *A&A*, **514**, A42

Brentjens, M. A. & de Bruyn, A. G. 2005, *A&A*, **441**, 1217

Broderick, A. E., Chang, P., & Pfrommer, C. 2012, *ApJ*, **752**, 22

Burlaga, L. F., Ness, N. F., Gurnett, D. A., & Kurth, W. S. 2013, *ApJL*, **778**, L3

Burn, B. J. 1966, *MNRAS*, **133**, 67

Caprini, C., Durrer, R., & Fenu, E. 2009, *J. Cosmol. Astropart. Phys.*, **11**, 1

Chamandy, L., Shukurov, A., & Subramanian, K. 2015, *MNRAS*, **446**, L6

Chandrasekhar, S. & Fermi, E. 1953, *ApJ*, **118**, 113

Cho, J. & Lazarian, A. 2005, *ApJ*, **631**, 361

Chyży, K. T. 2008, *A&A*, **482**, 755

Chyży, K. T. & Beck, R. 2004, *A&A*, **417**, 541

Chyży, K. T. & Buta, R. J. 2008, *ApJL*, **677**, L17

Chyży, K. T., Beck, R., Kohle, S., Klein, U., & Urbanik, M. 2000, *A&A*, **355**, 128

Chyży, K. T., Weżgowiec, M., Beck, R., & Bomans, D. J. 2011, *A&A*, **529**, A94

Colafrancesco, S., Profumo, S., & Ullio, P. 2007, *Phys. Rev. D*, **75**, 023513

Crocker, R. M., Jones, D. I., Melia, F., Ott, J., & Protheroe, R. J. 2010, *Nat.*, **463**, 65

Crutcher, R. M., Wandelt, B., Heiles, C., Falgarone, E., & Troland, T. H. 2010, *ApJ*, **725**, 466

Damas-Segovia, A., Beck, R., Vollmer, B., et al. 2016, *ApJ*, **824**, 30

de Avillez, M. A. & Breitschwerdt, D. 2005, *A&A*, **436**, 585

Dobbs, C. L. & Price, D. J. 2008, *MNRAS*, **383**, 497

Dolag, K., Kachelriess, M., Ostapchenko, S., & Tomàs, R. 2011, *ApJL*, **727**, L4

Dumas, G., Schinnerer, E., Tabatabaei, F. S., et al. 2011, *AJ*, **141**, 41

Dumke, M., Krause, M., Wielebinski, R., & Klein, U. 1995, *A&A*, **302**, 691

Ehle, M. & Beck, R. 1993, *A&A*, **273**, 45

Ellis, R. S. & Axon, D. J. 1978, *Astrophys. Space Sci.*, **54**, 425

Fendt, C., Beck, R., & Neininger, N. 1998, *A&A*, **335**, 123

Ferrière, K. 1996, *A&A*, **310**, 438

Ferrière, K. 2009, *A&A*, **505**, 1183

Ferrière, K. & Terral, P. 2014, *A&A*, **561**, A100

Fletcher, A. 2010, in *Astronomical Society of the Pacific Conference Series*, Vol. 438, ed. R. Kothes, T. L. Landecker, & A. G. Willis, 197

Fletcher, A. & Shukurov, A. 2001, *MNRAS*, **325**, 312

Fletcher, A., Berkhuijsen, E. M., Beck, R., & Shukurov, A. 2004, *A&A*, **414**, 53

Fletcher, A., Beck, R., Shukurov, A., Berkhuijsen, E. M., & Horellou, C. 2011, *MNRAS*, **412**, 2396

Fosalba, P., Lazarian, A., Prunet, S., & Tauber, J. A. 2002, *ApJ*, **564**, 762

Frick, P., Sokoloff, D., Stepanov, R., & Beck, R. 2011, *MNRAS*, **414**, 2540

Frick, P., Stepanov, R., Beck, R., et al. 2016, *A&A*, **585**, A21

Frisch, P. C., Andersson, B.-G., Berdyugin, A., et al. 2012, *ApJ*, **760**, 106

Gaensler, B. M., Haverkorn, M., Staveley-Smith, L., et al. 2005, *Science*, **307**, 1610

Gent, F. A., Shukurov, A., Sarson, G. R., Fletcher, A., & Mantere, M. J. 2013, *MNRAS*, **430**, L40

Gießübel, R. 2012, PhD dissertation, Universität zu Köln, Cuvillier Verlag Göttingen, ISBN 978-3-95404-308-8

Gießübel, R., Heald, G., Beck, R., & Arshakian, T. G. 2013, *A&A*, **559**, A27

Gissinger, C., Fromang, S., & Dormy, E. 2009, *MNRAS*, **394**, L84

Goldreich, P. & Kylafis, N. D. 1981, *ApJL*, **243**, L75

Golla, G. & Hummel, E. 1994, *A&A*, **284**, 777

Gómez, G. C. & Cox, D. P. 2002, *ApJ*, **580**, 235

Gräve, R. & Beck, R. 1988, *A&A*, **192**, 66

Greaves, J. S., Holland, W. S., Jenness, T., & Hawarden, T. G. 2000, *Nat.*, **404**, 732

Gressel, O., Elstner, D., Ziegler, U., & Rüdiger, G. 2008, *A&A*, **486**, L35

Han, J. L., Ferriere, K., & Manchester, R. N. 2004, *ApJ*, **610**, 820

Han, J. L., Manchester, R. N., Lyne, A. G., Qiao, G. J., & van Straten, W. 2006, *ApJ*, **642**, 868

Hanasz, M., Otmianowska-Mazur, K., & Lesch, H. 2002, *A&A*, **386**, 347

Hanasz, M., Wóltański, D., & Kowalik, K. 2009, *ApJL*, **706**, L155

Hanayama, H., Takahashi, K., Kotake, K., et al. 2005, *ApJ*, **633**, 941

Haverkorn, M., Gaensler, B. M., Brown, J. C., et al. 2006, *ApJL*, **637**, L33

Heald, G. 2009, in *IAU Symposium*, Vol. 259, 591–602

Heald, G. H. 2012, *ApJL*, **754**, L35

Heesen, V., Beck, R., Krause, M., & Dettmar, R.-J. 2009a, *A&A*, **494**, 563

Heesen, V., Krause, M., Beck, R., & Dettmar, R.-J. 2009b, *A&A*, **506**, 1123

Heesen, V., Beck, R., Krause, M., & Dettmar, R.-J. 2011a, *A&A*, **535**, A79

Heesen, V., Rau, U., Rupen, M. P., Brinks, E., & Hunter, D. A. 2011b, *ApJL*, **739**, L23

Heesen, V., Brinks, E., Leroy, A. K., et al. 2014, *AJ*, **147**, 103

Heiles, C. & Crutcher, R. 2005, in *Lecture Notes in Physics*, Berlin Springer Verlag, Vol. 664, Cosmic Magnetic Fields, ed. R. Wielebinski & R. Beck, 137

Heitsch, F., Zweibel, E. G., Slyz, A. D., & Devriendt, J. E. G. 2004, *ApJ*, **603**, 165

Helfer, T. T., Thornley, M. D., Regan, M. W., et al. 2003, *Astrophys. J. Suppl.*, **145**, 259

Hildebrand, R. H., Kirby, L., Dotson, J. L., Houde, M., & Vaillancourt, J. E. 2009, *ApJ*, **696**, 567

Hoang, T. & Lazarian, A. 2008, *MNRAS*, **388**, 117

Horellou, C. & Fletcher, A. 2014, *MNRAS*, **441**, 2049

Houde, M., Vaillancourt, J. E., Hildebrand, R. H., Chitsazzadeh, S., & Kirby, L. 2009, *ApJ*, **706**, 1504

Houde, M., Rao, R., Vaillancourt, J. E., & Hildebrand, R. H. 2011, *ApJ*, **733**, 109

Houde, M., Fletcher, A., Beck, R., et al. 2013, *ApJ*, **766**, 49

Hughes, A., Wong, T., Ekers, R., et al. 2006, *MNRAS*, **370**, 363

Ideguchi, S., Tashiro, Y., Akahori, T., et al. 2017, *ApJ*, **843**, 146

Jałocha, J., Bratek, Ł., Pekala, J., & Kutschera, M. 2012, *MNRAS*, **421**, 2155

Jansson, R. & Farrar, G. R. 2012a, *ApJ*, **757**, 14

Jansson, R. & Farrar, G. R. 2012b, *ApJL*, **761**, L11

Johnston-Hollitt, M., Govoni, F., Beck, R., et al. 2015, *Advancing Astrophysics with the Square Kilometre Array (AASKA14)*, 92

Kepley, A. A., Mühle, S., Everett, J., et al. 2010, *ApJ*, **712**, 536

Kim, W.-T. & Stone, J. M. 2012, *ApJ*, **751**, 124

Klein, U., Weiland, H., & Brinks, E. 1991, *A&A*, **246**, 323

Knapik, J., Soida, M., Dettmar, R.-J., Beck, R., & Urbanik, M. 2000, *A&A*, **362**, 910

Kotarba, H., Lesch, H., Dolag, K., et al. 2009, *MNRAS*, **397**, 733

Kotarba, H., Karl, S. J., Naab, T., et al. 2010, *ApJ*, **716**, 1438

Krause, F. & Beck, R. 1998, *A&A*, **335**, 789

Krause, M. 1990, in *IAU Symposium, Vol. 140, Galactic and Intergalactic Magnetic Fields*, ed. R. Beck, R. Wielebinski, & P. P. Kronberg, 187

Krause, M. 1993, in *IAU Symposium, Vol. 157, The Cosmic Dynamo*, ed. F. Krause, K. H. Rädler, & G. Rüdiger, 305

Krause, M. 2009, in *Revista Mexicana de Astronomia y Astrofisica Conference Series*, Vol. 36, 25

Krause, M. 2014, ArXiv e-prints

Krause, M., Beck, R., & Hummel, E. 1989a, *A&A*, **217**, 17

Krause, M., Hummel, E., & Beck, R. 1989b, *A&A*, **217**, 4

Krause, M., Wielebinski, R., & Dumke, M. 2006, *A&A*, **448**, 133

Kronberg, P. P., Lesch, H., & Hopp, U. 1999, *ApJ*, **511**, 56

Kronberg, P. P., Perry, J. J., & Zukowski, E. L. H. 1992, *ApJ*, **387**, 528

Kulpa-Dybeł, K., Otmianowska-Mazur, K., Kulesza-Żydzik, B., et al. 2011, *ApJL*, **733**, L18

Lacki, B. C. & Beck, R. 2013, *MNRAS*, **430**, 3171

Lacki, B. C., Thompson, T. A., & Quataert, E. 2010, *ApJ*, **717**, 1

Laine, S. & Beck, R. 2008, *ApJ*, **673**, 128

Lazar, M., Schlickeiser, R., Wielebinski, R., & Poedts, S. 2009, *ApJ*, **693**, 1133

Li, H.-B. & Henning, T. 2011, *Nat.*, **479**, 499

Mao, S. A., Gaensler, B. M., Haverkorn, M., et al. 2010, *ApJ*, **714**, 1170

Mao, S. A., McClure-Griffiths, N. M., Gaensler, B. M., et al. 2012, *ApJ*, **759**, 25

Mao, S. A., Zweibel, E., Fletcher, A., et al. 2015, *ApJ*, **800**, 92

Mao, S. A., Carilli, C., Gaensler, B. M., et al. 2017, *Nature Astronomy*, **1**, 621

Men, H., Ferrière, K., & Han, J. L. 2008, *A&A*, **486**, 819

Mestel, L. 1990, in *IAU Symposium, Vol. 140, Galactic and Intergalactic Magnetic Fields*, ed. R. Beck, R. Wielebinski, & P. P. Kronberg, 259

Mora, S. C. & Krause, M. 2013, *A&A*, **560**, A42

Moss, D., Sokoloff, D., Beck, R., & Krause, M. 2010, *A&A*, **512**, A61

Moss, D., Stepanov, R., Arshakian, T. G., et al. 2012, *A&A*, **537**, A68

Moss, D., Stepanov, R., Krause, M., Beck, R., & Sokoloff, D. 2015, *A&A*, **578**, A94

Mulcahy, D. D., Horneffer, A., Beck, R., et al. 2014, *A&A*, **568**, A74

Murphy, E. J. 2009, *ApJ*, **706**, 482

Nota, T. & Katgert, P. 2010, *A&A*, **513**, A65

Noutsos, A. 2009, in *IAU Symposium*, Vol. 259, ed. K. G. Strassmeier, A. G. Kosovichev, & J. E. Beckman, 15

Noutsos, A. 2012, *Space Sci. Rev.*, **166**, 307

Nyland, K., Young, L. M., Wrobel, J. M., et al. 2017, *MNRAS*, **464**, 1029

Ohno, H. & Shibata, S. 1993, *MNRAS*, **262**, 953

Opher, M., Bibi, F. A., Toth, G., et al. 2009, *Nat.*, **462**, 1036

Oppermann, N., Junklewitz, H., Robbers, G., et al. 2012, *A&A*, **542**, A93

Orlando, E. & Strong, A. 2013, *MNRAS*, **436**, 2127

Pakmor, R. & Springel, V. 2013, *MNRAS*, **432**, 176

Parrish, I. J. & Quataert, E. 2008, *ApJL*, **677**, L9

Patrikeev, I., Fletcher, A., Stepanov, R., et al. 2006, *A&A*, **458**, 441

Price, D. J. & Bate, M. R. 2008, *MNRAS*, **385**, 1820

Rand, R. J. & Kulkarni, S. R. 1989, *ApJ*, **343**, 760

Rees, M. J. 2005, in *Lecture Notes in Physics*, Berlin Springer Verlag, Vol. 664, *Cosmic Magnetic Fields*, ed. R. Wielebinski & R. Beck, 1

Reuter, H.-P., Klein, U., Lesch, H., Wielebinski, R., & Kronberg, P. P. 1994, *A&A*, **282**, 724

Robishaw, T., Quataert, E., & Heiles, C. 2008, *ApJ*, **680**, 981

Rodrigues, L. F. S., Shukurov, A., Fletcher, A., & Baugh, C. M. 2015, *MNRAS*, **450**, 3472

Rohde, R., Beck, R., & Elstner, D. 1999, *A&A*, **350**, 423

Ruiz-Granados, B., Rubiño-Martín, J. A., Florido, E., & Battaner, E. 2010, *ApJL*, **723**, L44

Scarrott, S. M., Ward-Thompson, D., & Warren-Smith, R. F. 1987, *MNRAS*, **224**, 299

Schleicher, D. R. G. & Beck, R. 2013, *A&A*, **556**, A142

Schleicher, D. R. G. & Beck, R. 2016, *A&A*, **593**, A77

Schleicher, D. R. G., Banerjee, R., Sur, S., et al. 2010, *A&A*, **522**, A115

Schlickeiser, R. 2012, *Phys. Rev. Lett.*, **109**, 261101

Schlickeiser, R. & Felten, T. 2013, *ApJ*, **778**, 39

Sellwood, J. A. & Balbus, S. A. 1999, *ApJ*, **511**, 660

Seymour, N., Dwelly, T., Moss, D., et al. 2008, *MNRAS*, **386**, 1695

Shukurov, A. & Berkhuijsen, E. M. 2003, *MNRAS*, **342**, 496

Shukurov, A., Sokoloff, D., Subramanian, K., & Brandenburg, A. 2006, *A&A*, **448**, L33

Siejkowski, H., Otmianowska-Mazur, K., Soida, M., Bomans, D. J., & Hanasz, M. 2014, *A&A*, **562**, A136

Soida, M., Beck, R., Urbanik, M., & Braine, J. 2002, *A&A*, **394**, 47

Soida, M., Krause, M., Dettmar, R.-J., & Urbanik, M. 2011, *A&A*, **531**, A127

Sokoloff, D., Shukurov, A., & Krause, M. 1992, *A&A*, **264**, 396

Sokoloff, D. D., Bykov, A. A., Shukurov, A., et al. 1998, *MNRAS*, **299**, 189

Stepanov, R., Arshakian, T. G., Beck, R., Frick, P., & Krause, M. 2008, *A&A*, **480**, 45

Stepanov, R., Shukurov, A., Fletcher, A., et al. 2014, *MNRAS*, **437**, 2201

Stil, J. M., Krause, M., Beck, R., & Taylor, A. R. 2009, *ApJ*, **693**, 1392

Strong, A. W., Moskalenko, I. V., & Reimer, O. 2000, *ApJ*, **537**, 763

Subramanian, K. 1998, *MNRAS*, **294**, 718

Sun, X.-H. & Reich, W. 2010, *Res. Astron. Astrophys.*, **10**, 1287

Sun, X. H., Reich, W., Waelkens, A., & Enßlin, T. A. 2008, *A&A*, **477**, 573

Tabatabaei, F. S., Beck, R., Krügel, E., et al. 2007, *A&A*, **475**, 133

Tabatabaei, F. S., Krause, M., Fletcher, A., & Beck, R. 2008, *A&A*, **490**, 1005

Tabatabaei, F. S., Berkhuijsen, E. M., Frick, P., Beck, R., & Schinnerer, E. 2013a, *A&A*, **557**, A129

Tabatabaei, F. S., Schinnerer, E., Murphy, E. J., et al. 2013b, *A&A*, **552**, A19

Tang, Y.-W., Ho, P. T. P., Koch, P. M., et al. 2009, *ApJ*, **700**, 251

Taylor, A. R., Stil, J. M., & Sunstrum, C. 2009, *ApJ*, **702**, 1230

Thompson, T. A., Quataert, E., Waxman, E., Murray, N., & Martin, C. L. 2006, *ApJ*, **645**, 186

Tingay, S. J., Kaplan, D. L., McKinley, B., et al. 2013, *AJ*, **146**, 103

Tüllmann, R., Dettmar, R.-J., Soida, M., Urbanik, M., & Rossa, J. 2000, *A&A*, **364**, L36

Urbanik, M., Elstner, D., & Beck, R. 1997, *A&A*, **326**, 465

Van Eck, C. L., Brown, J. C., Stil, J. M., et al. 2011, *ApJ*, **728**, 97

van Haarlem, M. P., Wise, M. W., Gunst, A. W., et al. 2013, *A&A*, **556**, A2

Vázquez-Semadeni, E., Kim, J., & Ballesteros-Paredes, J. 2005, *ApJL*, **630**, L49

Vollmer, B., Soida, M., Beck, R., et al. 2007, *A&A*, **464**, L37

Vollmer, B., Soida, M., Beck, R., et al. 2013, *A&A*, **553**, A116

Wang, P. & Abel, T. 2009, *ApJ*, **696**, 96

Weżgowiec, M., Urbanik, M., Vollmer, B., et al. 2007, *A&A*, **471**, 93

Weżgowiec, M., Urbanik, M., Beck, R., Chyży, K. T., & Soida, M. 2012, *A&A*, **545**, A69

Widrow, L. M. 2002, *Rev. Mod. Phys.*, **74**, 775

Wolleben, M., Fletcher, A., Landecker, T. L., et al. 2010, *ApJL*, **724**, L48

Yoast-Hull, T. M., Everett, J. E., Gallagher, III, J. S., & Zweibel, E. G. 2013, *ApJ*, **768**, 53

6. Primordial Magnetic Fields in the Early Universe and Cosmic Microwave Background Anisotropies

FABIO FINELLI AND DANIELA PAOLETTI

6.1 Introductory Remarks

Our aim in the talks given at the XXV Canary Islands Winter School of Astrophysics in 2013 was to present models for the generation of primordial magnetic fields in the early Universe and the imprints that these leave in cosmological observables, with an emphasis on Cosmic Microwave Background (CMB) anisotropies.

6.2 Introduction

The origin of the large-scale magnetic fields (LSMF) observed in galaxies, clusters of galaxies, with also hints of their presence in cosmic voids and on even larger scales in filaments, is an open issue of great importance in modern astrophysics (see Widrow, 2002, for reviews). Both large-scale and stochastic components are present in magnetic fields observed in galaxies (usually dependent on the morphology of the host) with magnitudes from a few to several microGauss (μG). In clusters of galaxies, stochastic magnetic fields from a few to several μG strength are present with a correlation scale of the order of magnitude of ten kiloparsec (Clarke et al., 2001; Enßlin & Vogt, 2006). More recently, the presence of intergalactic magnetic fields even in cosmic voids of the large-scale structure was proposed as a possible explanation for the gamma-ray observations of a couple of blazars (Neronov & Vovk, 2010; Taylor et al., 2011; Vovk et al., 2012; Tavecchio et al., 2010, 2011; Dolag et al., 2011). As a possible explanation for the lack of TeV photons observed in the high-energy spectrum from these blazars a lower bound of 10^{-18}–10^{-15} G was derived for such pervasive intergalactic magnetic fields.

In light of this multitude of observations of LSMF of different magnitudes and different coherence lengths, a primordial hypothesis for generating the seed magnetic fields, which are amplified afterwards by adiabatic compression and dynamo during structure formation, is a viable possibility (Widrow, 2002), also taking in consideration the recent observations of strong magnetic fields in galaxies at high redshift (Bernet et al., 2008; Wolfe et al., 2008). Recently, initial seeds motivated as primordial magnetic fields (PMF) have been used in N-body simulations to reproduce magnetic fields in clusters of galaxies (Govoni et al., 2013; Xu et al., 2012). What favours even more strongly a primordial component for the origin of LSMF is the possibility of the presence of intergalactic magnetic fields in the voids of large-scale structure; such fields would be more difficult to generate by astrophysical mechanisms.

Such a stochastic background can leave footprints in several cosmological observables that can be used to constrain the PMF amplitude. In particular, a limit of the order of 0.1 μG is set by the impact of PMF to the Big Bang Nucleosynthesis (Grasso & Rubinstein, 1995). Instead, an amplitude larger than 30 nG would have left observable distortions in the spectrum of the CMB (Jedamzik et al., 2000).

One of the most important cosmological observables where we can investigate footprints of the PMF is the Cosmic Microwave Background radiation (CMB). In fact, a stochastic background of PMF modelled as a fully inhomogeneous source to the Einstein

TABLE 6.1. Current constraints on primordial magnetic fields Part I.

Probe	Data	Constraint	Reference
BBN		$\sqrt{\langle B^2 \rangle} \leq 300$ nG	Grasso & Rubinstein, 1995
CMB SD	COBE FIRAS	$\sqrt{\langle B^2 \rangle} \leq 30$ nG	Jedmazik et al., 2000
CMB PS	WMAP 7 + SPT	$B_{1\,\mathrm{Mpc}} \leq 3.5$ nG	Paoletti & Finelli, 2013
CMB PS ($n_B = 2$)	WMAP 7 + SPT	$B_{1\mathrm{Mpc}} \leq 6.6 \times 10^{-3}$ nG	Paoletti & Finelli, 2013
CMB PS	*Planck* 2013 + ACT/SPT	$B_{1\,\mathrm{Mpc}} \leq 3.5$ nG	Planck Collaboration XVI, 2014
CMB PS	*Planck* 2015	$B_{1\,\mathrm{Mpc}} \leq 4.4$ nG	Planck Collaboration XIX, 2016
CMB PS	*Planck* 2015 + Helical fields	$B_{1\,\mathrm{Mpc}} \leq 5.8$ nG	Planck Collaboration XIX, 2016
CMB PS ($n_B = 2$)	*Planck* 2015	$B_{1\,\mathrm{Mpc}} \leq 0.011$ nG	Planck Collaboration XIX, 2016
CMB PS ($n_B = -2.9$)	*Planck* 2015	$B_{1\,\mathrm{Mpc}} \leq 2.1$ nG	Planck Collaboration XIX, 2016
CMB Faraday rotation	WMAP 7	$\sqrt{\langle B^2 \rangle} \leq 31$ nG	Kahniashvili et al., 2010
CMB Faraday rotation	*Planck* 2015	$B_{1\mathrm{Mpc}} \leq 1380$ nG	Planck Collaboration XIX, 2016
Gev-TeV from blazars	FERMI-LAT	$B_{1\mathrm{Mpc}} \geq 3 \times 10^{-7}$ nG	Neronov & Vovk, 2010

TABLE 6.2. Current constraints on primordial magnetic fields Part II.

Probe	Data	Constraint	Reference
CMB BS ($n_B = -2.9$) (compensated scalar mode on large angular scales)	*Planck* 2015	$B_{1\mathrm{Mpc}} \leq 3$ nG	Planck Collaboration XIX, 2016
CMB BS ($n_B = -2.9$) (bimodal decomposition, tensor passive mode)	*Planck* 2015	$B_{1\mathrm{Mpc}} \leq 2.9$ nG	Planck Collaboration XIX, 2016
CMB BS ($n_B = -2.9$) (directional bispectrum, passive modes)	*Planck* 2015	$B_{1\mathrm{Mpc}} \leq 4.5$ nG	Planck Collaboration XIX, 2016
CMB BS (compensated scalar mode on large angular scales)	*Planck* 2013	$\sqrt{\langle B^2 \rangle} \leq 22$ nG	Trivedi et al., 2013
CMB BS (passive scalar mode)	WMAP 7	$\sqrt{\langle B^2 \rangle} \leq 2.4$ nG	Trivedi et al., 2010
CMB BS (vector mode)	WMAP 7	$\sqrt{\langle B^2 \rangle} \leq 10$ nG	Shiraishi et al., 2010
CMB BS (tensor mode)	WMAP 7	$\sqrt{\langle B^2 \rangle} \leq 3.2$ nG	Shiraishi & Sekiguchi, 2014
CMB TS (compensated scalar mode on large angular scales)	*Planck* 2013	$\sqrt{\langle B^2 \rangle} \leq 19$ nG	Trivedi et al., 2014
CMB TS (passive scalar mode)	*Planck* 2013	$\sqrt{\langle B^2 \rangle} \leq 0.6$ nG	Trivedi et al., 2014

equations leaves several different footprints on the anisotropy pattern of the CMB, which need careful theoretical predictions. A stochastic background of PMF generates scalar, vector and tensor perturbations (Paoletti et al., 2009, PFP09 hereafter; Kahniashvili & Ratra, 2007; Lewis, 2004a,b; Mack et al., 2002) – whose contribution to CMB anisotropies in temperature and polarization is not suppressed by Silk damping – with different initial conditions – compensated (Finelli et al., 2008, FPP08 hereafter), passive (Shaw & Lewis, 2010; Lewis, 2004a,b) and inflationary (Bonvin et al., 2013) – which could depend on the generation epoch of PMF. The χ^2 statistics of magnetic source terms to cosmological perturbations, given by the stochastic background (Brown & Crittenden, 2005), makes the contribution to CMB anisotropies totally non-Gaussian: the CMB bispectrum is therefore a useful probe for PMF that provides constraints independent from those based on the CMB power spectrum (see for example Caprini et al., 2009; Seshadri & Subramanian, 2009). A stochastic background of PMF has distinctive predictions for the CMB polarization pattern. Owing to the generation of vector perturbations, a peculiar BB power spectrum is predicted with a shallow maximum at high multipoles $\ell \gtrsim 1000$; such a BB signal is not degenerate with the one predicted by gravitational waves from inflation. Moreover, if a helical component is present in the PMF spectrum, a stochastic background of PMF could also generate parity-violating correlators such as TB and EB (Kahniashvili & Ratra, 2007; Caprini et al., 2004; Ballardini et al., 2014, hereafter BFP14). A stochastic background of PMF can also modify the CMB polarization pattern by the Faraday rotation effect, which has the characteristic dependence on the observed frequency $\propto 1/\nu_{\rm obs}^4$ (Kahniashvili et al., 2009, 2014; Kosowsky et al., 2005; Pogosian et al., 2011; Planck Collaboration et al., 2016). A list of the existing constraints is summarized in Table 6.1 and Table 6.2.

6.3 Generation Mechanisms

PMF, which could have provided the initial seed magnetic fields for the multitude of observed LSMF described above, can be described as a stochastic inhomogeneous background. Such a stochastic background of PMF could be generated by very different mechanisms operating at very different epochs. Below is a very coincise summary.

- PMF can be generated by electric currents induced by the difference in motion of protons and electrons. Vorticity of the plasma can produce such electric currents (Harrison, 1970). However, since linear vector perturbations decay in a perfect fluid dominated Universe, the second-order perturbation theory of cosmological fluctuations keeping separated electrons, protons and photons in the Einstein–Boltzmann system is needed. Takahashi et al. (2005) concluded that $B \sim \mathcal{O}(10^{-25})$ G on scales of 10 Mpc can be obtained (see also Ichiki et al., 2006; Matarrese et al., 2005).
- PMF can be generated during phase transitions motivated by particle physics (see for instance Vachaspati, 1991; Grasso & Riotto, 1998).
- Generation mechanisms based on inflation can amplify $U(1)$ 'electromagnetic' fields (or hyper-electromagnectic fields, which decay later into the $U(1)$ of the Standard Model of particle physics) by breaking conformal invariance. The Maxwell equations in four dimensions in a Robertson–Walker metric are indeed equivalent to those in Minkowski metric up to a rescaling factor function of the cosmological scale factor (the same applies to massless conformally coupled scalar fields and massless spin-1/2 fields, see e.g. Finelli et al., 1999). Therefore the geometrical amplification of quantum fluctuations at the core of inflation that operates for massless minimally coupled scalar fields and for gravitational waves does not apply to electromagnetic fields. Breaking of conformal invariance for the electromagnetic fields can occur in

many ways during inflation, as first explored and studied by Turner & Widrow (1988).

- Conformal invariance can be broken after inflation during (p)reheating through the coupling with the inflaton or other charged fields (Finelli & Gruppuso, 2001; Calzetta & Kandus, 2002; Byrnes et al., 2012).
- Conformal invariance can be broken in the presence of extra-dimensions, which appear as a coupling with the dilaton after dimensional reduction (Ratra, 1992; Gasperini et al., 1995).

It is important to say that both electric and magnetic fields are amplified during the stage in which conformal invariance is broken. However, in the subsequent stage in which charged particles are created and conductivity grows to a large value, electric fields are dissipated and magnetic fields remain frozen in the plasma (Ratra, 1992).

These different scenarios can be potentially discriminated since they lead to different predictions for the spectral index n_B of the stochastic background. Durrer & Caprini (2003) argued that causal mechanisms after inflation predict $n_B \geq 2$ ($n_H \geq 3$) for non-helical (helical) PMF, whereas inflationary mechanisms can produce any spectral index with a large coherence length.

6.4 A Stochastic Background of Primordial Magnetic Fields

The most general two-point correlation function for a stochastic background that preserves stochastic homogeneity and isotropy is:

$$\langle B_i(\mathbf{k})B_j^*(\mathbf{h})\rangle = \frac{(2\pi)^3}{2}\delta(\mathbf{k}-\mathbf{h})\Big[P_{ij}(k)P_B(k) + \imath\epsilon_{ijl}\hat{k}_l P_H(k)\Big], \tag{6.1}$$

where $P_{ij}(k) = \delta_{ij} - \hat{k}_i\hat{k}_j$ ensures the solenoidal condition on the magnetic field (i.e. Maxwell's equations $\nabla \cdot \mathbf{B} = 0$), ϵ_{ijl} is the totally antisymmetric tensor related to the P-violation under the translation $\mathbf{k} \to -\mathbf{k}$, and P_B and P_H are the non-helical and helical part of the spectrum for the stochastic background, respectively. The symmetric part of the power spectrum represents the averaged magnetic field energy density, and the antisymmetric part represents instead the averaged helical component (Malyshkin & Boldyrev, 2007):

$$\langle B_i(\mathbf{k})B_i^*(\mathbf{h})\rangle = (2\pi)^3\delta(\mathbf{k}-\mathbf{h})P_B(k)\,, \tag{6.2}$$

$$\langle(\widehat{\nabla\times\mathbf{B}})_i(\mathbf{k})B_i^*(\mathbf{h})\rangle = (2\pi)^3\delta(\mathbf{k}-\mathbf{h})P_H(k)\,. \tag{6.3}$$

More exactly the antisymmetric part represents the current helicity associated with the effects of turbulence of the magnetic field (Malyshkin & Boldyrev, 2007). An alternative definition is the helicity connected to the gauge field:

$$\langle(\widehat{\nabla\times\mathbf{A}})_i(\mathbf{k})A_i^*(\mathbf{h})\rangle, \tag{6.4}$$

which is not gauge-invariant in general. In the following we will use just the definition of helicity as in Eq. (6.3).

We assume that both non-helical and helical terms of the PMF power spectrum can be described by a simple power law:

$$P_B(k) = A_B\left(\frac{k}{k_*}\right)^{n_B}, \tag{6.5}$$

$$P_H(k) = A_H\left(\frac{k}{k_*}\right)^{n_H}, \tag{6.6}$$

where A_B, A_H are the amplitudes, n_B, n_H are the spectral indices of the non-helical and helical parts respectively and k_* is the pivot scale.

Even though magnetically induced perturbations do not suffer from Silk damping, PMFs are nevertheless suppressed on small scales by radiation viscosity (Jedamzik et al., 1998; Kahniashvili et al., 2001). We model this damping by introducing a sharp cut-off in the PMF power spectra at the damping scale k_D, which represents the dissipation scale of the fields, and we adopt the one proposed in Subramanian & Barrow (1998). With this model the power spectra (6.5) and (6.6) are assumed to be zero for scales $k \geq k_D$.

In this description of a stochastic background, the PMF are characterized by two quantities, an amplitude of the power spectra, A_X, and a spectral index n_X. The latter is one of the major discriminating factors between generation mechanisms, since different mechanisms generate fields with different spectral indices (for example, causal mechanisms generate fields with $n_B > 2$ for the non-helical part and $n_H > 3$ for the helical part; Durrer & Caprini, 2003).

We can express the amplitudes A_B and A_H in terms of mean-square values of the magnetic field and of the absolute value of the helicity as:

$$\langle B^2 \rangle = \int \frac{d^3k}{(2\pi)^3} P_B(k) = \frac{A_B}{2\pi^2} \frac{k_D^{n_B+3}}{n_B+3}, \tag{6.7}$$

$$\langle \mathcal{B}^2 \rangle = \frac{1}{k_D} \int \frac{d^3k}{(2\pi)^3} k |P_H(k)| = \frac{|A_H|}{2\pi^2} \frac{k_D^{n_H+3}}{n_H+4}, \tag{6.8}$$

where the integrals over momentum run from 0 to k_D. Conventionally the fields can be parametrized convolving the PMF amplitude and its helicity with a 3D Gaussian window function, which smooths the field over a sphere of comoving radius λ (where usually we assume $\lambda = 1$ Mpc) like:

$$\langle B_\lambda^2 \rangle = \int_{\mathbb{R}} \frac{d^3k}{(2\pi)^3} P_B(k)\, e^{-\lambda^2 k^2} = \frac{A_B}{(2\pi)^2} \frac{1}{\lambda^{n_B+3}} \Gamma\Big(\frac{n_B+3}{2}\Big), \tag{6.9}$$

$$\langle \mathcal{B}_\lambda^2 \rangle = \lambda \int_{\mathbb{R}} \frac{d^3k}{(2\pi)^3} k |P_H(k)|\, e^{-\lambda^2 k^2} = \frac{|A_H|}{(2\pi)^2} \frac{1}{\lambda^{n_H+3}} \Gamma\Big(\frac{n_H+4}{2}\Big), \tag{6.10}$$

where for the convergence of the integrals above it is requested $n_B > -3$ and $n_H > -4$.

PMF evolved as a stiff source, therefore at the lowest order we can neglect back-reactions of the fluid or gravity onto the fields. Before the decoupling epoch, the electric conductivity of the primordial plasma is very large, therefore it is possible at lowest order to consider the infinite conductivity limit, in which the induced electric field is zero. Within this limit, the magnetic field amplitude scales simply as $B(\mathbf{x}, \tau) = B(\mathbf{x})/a(\tau)^2$, where τ is the conformal time and $a(\tau)$ is the scale factor usually normalized at $a_0 = 1$. Therefore the energy–momentum tensor (EMT) of a stochastic background of PMF is:

$$\tau^0_{\ 0}(\mathbf{x}) = -\frac{1}{8\pi a^4} |\mathbf{B}(\mathbf{x})|^2, \tag{6.11}$$

$$\tau^0_{\ i}(\mathbf{x}) = 0, \tag{6.12}$$

$$\tau^i_{\ j}(\mathbf{x}) = \frac{1}{4\pi a^4} \Big(\delta^i_{\ j} \frac{|\mathbf{B}(\mathbf{x})|^2}{2} - B^i(\mathbf{x}) B_j(\mathbf{x})\Big). \tag{6.13}$$

The spatial part of magnetic field EMT in Fourier space is given by:

$$\tau_{ij}(\mathbf{k}) = \frac{1}{32\pi^4} \int d^3p \Big[B_i(\mathbf{p}) B_j(\mathbf{k}-\mathbf{p}) - \frac{\delta_{ij}}{2} B_l(\mathbf{p}) B_l(\mathbf{k}-\mathbf{p})\Big]. \tag{6.14}$$

The Fourier transforms of the various components of the EMT are all convolutions. It was understood in Paoletti (2007) and FPP08 that these convolutions could be *exactly* performed analytically by assuming a sharp cut-off at $k = k_D$ in the power spectra of the fields. In particular, since $P_B(k) = 0$ and $P_H(k) = 0$ for $k > k_D$, two conditions need to be taken into account in the convolution integrals: $p < k_D$ and $|\mathbf{k} - \mathbf{p}| < k_D$. The second condition introduces a k-dependence on the angular integration domain and the two allow the spectra for the EMT coefficients to be non-zero only for $0 < k < 2k_D$. Such conditions split the double integral (over $\gamma = \hat{\mathbf{k}} \cdot \hat{\mathbf{p}}$ and over p) in three parts depending on the γ and p lower and upper limit of integration. For more details see FPP08, PFP09 and BFP14.

6.4.1 *Scalar Sector*

Scalar magnetized perturbations are sourced by the energy density and the scalar part of the anisotropic stress of the stochastic background of PMF, together with the induced scalar part of the Lorentz force acting on baryons. Owing to the inhomogeneous nature of the stochastic background, the conservation law for the EMT of PMF implies that only two of the above quantities are independent and the following relation holds:

$$\sigma = \frac{\rho}{3} + L\,. \tag{6.15}$$

We will limit our detailed description to the ρ_B treatment, we refer to FPP08, PFP09 and BFP14 for more details and the other correlators. The two-point correlation function of the energy density can be written in the Fourier space as:

$$\begin{aligned} \langle \rho_B(\mathbf{k}) \rho_B^*(\mathbf{h}) \rangle &\equiv (2\pi)^3 \delta^{(3)}(\mathbf{k} - \mathbf{h}) \left|\rho_B(k)\right|^2 , \\ &= \delta_{ab}\delta_{cd} \langle \tau_{ab}(\mathbf{k}) \tau_{cd}^*(\mathbf{h}) \rangle\,. \end{aligned} \tag{6.16}$$

With a little algebra (FPP08, BFP14) the energy density spectrum can be written as:

$$\begin{aligned} \left|\rho_B(k)\right|^2 \equiv \left|\rho_B(k)\right|^2 - 2\rho_H(k) = \int_\Omega \frac{d^3p}{(4\pi)^5} \Big[P_B(p) P_B(|\mathbf{k} - \mathbf{p}|)(1 + \mu^2) \\ - 2P_H(p) P_H(|\mathbf{k} - \mathbf{p}|)\mu \Big] , \end{aligned} \tag{6.17}$$

where $\mu \equiv \hat{p} \cdot \widehat{(\mathbf{k} - \mathbf{p})} = \frac{k\gamma - p}{\sqrt{k^2 - 2kp\gamma + p^2}}$ and $\gamma \equiv \hat{k} \cdot \hat{p}$. There is therefore a non-negligible helical contribution to the scalar mode.

We give here the explicit values for the non-helical part when $n_B = 2$:

$$|\rho_B(k)|^2 = \frac{A_B^2\, k_D^7}{512\, \pi^4\, k_*^4} \left[\frac{4}{7} - \tilde{k} + \frac{8\tilde{k}^2}{15} - \frac{\tilde{k}^5}{24} + \frac{11\tilde{k}^7}{2240} \right] ,$$

and for the helical part when $n_H = 2$:

$$\rho_H(k) = \frac{A_H^2\, k_D^7}{512\, \pi^4\, k_*^4} \begin{cases} -\frac{2}{7} + \frac{\tilde{k}}{2} - \frac{4\tilde{k}^2}{15} + \frac{2\tilde{k}^4}{45} - \frac{\tilde{k}^7}{210} & \text{for} \quad 0 \le \tilde{k} \le 1 \\ \frac{2}{7} - \frac{2}{15\tilde{k}} - \frac{7\tilde{k}}{18} + \frac{4\tilde{k}^2}{15} - \frac{2\tilde{k}^4}{45} + \frac{\tilde{k}^7}{630} & \text{for} \quad 1 \le \tilde{k} \le 2 \end{cases} .$$

For $k \ll k_D$ and $n_{B,\,H} > -3/2$ the energy density spectrum is:

$$\left|\rho(k)\right|^2 \simeq \frac{A_B^2\, k_D^{2n_B+3}}{128\pi^4 k_*^{2n_B}(2n_B + 3)} + \frac{A_H^2\, k_D^{2n_H+3}}{128\pi^4 k_*^{2n_H}(2n_H + 3)}\,. \tag{6.18}$$

For $n_{B,\,H} = -3/2$ we have a removable parametric divergence that is replaced by a logarithmic divergence in k.

For $-3 < n_{B,\,H} < -3/2$ we obtain a behaviour $|\rho_B|^2 \propto k^{2n_B+3}$ ($|\rho_B|^2 \propto k^{2n_H+3}$). For any value of $n_{B,\,H}$, the exact integration is important to obtain the correct amplitudes.

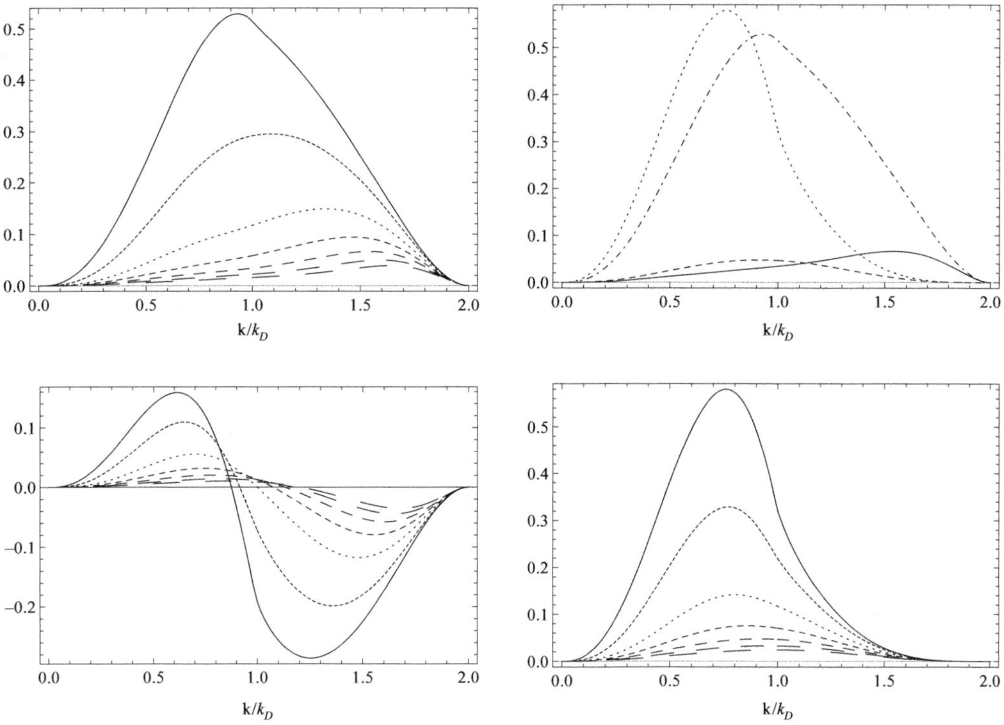

FIGURE 6.1. Non-helical (helical) contribution to $k^3|\rho(k)|^2$ in units of $\langle B^2\rangle^2/(4\pi)^4$ $(\langle \mathcal{B}^2\rangle^2/(4\pi)^4)$ versus k/k_D is plotted in the upper left (bottom left) panel. The total contribution is displayed in the bottom right panel for $\langle B^2\rangle^2 = \langle \mathcal{B}^2\rangle^2$ and $n_B = n_H$. The different lines are for n_B $(n_H) = -3/2, -1, 0, 1, 2, 3, 4$ ranging from the solid to the longest dashed. The panel in the upper right displays the comparison between the non-helical case and the maximal helical case for $n_{B,\,H} = 1$ (solid vs dashed) and $n_{B,\,H} = -3/2$ (dot-dashed vs dotted).

FPP08, PFP09 and BFP14 improve on previous results, which simplified the integration over the angles in an incorrect way (Mack et al., 2002; Kahniashvili & Ratra, 2007).

See the panels in the left in Fig. 6.1 for the shape of $|\rho_B(k)|^2$ and $-2|\rho_H(k)|^2$ for different spectral indices. See the panel in the bottom right of Fig. 6.1 for the total contribution $|\rho(k)|^2$ in the maximal helical case, $A_B = A_H$, and $n_B = n_H$. The panel in the upper right of Fig. 6.1 displays the comparison of $|\rho(k)|^2$ between the non-helical case, $A_H = 0$, and the maximal helical one.

6.4.2 *Vector Sector*

The vector contribution to τ_{ab} is given by:

$$\Pi^{(V)}_{(B)i} \equiv \hat{k}_a P_{ib}(k)\tau_{ab}\,, \tag{6.19}$$

and the two-point correlation function for the vector source in the Fourier space is:

$$\left\langle \Pi^{(V)}_{(B)i}(\mathbf{k})\Pi^{(V)*}_{(B)j}(\mathbf{h})\right\rangle \equiv \frac{(2\pi)^3}{2}\delta(\mathbf{k}-\mathbf{h})\left[P_{ij}(k)|\Pi^{(V)}_{(B)}(k)|^2 + \imath\epsilon_{ijl}\hat{k}_l X^{(V)}_{(B)}(k)\right], \tag{6.20}$$

where, differently from the scalar case, we have both a symmetric part $|\Pi^{(V)}_{(B)}(k)|^2$ and and antisymmetric one $X^{(V)}_{(B)}$.

Vector perturbations are sourced by $\Pi^{(V)}_{(B)}$ with the additional contribution of the Lorentz force on baryons. Again, as in the scalar sector, the conservation equations for the fields give a relation between the vector EMT component and the Lorentz force: $\Pi^{(V)}_{(B)}(k) = -k_i L^{(V)i}_{(B)}$, therefore only one of the two correlators needs to be taken into account. Again, as for the scalar sector, the vector anisotropic stress in Fourier space is given by convolutions of the fields. We refer to PFP09 for the exact solutions; for the case $n_B = 2$ we have:

$$|\Pi^{(V)}_{(B)}(k)|^2_{n_B=2} = \frac{A^2 k_D^7}{256\pi^4 k_*^4}\left[\frac{4}{15} - \frac{5\tilde{k}}{12} + \frac{4\tilde{k}^2}{15} - \frac{\tilde{k}^3}{12} + \frac{7\tilde{k}^5}{960} - \frac{\tilde{k}^7}{1920}\right].$$

6.4.3 *Tensor Sector*

The stochastic background of PMF source gravitational waves comes through the tensor part of the anisotropic pressure.

The tensor part of EMT is given by:

$$\Pi^{(T)}_{(B)ij}(\mathbf{k}) \equiv \mathcal{P}_{ijab}(k)\tau_{ab}(\mathbf{k}), \tag{6.21}$$

where the tensor projector $\mathcal{P}_{ijab}$ is:

$$\mathcal{P}_{ijab}(k) = P_{ia}(k)P_{jb}(k) - \frac{1}{2}P_{ij}(k)P_{ab}(k). \tag{6.22}$$

The two-point correlation function for the tensor part is:

$$\left\langle \Pi^{(T)}_{(B)ij}(\mathbf{k})\Pi^{(T)*}_{(B)lm}(\mathbf{h})\right\rangle \equiv \tfrac{(2\pi)^3}{4}\delta(\mathbf{k}-\mathbf{h}) \times \times\left[\mathcal{M}_{ijlm}|\Pi^{(T)}_{(B)}|^2(k) + \imath\mathcal{A}_{ijlm}X^{(T)}_{(B)}(k)\right], \tag{6.23}$$

where as in the vector sector we have both symmetric, $|\Pi^{(T)}_{(B)}|^2$, and antisymmetric, $X^{(T)}_{(B)}$, parts. The tensors $\mathcal{M}_{ijlm}$ and $\mathcal{A}_{ijlm}$ are

$$\mathcal{M}_{ijlm} \equiv P_{il}P_{jm} + P_{im}P_{jl} - P_{ij}P_{lm}, \tag{6.24}$$

$$\mathcal{A}_{ijlm} \equiv \frac{\hat{k}_t}{2}\left(P_{il}\epsilon_{jmt} + P_{im}\epsilon_{jlt} + P_{jl}\epsilon_{imt} + P_{jm}\epsilon_{ilt}\right). \tag{6.25}$$

The tensor component of the EMT relevant for the magnetically induced perturbations is again a convolution in the Fourier space; the analytical solutions for specific spectral index are given in PFP09, and as an example we report the case for $n_B = 2$:

$$|\Pi^{(T)}_{(B)}(k)|^2_{n_B=2} = \frac{A^2 k_D^7}{256\pi^4 k_*^4}\left[\frac{8}{15} - \frac{7\tilde{k}}{6} + \frac{16\tilde{k}^2}{15} - \frac{7\tilde{k}^3}{24} - \frac{13\tilde{k}^5}{480} + \frac{11\tilde{k}^7}{1920}\right].$$

Let us end this section by assessing possible modifications to the integration scheme with a sharp cut-off in $k = k_D$, which allows us to derive simple polynomial expressions for most of the correlators as in FPP08, PFP09 and BFP14. By substituting a sharp cut-off with a Gaussian filter $W(k) = e^{-k^2/k_D^2}$, the result slightly changes for the non-helical part of $|\rho_B(k)|^2$ with $n_B = 2$, as shown in Fig. 6.2. This means that the physics at the damping scale has little impact on the EMT correlators at the cosmological scales relevant for the CMB and LSS power spectra. Larger differences might arise in other contexts.

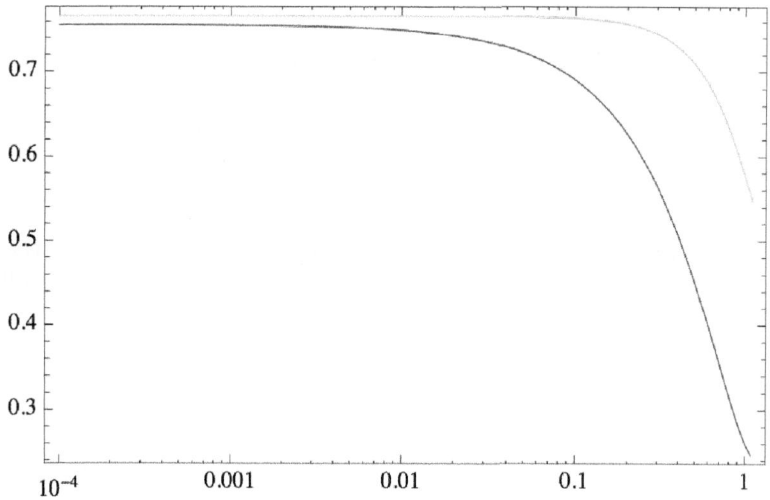

FIGURE 6.2. Comparison of the non-helical part of $|\rho_B(k)|^2$ with $n = 2$ obtained by the sharp cut-off (dark grey line) and the exponential damping (light grey line) versus k/k_D.

6.5 Cosmological Perturbations with PMF

A stochastic background of PMF EMT acts as a stiff source to the Einstein–Boltzmann system of cosmological fluctuations. As a consequence, the Einstein–Boltzmann system of differential equations becomes *inhomogeneous* under the assumption of negligible contribution at the homogeneous level by the stochastic background of PMF. This generates a mode completely sourced by PMF that is basically the special solution of the inhomogeneous system of Einstein–Boltzmann equations. As described in the previous section, the EMT of PMF has scalar, vector and tensor components and therefore PMF source scalar, vector and tensor perturbations. We will now describe the case for non-helical PMF, for the treatment considering also the helical component we refer to BFP14.

6.5.1 *The Scalar Sector*

Magnetically induced metric and matter fluctuations in the scalar sector (see Ma & Bertschinger, 1995, for a treatment in the Netwonian and in the synchronous gauge of primary cosmological perturbations) are driven by the scalar components of the EMT of the PMFs. Using the notation of Ma & Bertschinger (1995), the Einstein equations with the contribution by PMF in the synchronous gauge are

$$\begin{aligned} k^2\eta - \frac{1}{2}\mathcal{H}\dot{h} &= 4\pi G a^2(\Sigma_n \rho_n\delta_n + \rho_B)\,, \\ k^2\dot{\eta} &= 4\pi G a^2 \Sigma_n(\rho_n + P_n)\theta_n\,, \\ \ddot{h} + 2\mathcal{H}\dot{h} - 2k^2\eta &= -8\pi G a^2(\Sigma_n c_{s\,n}^2\rho_n\delta_n + \frac{\rho_B}{3})\,, \\ \ddot{h} + 6\ddot{\eta} + 2\mathcal{H}(\dot{h} + 6\dot{\eta}) - 2k^2\eta &= -24\pi G a^2[\Sigma_n(\rho_n + P_n)\sigma_n + \sigma_B], \end{aligned} \tag{6.26}$$

where by n we label the components, i.e. baryons, cold dark matter (CDM), photons and neutrinos. We note the contributions by magnetic energy density and anisotropic stress, whereas there is no magnetic contribution to the source term of the momentum constraint. To the gravitational contribution we also need to add the contribution of the Lorentz force to the charged particles of the plasma, the baryons. The scalar part of the

Lorentz force can be written as

$$\nabla^2 L^{(S)} = \frac{1}{4\pi}\left[(\nabla_i B_j(\mathbf{x}))\nabla_j B_i(\mathbf{x}) - \frac{1}{2}\nabla^2 B^2(\mathbf{x})\right]. \tag{6.27}$$

The presence of the Lorentz force enters in the Euler equations for baryons:

$$\dot\theta_b = -\mathcal{H}\theta_b + k^2 c_{sb}^2 \delta_b - k^2 \frac{L}{\rho_b}. \tag{6.28}$$

Prior to the decoupling, baryons are tightly coupled with photons through the Thomson scattering. Therefore baryon and photons may be treated as a single coupled fluid in the approximation called 'tight coupling'. The most important consequence for PMF of the tight coupling is that, because of the coupling of the fluids, there is also an indirect effect of the Loretnz force on photons, which can be seen in the Euler equation during the tight coupling:

$$\dot\theta_\gamma = -R^{-1}\Big(\dot\theta_b + \mathcal{H}\theta_b - c_s^2 k^2 \delta_b + k^2 \frac{L}{\rho_b}\Big) + k^2\Big(\frac{\delta_\gamma}{4} - \sigma_\gamma\Big), \tag{6.29}$$

which, when the tight coupling ends, reduces to the regular Euler equation:

$$\dot\theta_\gamma = k^2\Big(\frac{\delta_\gamma}{4} - \sigma_\gamma\Big). \tag{6.30}$$

To complete the analysis it is necessary to fix the initial conditions for magnetically induced perturbations. The initial conditions are given by the solutions of the (now inhomogeneous) Einstein–Boltzmann system at early times and on large wavelengths, namely $k\tau \ll 1$. By assuming only *regular* modes deep in the radiation era for $\tau \to 0$, we obtained the *compensated* initial conditions derived in FPP08 and PFP09 and then extended to the case where also the matter contribution is included in Paoletti & Finelli (2011), hereafter PF11. The term *compensated* derives from the compensation between the magnetic source terms and the fluid terms that is necessary for the solution of the system. This compensation is caused by the fact that PMF modelled as a stochastic background have a negligible contribution at the homogeneous level. In PF11 the initial conditions for the magnetically induced scalar mode are:

$$\begin{aligned}
h(k,\tau) &= -\frac{3}{4}\Omega_B\omega\tau + \frac{9}{32}\Omega_B\omega^2\tau^2\\
\eta(k,\tau) &= \frac{1}{8}\Omega_B\omega\tau - \frac{3\Omega_B\omega^2\tau^2}{64} + \frac{(-165L_B - 55\Omega_B + 28R_\nu\Omega_B)}{168(15+4R_\nu)}k^2\tau^2\\
\delta_\gamma(k,\tau) &= -\Omega_B + \frac{\Omega_B\omega\tau}{2} - \frac{3\Omega_B\omega^2\tau^2}{16} - \frac{(3L_B + \Omega_B - R_\nu\Omega_B)}{6(-1+R_\nu)}k^2\tau^2\\
\delta_\nu(k,\tau) &= -\Omega_B + \frac{\Omega_B\omega\tau}{2} - \frac{3\Omega_B\omega^2\tau^2}{16} - \frac{(3L_B + \Omega_B - R_\nu\Omega_B)}{6R_\nu}k^2\tau^2\\
\delta_b(k,\tau) &= -3\frac{\Omega_B}{4} + \frac{3\Omega_B\omega\tau}{8} + \frac{1}{8}k^2\tau^2\Omega_B - \frac{9}{64}\Omega_B\omega^2\tau^2\omega^2 - \frac{3L_Bk^2\tau^2}{8(-1+R_\nu)}\\
\delta_c(k,\tau) &= -\frac{3\Omega_B}{4} + \frac{3\Omega_B\omega\tau}{8} - \frac{9}{64}\Omega_B\omega^2\tau^2\\
\theta_\gamma(k,\tau) &= \frac{3L_Bk^2\tau}{4(-1+R_\nu)} - \frac{\Omega_B}{4}k^2\tau + k^2\tau^2\Big(-\frac{9L_B(-1+R_c)\omega}{16(-1+R_\nu)^2} + \frac{(-4+R_\nu+3R_c)\omega\Omega_B}{16(-1+R_\nu)}\Big)\\
\theta_\nu(k,\tau) &= \frac{3L_Bk^2\tau}{4R_\nu} - \frac{k^2(-1+R_\nu)\Omega_B\tau}{4R_\nu} + \frac{1}{16}k^2\tau^2\omega\Omega_B
\end{aligned}$$

$$\theta_b(k,\tau) = \frac{3L_B k^2\tau}{4(-1+R_\nu)} - \frac{1}{4}\Omega_B k^2\tau + k^2\tau^2\Big(- \frac{9L_B(-1+R_c)\omega}{16(-1+R_\nu)^2} + \frac{(-4+R_\nu+3R_c)\omega\Omega]}{16(-1+R_\nu)}$$

$$\theta_c(k,\tau) = 0$$

$$\sigma_\nu(k,\tau) = -\frac{3L_B+\Omega_B}{4R_\nu} + \frac{\Omega_B k^2(55-28R_\nu)\tau^2}{56R_\nu(15+4R_\nu)} + \frac{165L_B k^2\tau^2}{56R_\nu(15+4R_\nu)}$$

$$F_3(k,\tau) = -\frac{3k\tau(3L_B+\Omega_B)}{14R_\nu} + \frac{165L_B+55\Omega_B-28R_\nu\Omega_B}{7(430R_\nu+112R_\nu^2)}, \tag{6.31}$$

where:

$$a(\tau) = \frac{\Omega_m H_0^2}{\omega^2}\Big(\omega\tau + \frac{1}{4}\omega^2\tau^2\Big) \tag{6.32}$$

$$\omega = \frac{\Omega_m H_0}{\sqrt{\Omega_\nu+\Omega_\gamma}} \tag{6.33}$$

$$R_\nu = \frac{\rho_\nu}{\rho_\nu+\rho_\gamma} \tag{6.34}$$

$$R_c = \frac{\Omega_c}{\Omega_m} \equiv \frac{\Omega_c}{\Omega_b+\Omega_c} \tag{6.35}$$

$$\Omega_B = \frac{\rho_B}{(\rho_\nu+\rho_\gamma)} \tag{6.36}$$

$$L_B = \frac{L_B}{(\rho_\nu+\rho_\gamma)}. \tag{6.37}$$

6.5.2 *The Vector Sector*

In the standard ΛCDM model vector modes decay with the expansion of the Universe and have no observational signature at any significant level. However the associated temperature fluctuations, once generated, do not decay but in this case they have to be sourced by some shear (Lewis, 2004b). PMF carrying vector anisotropic stress generate a fully magnetized vector mode (Lewis, 2004a) that is the dominant PMF compensated contribution to the CMB angular power spectra on small angular scales. On these scales the primary CMB contribution is suppressed by Silk damping therefore magnetic vector mode dominates over CMB angular power spectrum as shown in PFP09.

The vector metric perturbation is described through a divergenceless vector:

$$h_{ij}^V = \partial_i h_j + \partial_j h_i \tag{6.38}$$

where we have $\partial_i h_i = 0$. For the magnetically induced perturbations we will have two independent metric perturbation modes which are sourced by $|\Pi_{(B)}^{(V)}|^2$:

$$\dot{h}_V^\pm + 2\mathcal{H}h_V^\pm = -16\pi G a^2 \frac{\Pi_\nu^{(V)} + \Pi_\gamma^{(V)} + \Pi_{(B)}^{(V)\pm}}{k}. \tag{6.39}$$

To the contribution to the metric perturbations as in the scalar sector we need to add the contribution of the vector Lorentz force to the baryon Euler equation:

$$\dot{v}_b + \mathcal{H}v_b = -\frac{\rho_\gamma}{\rho_b}\Big(\frac{4}{3}n_e a\sigma_T(v_b - v_\gamma) - \frac{L_{(B)}^{(V)}}{\rho_\gamma}\Big). \tag{6.40}$$

6.5.3 *The Tensor Sector*

The evolution of tensor metric perturbations is described by Einstein equations where PMF contribution is again an additional source term, given by PMF stress tensor:

$$\ddot{h}_{ij} + 2\mathcal{H}\dot{h}_{ij} + k^2 h_{ij} = 16\pi G a^2 \big(\rho_\nu \pi^\nu_{ij} + \Pi^{(T)}_{(B)ij}\big) \,. \tag{6.41}$$

Also in this case the regular solution requires compensation between PMF and the neutrino sector. The solution can be found in PFP09:

$$\begin{aligned} h_k &= A\Big[1 - \frac{5(k\tau)^2}{2(15+4R_\nu)}\Big] + \frac{15(1-R_\nu)\tilde{\Pi}^{(T)}_{(B)}(k\tau)^2}{56(15+4R_\nu)} \,, \\ \sigma^{(T)}_\nu &= -\frac{(1-R_\nu)}{R_\nu}\tilde{\Pi}^{(T)}_{(B)}\Big[1 - \frac{15(k\tau)^2}{14(15+4R_\nu)}\Big] + A\frac{4(k\tau)^2}{3(15+4R_\nu)} \,, \end{aligned} \tag{6.42}$$

where the A represents the amplitude and spectrum of the primary pertubations.

6.5.4 *Passive Modes*

In the previous sections we have discussed the PMF contribution in the case of initial conditions that are non-singular close to the Big Bang (i.e. for $\tau \to 0$). These are *regular* initial conditions, which are unavoidably generated and are analogous to the non-singular initial conditions for isocurvature initial conditions (Bucher et al., 2001). For the scalar (tensor) modes, a compensation is required between PMF and free-streaming neutrinos in energy-densities, velocity and pressure (tensor anisotropic stress) and therefore are often referred to in the literature as *compensated* modes. For a cosmological scenario in which PMF are present before neutrino decoupling (i.e. when neutrinos do not yet free-stream and behave as a perfect fluid), *passive* modes are also considered as generated by *singular* modes (i.e. not regular for $\tau \to 0$) driven by the anisotropic stress of PMF which is not compensated by neutrinos (since are still not free-streaming they do not carry anisotropic stress).[1] These passive modes correspond to homogeneous solutions (i.e. without magnetic contributions) of the Einstein–Boltzmann system in the scalar and tensor sector (Shaw & Lewis, 2010) where the amplitude is derived by the matching of the initial conditions before and after neutrino decoupling, and it is dependent on the PMF stochastic background:

$$\mathcal{R} \simeq \mathcal{R}(\tau_B) - \frac{3}{2}(1-R_\nu)\sigma_B\left[\ln\left(\frac{\tau_\nu}{\tau_B}\right) + \left(\frac{5}{8R_\nu} - 1\right)\right] \tag{6.43}$$

$$h \simeq (1-R_\nu)\Pi^{(T)}_B\left[\ln\left(\frac{\tau_\nu}{\tau_B}\right) + \left(\frac{5}{8R_\nu} - 1\right)\right] . \tag{6.44}$$

6.6 Contributions to the CMB Anisotropies Power Spectrum

The magnetically induced scalar, vector and tensor contributions to CMB anisotropy angular power spectra for compensated modes in the non-helical case are shown in Fig. 6.3. These theoretical predictions have been obtained through a modified version of the publicly available Einstein–Boltzmann code CAMB (Lewis et al., 2000), which evolves the equation system and derives the CMB anisotropy angular power spectra; the modified

[1] Note that the non-regular curvature solution after neutrino decoupling has been introduced and characterized in terms of CMB anisotropies and matter power spectrum by Amendola & Finelli (2005).

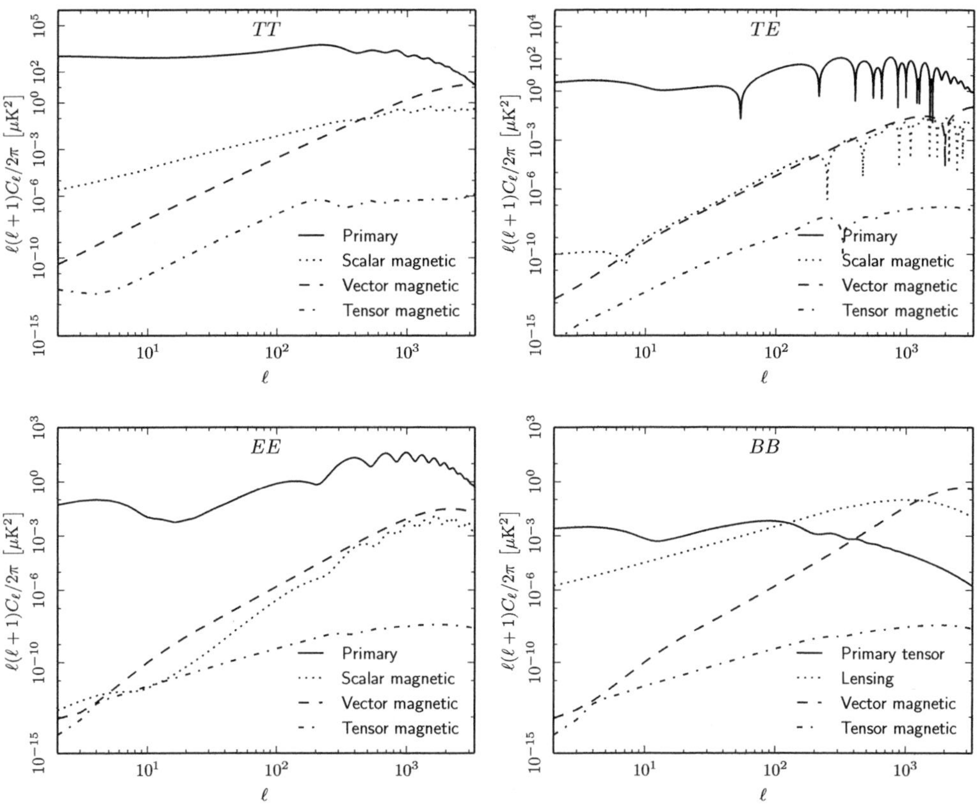

FIGURE 6.3. Magnetically induced CMB TT (*top left*), TE (*top right*), EE (*bottom left*), and BB (*bottom right*) power spectra. The solid lines represent primary CMB anisotropies, the dotted lines represent magnetically induced compensated scalar modes (except for the BB panel, where it represents the lensing contributions and the solid line represents primary tensor modes with a tensor-to-scalar ratio of $r = 0.1$), the dashed lines represent vector modes, whereas the dot-dashed lines represent magnetically induced compensated tensor modes. We consider PMFs with $B_{1\,\mathrm{Mpc}} = 4.1\,\mathrm{nG}$ and $n_B = -1$. Reproduced with permission from *Astronomy & Astrophysics*, © ESO; original source ESA and the Planck Collaboration (2016).

version includes the PMF contributions to the perturbations equations and the initial conditions. Analytic treatments are also available in Caprini et al. (2004), although often with evaluations of the PMF EMT components obtained by simplifying the integration on the angular part, and therefore need to be upgraded as described in the above sections.

The values of the parameters chosen are $B_{1\,\mathrm{Mpc}} = 4.1$ nG and $n_B = -1$. The remaining parameter k_D is fixed as (Subramanian & Barrow, 1998; Mack et al., 2002)

$$k_D = (5.5 \times 10^4)^{\frac{1}{n_B+5}} \left(\frac{B_{1\,\mathrm{Mpc}}}{\mathrm{nG}}\right)^{-\frac{2}{n_B+5}} \left(\frac{2\pi}{\lambda/\mathrm{Mpc}}\right)^{\frac{n_B+3}{n_B+5}} \times$$
$$h^{\frac{1}{n_B+5}} \left(\frac{\Omega_b h^2}{0.022}\right)^{\frac{1}{n_B+5}} \mathrm{Mpc}^{-1}, \tag{6.45}$$

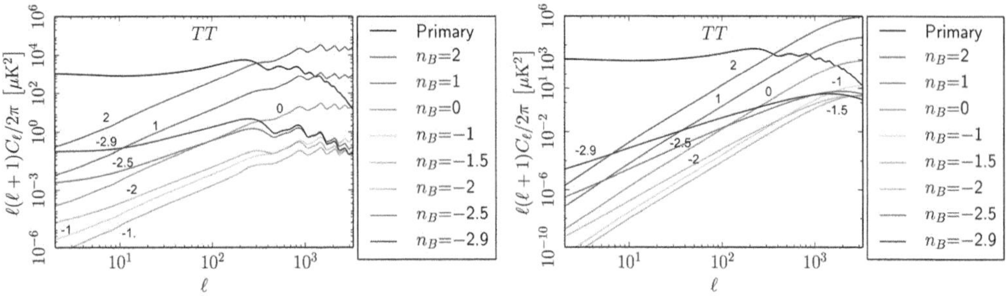

FIGURE 6.4. Dependence of the magnetically induced CMB power spectrum on the spectral index. For all plotted cases, the amplitude is $B_{1\,\mathrm{Mpc}} = 4.1\,\mathrm{nG}$. The black lines show primary CMB anisotropies; for the other grey lines, we refer to the legend. *Left*: scalar contributions, *right*: vector contributions. Reproduced with permission from *Astronomy & Astrophysics*, © ESO; original source ESA and the Planck Collaboration (2016).

where h is the reduced Hubble constant, $H_0 = 100\,h\,\mathrm{km\,s^{-1}\,Mpc^{-1}}$, and Ω_b is the baryon density parameter. For other choices of the damping physics see, for example, Kunze & Komatsu (2014, 2015).

The magnetically induced angular power spectrum is strongly dependent on the spectral shape of the components of the PMF EMT and is especially sensitive to the infrared behaviour of the EMT spectra, i.e. $k \to 0$. In particular, Fig. 6.4 shows the dependence on n_B of the compensated scalar and vector contributions to CMB anisotropies. We note how for spectral indices $n_B > -3/2$ we have a behaviour driven by the white noise shape of the EMT, whereas for indices $n_B < -3/2$ we note that the CMB angular power spectrum tilts changing its spectral shape accordingly to the behaviour of the EMT, which for these indices is infrared dominated as k^{2n_B+3}.

6.6.1 *The Impact of Helicity*

At the CMB scales, the helical even-correlators contribute *negatively*, with respect to the non-helical part, to the various EMT components. This negative contribution has been first discussed analytically in Caprini et al. (2004) for the tensor mode at low multipoles, and then shown at all multipoles for scalar, vector and tensor contribution to any of the TT, TE, EE, BB spectra in BFP14. See Fig. 6.5 for the comparison between the CMB anisotropy angular power spectra induced by non-helical PMF and by maximally helical PMF with $B_\lambda = 5$ nG and $n_B = n_H = -1$.

Unlike in the scalar case, the two-point correlation functions for the vector and the tensor parts present a parity-odd part, given by $X^{(V)}_{(B)}(k)$ and $X^{(T)}_{(B)}(k)$, respectively, in Eqs. (6.20) and (6.23). For example, the vector part is given by (see BFP14 for the tensor part and analytical solutions):

$$X^{(V)}_{(B)}(k) = \int \frac{d^3p}{(4\pi)^5} \Big\{ P_B(p) P_H(|\mathbf{k}-\mathbf{p}|) \big[\beta(1-\gamma^2) - (\gamma\beta - \mu)\gamma\big] + P_B(p) P_H(|\mathbf{k}-\mathbf{p}|) \big[\gamma(1-\beta^2) - (\gamma\beta-\mu)\beta\big] \Big\}. \tag{6.46}$$

In absence of parity violating interactions the odd-parity correlators C_ℓ^{TB} and C_ℓ^{EB} would vanish for symmetry. In presence of non-zero helicity, PMF source $C_\ell^{\mathrm{TB}} \neq 0$ and $C_\ell^{\mathrm{EB}} \neq 0$, thanks to the contribution of the helical part of tensor and vector perturbations.

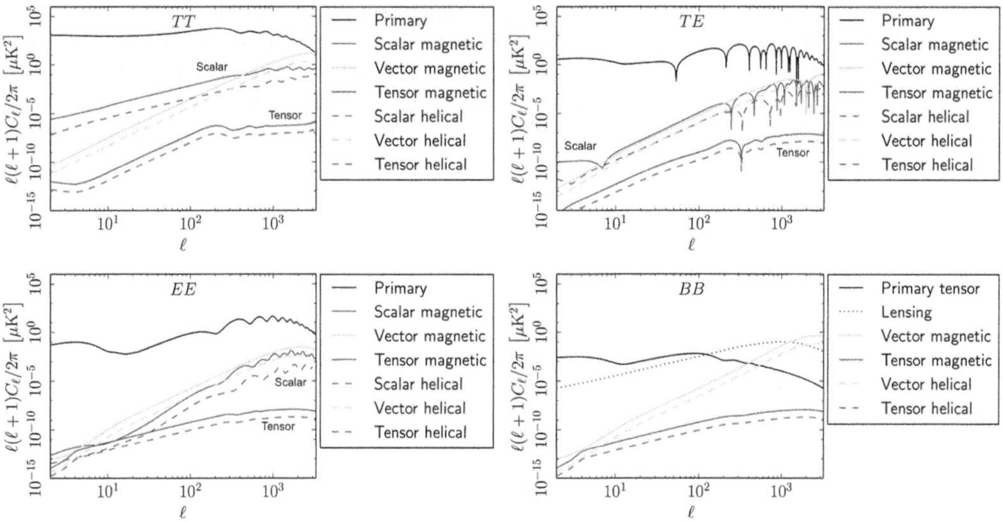

FIGURE 6.5. CMB TT (*top left*), TE (*top right*), EE (*bottom left*), and BB (*bottom right*) power spectra due to helical PMFs compared with the ones due to non-helical PMFs. Solid lines are non-helical predictions, while dashed lines are helical predictions. We consider PMFs with $B_{1\,\mathrm{Mpc}} = 5\,\mathrm{nG}$ and $n_B = -1$. Reproduced with permission from *Astronomy & Astrophysics*, © ESO; original source ESA and the Planck Collaboration (2016).

See Fig. 6.6 for the expected signal in the maximal helical case with $B_{1\,\mathrm{Mpc}} = 5\,\mathrm{nG}$ and $n_B = -1$.

6.6.2 *The Impact on the Ionization History*

Primordial magnetic fields are damped on scales smaller than the photon diffusion and free-streaming scale. This leads to heating of the primordial plasma before the decoupling of matter and radiation and of ordinary matter (electrons and baryons) afterwards. The

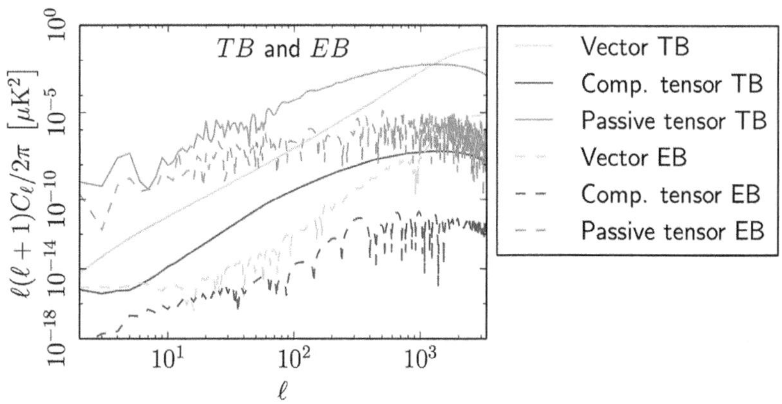

FIGURE 6.6. Magnetically induced odd-parity CMB cross-correlations for a helical PMF. The solid lines show TB, the dashed lines show EB. Light grey are the vector modes, dark grey are the compensated tensor modes, and mid grey are the passive tensor modes. We consider a PMF with $B_{1\,\mathrm{Mpc}} = 5\,\mathrm{nG}$ and $n_B = -1$. Reproduced with permission from *Astronomy & Astrophysics*, © ESO; original source ESA and the Planck Collaboration (2016).

heating by PMF affects both the thermal and ionization history of the Universe (Subramanian & Barrow, 1998; Jedamzik et al., 2000; Sethi & Subramanian, 2005; Schleicher et al., 2008; Kunze & Komatsu, 2014), leading to a Compton-y distortion of the CMB spectrum and changes in the CMB anisotropies power spectra through modifications of the Thomson visibility function around decoupling. In the following we summarize the heating effect described in Chluba et al. (2015).

The heating rate Γ by PMF has to be added to the electron temperature evolution equation with

$$\frac{dT_e}{dt} = -2HT_e + \frac{8\sigma_T N_e \rho_\gamma}{3m_e c N_{\rm tot}}(T_\gamma - T_e) + \frac{\Gamma}{(3/2)kN_{\rm tot}}. \tag{6.47}$$

Here, $H(z)$ denotes Hubble rate, $N_{\rm tot} = N_{\rm H}(1 + f_{\rm He} + X_{\rm e})$ the number density of all ordinary matter particles that share the thermal energy, beginning tightly coupled by Coulomb interactions; $N_{\rm H}$ is the number density of hydrogen nuclei, $f_{\rm He} \approx Y_p/4(1 - Y_p) \approx 0.079$ for helium mass fraction $Y_p = 0.24$; $X_{\rm e} = N_{\rm e}/N_{\rm H}$ denotes the free electron fraction and ρ_γ the CMB energy density. The first term in Eq. (6.47) describes the adiabatic cooling of matter due to the Hubble expansion, while the second term is caused by Compton cooling and heating. The last term accounts for the PMF heating. Notice that the last two terms in Eq. (6.47) differ slightly from those presented in earlier works (e.g., Sethi & Subramanian, 2005).

The heating rate Γ is the sum of a decaying magnetic turbulent term and an ambipolar diffusion term:

$$\Gamma = \Gamma_{\rm turb} + \Gamma_{\rm ambi}\,. \tag{6.48}$$

The heating rate caused by decaying magnetic turbulence can be approximated as (Sethi & Subramanian, 2005)

$$\Gamma_{\rm turb} = \frac{3m}{2}\frac{\left[\ln\left(1+\frac{t_i}{t_{\rm d}}\right)\right]^m}{\left[\ln\left(1+\frac{t_i}{t_{\rm d}}\right)+\frac{3}{2}\ln\left(\frac{1+z_i}{1+z}\right)\right]^{m+1}} H(z)\,\rho_{\rm B}(z), \tag{6.49}$$

with $m = 2(n_B + 3)/(n_B + 5)$, $t_i/t_{\rm d} \approx 14.8(B/{\rm nG})^{-1}(k_D/{\rm Mpc}^{-1})^{-1}$, k_D is the damping scale and $\rho_B(z)$ is the energy density of PMF.

The heating rate caused by ambipolar diffusion can be approximated as (Sethi & Subramanian, 2005; Schleicher et al., 2008):

$$\Gamma_{\rm am} \approx \frac{(1 - X_{\rm p})}{\gamma X_{\rm p}\,\rho_{\rm b}^2}\frac{\left\langle|(\nabla \times B) \times B|^2\right\rangle}{16\pi^2}, \tag{6.50}$$

where $\langle L^2\rangle = \left\langle|(\nabla \times B) \times B|^2\right\rangle/(4\pi)^2$ denotes the average square of the Lorentz force and $\rho_{\rm b} = m_{\rm H} N_{\rm b}$ the baryon *mass* density with baryon number density $N_{\rm b}$. Neglecting corrections from helium, we only need the free proton fraction, $X_{\rm p} = N_{\rm p}/N_{\rm H}$, to describe the coupling between the ionized and neutral component. The coupling coefficient is given by $\gamma = \langle\sigma v\rangle_{H\,H^+}/2m_{\rm H}$ with $\langle\sigma v\rangle_{H\,H^+} \approx 6.49 \times 10^{-10}(T/{\rm K})^{0.375}{\rm cm}^3\,{\rm s}^{-1}$.

As explained in Chluba et al. (2015), the effect on the ionization history described in Eq. (6.48) was previously overestimated (Kunze & Komatsu, 2014, 2015). The reason is that in RECFAST (Seager et al., 2000) the photoionization rates are evaluated assuming temperature $T \equiv T_{\rm e}$, although from a physical point of view the radiation temperature should be used. This reduces the effect on the ionization history by up to one order of magnitude and a consistent treatment is implemented in both RECFAST++ and COSMOREC (Chluba & Thomas, 2011).

This ionization effect *alone* modelled as described above leads to constraints on the amplitude of PMF of the order of nG for $n_B = -2.9$, see for instance (Chluba et al., 2015; Planck Collaboration et al., 2016). These results show how the effect of PMF on the ionizing history needs further investigation since its magnitude is comparable to the direct effect on the Einstein–Boltzmann system of cosmological perturbations.

It is interesting to close this subsection by noticing that the heating by PMF is estimated through an *average* effect. The ratio of the PMF energy density and radiation energy density ρ_γ is:

$$\rho_B \simeq 9.5 \times 10^{-8} \left(\frac{B}{nG}\right)^2 \rho_\gamma \qquad (6.51)$$

and therefore neglecting to this contribution in the Friedmann equations is a good approximation.

However, neglecting the average effect of PMF could not be a good approximation in the thermal and ionization history. Since the baryon to photon ratio is quite small ($n_b/n_\gamma \simeq 2.75 \times 10^{-8}\Omega_b h^2$), the magnetic energy density is then large compared with the electron thermal energy density and can therefore alter the thermal and ionization history.

6.7 Constraints on a Stochastic Background of PMF with the CMB Anisotropies Power Spectrum

We now present the analysis described in details in PF11 and Paoletti & Finelli (2013), hereafter PF13. In order to constrain PMF, we developed an extension of the Markov Chain Monte Carlo code `CosmoMC` (Lewis & Bridle, 2002, 2011), which computes the Bayesian probability distribution of the basic cosmological parameter and the PMF parameters: the amplitude smoothed on 1 Mpc length B_{1Mpc} and the spectral index n_B. We keep the damping scale k_D as a derived parameter depending on the amplitude of the fields, the spectral index and the background cosmology (in units of Mpc^{-1}) (Jedamzik et al., 1998; Subramanian & Barrow, 1998) as in Eq. (6.45).[2] Since we vary as a free parameter the spectral index of the fields it is necessary to include in the treatment the Fourier spectra of the EMT components and scalar Lorentz force of PMF for generic spectral index. The analytic solutions presented in PFP09 and FPP08 present special functions that are not of easy implementation on a numerical code, therefore we used the empirical fits provided in PF11. The dominant compensated contributions are the scalar and vector compensated magnetized mode, these are the modes considered in every analysis, for the Planck 2016 constraints it has been considered also the contribution by the tensor passive mode (Planck Collaboration et al., 2016), for the pre-Planck constraints including also passive tensor modes we refer to (Shaw & Lewis, 2012). We exclude from the analysis the compensated tensor mode because, as previously shown, it is strongly subdominant.

We vary the baryon density $\omega_b = \Omega_b h^2$, the cold dark matter density $\omega_c = \Omega_c h^2$ (with h being $H_0/100\mathrm{km\,s^{-1}Mpc^{-1}}$), the reionization optical depth τ (not to be confused with the conformal time τ), the ratio of the sound horizon to the angular diameter distance at decoupling θ (leaving H_0 as a derived parameter), $\ln(10^{10}A_S)$, n_S and the magnetic parameters $B_{1\mathrm{Mpc}}$ (in units of 10 nG) and n_B. As priors we use $[0\,,1]$ for $B_{1\mathrm{Mpc}}/(10\mathrm{nG})$ and $[-2.9\,,3]$ for n_B (> -3 in order to avoid infrared divergencies in the PMF EMT

[2] Note that in PF11 and PF13 we did not take into account the dependence of k_D on the baryon density as in Eq. (6.45), but this has been demonstrated to be a negligible effect in terms of the constraints on PMF amplitude.

correlators). We assume a flat universe and a CMB temperature $T_{\rm CMB} = 2.725$ K. The pivot scale of the primordial scalar was set to $k_* = 0.05$ Mpc^{-1} and we restrict our analysis to three massless neutrinos (a non-vanishing neutrino mass leads to a large-scale enhancement in the power spectrum of CMB anisotropies in the presence of PMF (Shaw & Lewis, 2010) and would not change our results). Here we set the primordial helium fraction to $y_{\rm He} = 0.24$. We sample the posterior using the Metropolis–Hastings algorithm (Hastings, 1970), generating four parallel chains and imposing a conservative Gelman-Rubin convergence criterion (Gelman & Rubin, 1992) of $R - 1 < 0.01$. We use the lensed CMB and matter power spectra.

6.7.1 *Analysis of Pre-Planck Datasets*

We begin our description of the constraints on PMF with CMB anisotropy angular power spectra with the results of pre-Planck datasets. The dataset considered in PF11 was the combination of WMAP 7 year (Jarosik et al., 2011; Larson et al., 2011) and ACBAR (Reichardt et al., 2009) data in temperature, and of the BICEP (Chiang et al., 2010) and QUaD (Brown et al., 2009) data in polarization. It followed the method implemented in `CosmoMC` consisting of varying a nuisance parameter $A_{\rm SZ}$, which accounts for the unknown amplitude of the thermal Sunyaev–Zeldovich (SZ) contribution to the small-scale CMB data points assuming the model of Komatsu & Seljak (2002).

In PF11 are derived the following constraints on the amplitude of PMF: $B_{1\rm Mpc} < 5.0$ nG at 95% confidence level (in agreement with $B_{1\,\rm Mpc} < 6.4$nG by Shaw & Lewis, 2012). The results show also that positive spectral indices are allowed with lower amplitudes with respect to the negative ones. In PF11 it is demonstrated that the dominant magnetic compensated contribution for the constraints is given by the vector mode. The magnetized vector mode peaks at small angular scales where the primary CMB anisotropies are suppressed by Silk damping, whereas the scalar mode is the dominant magnetic compensated contribution on small and intermediate multipoles, where the primary CMB remains dominant.

The CMB is not the only cosmological observable that can be used to constrain PMF. In FPP08 and Shaw & Lewis (2012) it is shown that PMF have a considerable impact also on the matter power spectrum. To investigate the possibility of the combination of CMB and Large Scale Structure (LSS) data, MCMC analysis is performed in PF11, adding the LSS data of SDSS LRG DR4 (Tegmark et al., 2006) to the pre-Planck data combination used in the previous analysis. The results give $B_{1\rm Mpc} < 5.0$ nG at 95% CL, showing that the addition of the SDSS data does not introduce any significant improvement on the constraints. This result is related with the scales on which PMF have an impact on the matter power spectrum. In particular, the impact of PMF is on very small scales that are far beyond the scales reached by the dataset used. Even though there are datasets that reach smaller scales, their use would require the introduction of nonlinear treatments of magnetized perturbations that have still not been fully developed (see Shaw & Lewis, 2012, for an approximated approach).

We have already discussed the importance of small-angular-scale data to constrain PMF. The analysis of the angular power spectra described shows how the constraining power for the gravitational contribution to CMB anisotropies comes from $\ell \sim \mathcal{O}(10^3)$: at these multipoles the primary CMB anisotropies sourced by adiabatic initial conditions are suppressed by Silk damping, whereas foreground residuals and CMB secondary anisotropies become more and more important going towards higher multipoles. A careful treatment of foreground residuals and CMB secondary anisotropies is therefore important to constrain PMF. In PF11 we limited ourselves to conservatively cut data below

$\ell = 2000$ and the only foreground nuisance parameter was the amplitude of the residual Sunyaev–Zeldovich contribution A_{SZ}.

With the release of data from CMB ground experiments such the South Pole Telescope (SPT) (Reichardt et al., 2009, 2012) and Atacama Cosmology Telescope (ACT) (Fowler et al., 2010), which probed the CMB temperature anisotropy pattern through the Silk damping tail, more refined (empirical) models of foreground residuals/secondary anisotropies were necessary. In particular, in PF13 are presented the constraints on PMF derived with the combination of WMAP 7 year data with the small-scale data from the SPT 2 year observations (Reichardt et al., 2012; Keisler et al., 2011). See Dvorkin et al. (2011) for the use of the same dataset to constrain another extended cosmological model of high interest as ΛCDM model plus cosmic strings. The contribution of cosmic strings to CMB anisotropies is mainly located at those high multipoles, i.e. $\ell \sim \mathcal{O}(10^3)$, in a similar way to PMF.

Before going on to the constraints we briefly describe the SPT model of foreground residuals/secondary anisotropies, which is characterized by three parameters accounting for the three different contributions that are considered. The dominant astrophysical contamination for the SPT data is given by residual extragalactic point sources (with fluxes below the detection threshold) and cluster of galaxies. Astrophysical contributions to the angular power spectrum, differently from the primary CMB, depend on the frequency and lead to different contributions which dominate over the others. In particular, lower frequencies are dominated by the Poissonian term of radio galaxies, growing in frequency the contribution of dusty galaxies becomes relevant up to the higher frequencies that are completely dominated by the clustering term of infrared galaxies; the contribution of galaxy clusters is through the SZ effect, which has an analytically known frequency dependence (Sunyaev & Zeldovich, 1970). The Poissonian term is simply due to the random distribution in the sky of the unresolved galaxies and is a flat spectrum that depends on the integration of the number counts (Serra et al., 2008; Planck Collaboration et al., 2011a,b). It affects both radio and infrared galaxies, but for the latter it is only a subdominant term. Infrared galaxies contribution is dominated by the clustering, much more complex with respect to the Poissonian (Serra et al., 2008; Millea et al., 2012; Paoletti et al., 2012; Addison et al., 2012; Xia et al., 2012). The SZ effect is a spectral distortion induced by the inverse Compton interaction of CMB photons with the intracluster gas (thermal) or by the peculiar motion of the cluster (kinetic) (Sunyaev & Zeldovich, 1970, 1972). In PF13 analysis we used the three templates provided by the SPT collaboration for the 150 GHz data (Reichardt et al., 2012), which are shown compared with magnetized modes in Fig. 6.7.

The flat shape of the clustering term on large scales is given in the template provided by the SPT collaboration, which considers a shape variation only on multipoles where the clustering contribution is relevant, assuming a flat contribution elsewhere, considering that for the SPT data release the lowest multipole used is $\ell \sim 650$.

In order to marginalize over the astrophysical contribution together with cosmological and magnetic parameters, the modified `cosmoMC` includes the parameters describing the astrophysical residual contributions: D^{SZ}_{3000}, D^{PS}_{3000}, D^{CL}_{3000}, which are the amplitudes of Poissonian, clustering and SZ signals at $\ell = 3000$. The prior used is $[0, 100]$ for all the three astrophysical parameters.

The one-dimensional (1D) posterior probabilities for the cosmological and magnetic parameters from the analysis performed with the combination of WMAP 7 and SPT data are shown as dashed lines in Fig. 6.9. The constraint is $B_{1\,\mathrm{Mpc}} < 3.5$ nG at 95% confidence level. As with the previous dataset, positive spectral indices are allowed with smaller amplitudes with repect to negative ones. In order to estimate the importance of

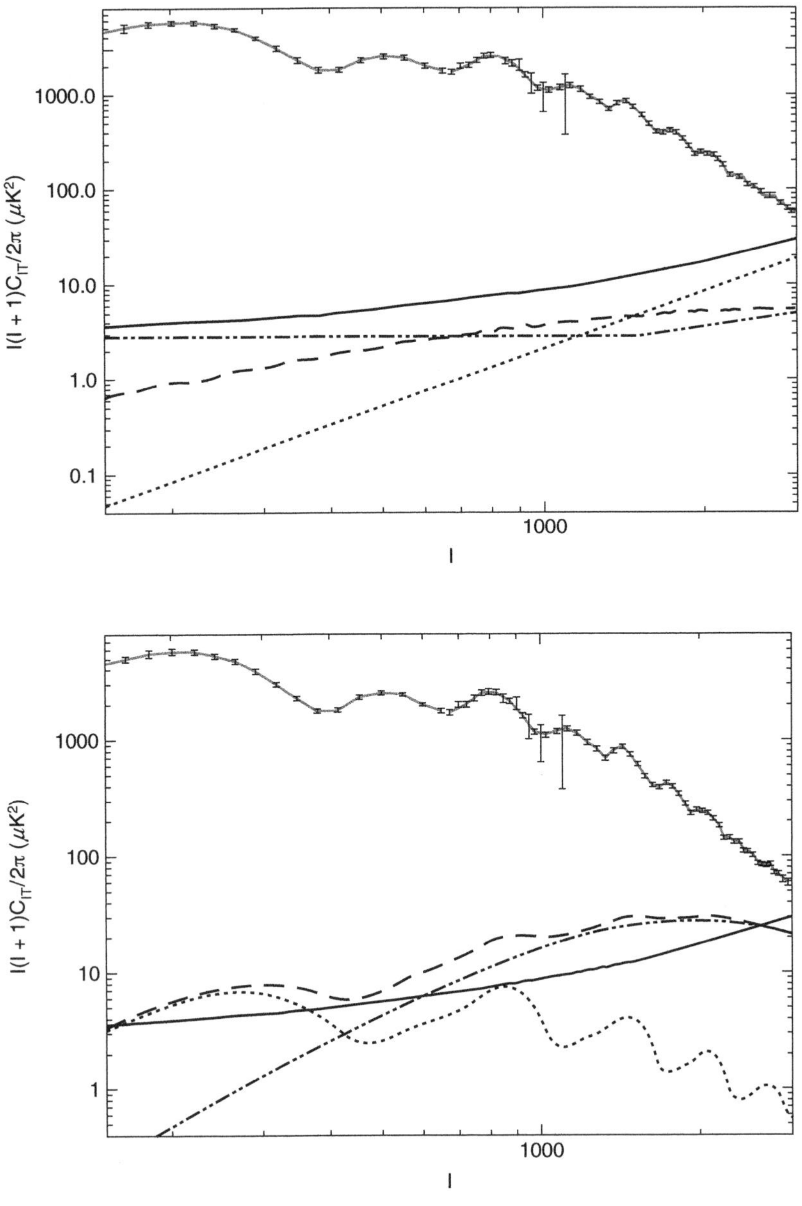

FIGURE 6.7. Comparison of magnetized modes for a field of $B_{1\,\mathrm{Mpc}} = 6$ nG and $n_B = -2.5$ with the astrophysical contributions from SPT 150 GHz data as in Reichardt et al. (2012). The top panel shows the astrophysical contributions: TSZ effect (dashed line), the Poissonian term (dotted line), the clustering term (triple dot-dashed line) and the solid line represents the sum of the three. In the lower panel the magnetic contributions including the uncorrelated sum of the two (dashed line) is compared with the total astrophysical contribution (solid line).

the astrophysical contamination of CMB data for the PMF constraints, the same analysis is repeated without accounting for the astrophysical residuals (all astrophysical parameters are set to zero). The results of this analysis show a tentative detection of a few nG amplitude with a $n_B \sim -1$ spectrum. In Fig. 6.9 we show the posterior distributions as solid lines; this false tentative detection of PMF is the result of the contamination of

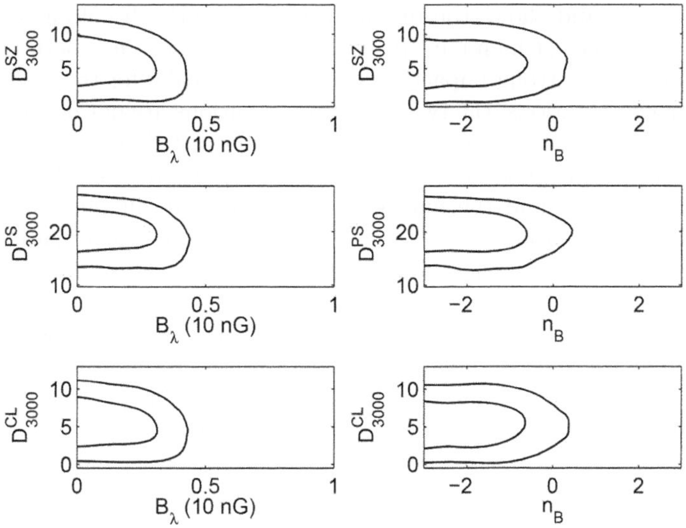

FIGURE 6.8. Constraints with WMAP 7 + SPT data for the two magnetic parameters, B_λ and n_B, versus the astrophysical ones: D^{SZ}_{3000}, D^{PS}_{3000}, D^{CL}_{3000}.

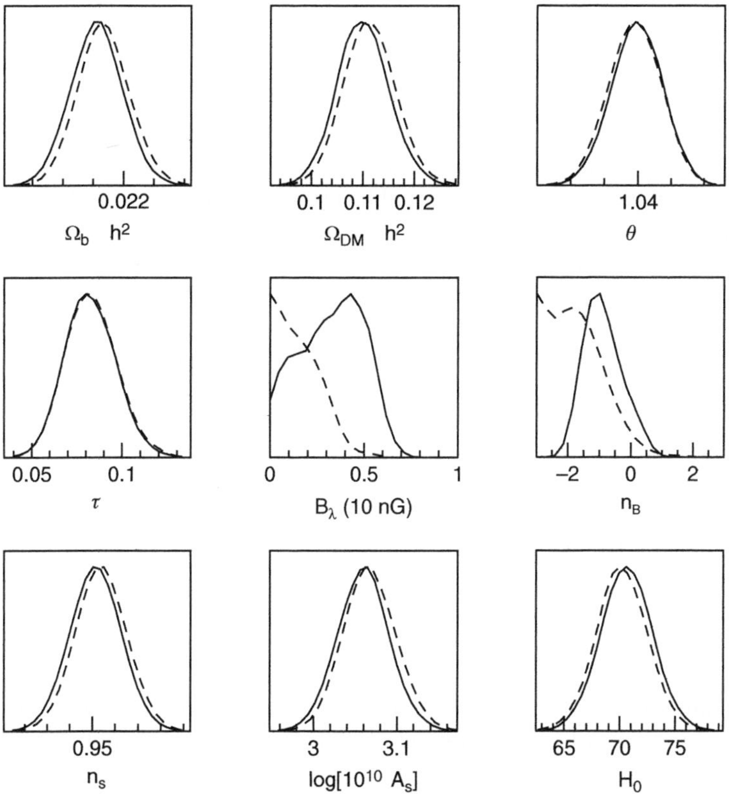

FIGURE 6.9. Constraints with WMAP 7 + SPT data, without the inclusion of astrophysical residual contributions. In the analysis we included only the six ΛCDM cosmological parameters and the two magnetic ones, B_λ and n_B. Dashed lines are for comparison with the results of the analysis that includes the astrophysical contribution.

astrophysical signals and shows the importance of correctly disentangling cosmological and astrophysical signals at high multipoles. Even though both magnetic and astrophysical contributions have a strong impact on small angular scales for the case of the WMAP 7 and SPT data combinations, the two sets of contributions are not degenerate with each other.[3] In Fig. 6.8 are shown the marginalized posterior distributions of the astrophysical parameters versus the magnetic ones, and we note that there is no degeneracy between the astrophysical models and the magnetic modes. This result holds for the simple 3-astrophysical contribution model used for WMAP 7 + SPT; the situation can be much more complex in other cases where it is necessary to use a more complicated model for the astrophysical contributions.

In PF13 are presented the results from analyses dedicated to specific models of PMF, realized by keeping the spectral index fixed to physical representative values. The results are: $B_{1\mathrm{Mpc}} < 5.6 \times 10^{-1}$ nG for $n_B = 0$, $B_{1\mathrm{Mpc}} < 6.6 \times 10^{-3}$ nG for $n_B = 2$ and $B_{1\mathrm{Mpc}} < 7 \times 10^{-4}$ nG for $n_B = 3$. As expected from the results of the analysis for variable spectral index the constraints on $B_{1\mathrm{Mpc}}$ for positive spectral indices are much stronger than the generic ones. These tight limits may be very important for their implications on PMF generation and evolution.

6.7.2 *The Planck Era*

The Planck mission data represent a milestone for CMB anisotropy in temperature and polarization. In particular, Planck data cover from $\ell = 2$ to approximatively $\ell = 2500$, and are essentially limited by the ability in subtracting foregrounds at high multipoles rather than from instrumental noise and angular resolution. In particular, with the single Planck dataset alone it is possible to avoid cross-correlation between different datasets, which can introduce additional uncertainties, if not biases.

The analysis with pre-Planck data described above shows the interplay between foreground residuals and imprints of PMF at high multipoles. For this reason it is useful to first forecast the Planck capabilities in constraining PMF in presence of foregrounds.

For Planck semi-realistic simulated data that include foreground residuals we follow the treatment in Paoletti et al. (2012): we consider Planck frequencies from 70 to 353 GHz with the same contributions studied in WMAP 7 plus SPT, i.e. the Poissonian, clustering and the TSZ terms. The three foreground fiducial contributions to the angular power spectrum are modelled with empirical parametrizations based on Planck early data results for the point sources and on theoretical models for the TSZ. In order to estimate the impact of foreground and astrophysical uncertainties on the constraints on PMF, in PF11 we performed an MCMC analysis with Planck simulated data with the inverse noise weighted combination of the five frequencies with inflight beams and noise characteristics from Planck HFI Core Team et al. (2011a,b) and Mennella et al. (2011); the analysis is performed with and without the astrophysical contamination. The results of the semi-realistic analysis show a degradation of the constraints as expected, the constraints from 2.4 nG are degraded to $B_{1\mathrm{Mpc}} < 3.6$ nG at the 95% confidence level.

The constraints on PMF from the Planck nominal mission data in temperature have been presented in Planck Collaboration et al. (2014). These constraints have been derived

[3] A degeneracy between two parameters takes place when different sets of the two parameters give the same observable making undistinguishable one set from the others with the data. They are evidenced by diagonal-shape 2D posterior distributions.

with the Planck 2013 likelihood, which contains 11 nuisance foreground parameters and is supplemented by the large angular polarization from WMAP 9 year data. The results show how the cosmological parameters are in agreement with those estimated assuming no PMF, confirming that the magnetic parameters are not degenerate with the cosmological ones. The constraints on PMF with Planck 2013 data are $B_{1\mathrm{Mpc}} < 4.1$ nG at the 95% confidence level. A powerful dataset in CMB temperature anisotropies comes from the combination of Planck data with very small angular-scales data from ACT and SPT (for the detailed description of the dataset used we refer the reader to Planck Collaboration et al., 2014). The Planck 2013 + highL combination results give the tight constraint $B_{1\mathrm{Mpc}} < 3.4$nG at the 95% confidence level. The posterior distributions for the amplitude and spectral index of the analyses with Planck 2013 and Planck 2013 + highL are shown in Fig. 6.10.

In Planck Collaboration et al. (2016) are presented the constraints on PMF using Planck 2015 data, derived using the new Planck likelihood which is fully based on Planck temperature and polarization data (i.e. without the WMAP 9 low-ℓ support for polarization) from the full mission. The analysis with the compensated scalar and vector modes, using both temperature and polarization for $2 < \ell < 50$ and temperature only for $50 < \ell < 2500$, provides the constraint $B_{1\,\mathrm{Mpc}} < 4.4\,\mathrm{nG}$, which is not sensitive to the inclusion of high-ℓ polarization data. The comparison between the 2015 and 2013 results is shown in Fig. 6.10; the slightly higher upper limits with respect to the 2013 Planck release are due to the changes in the 2015 data and are described in Planck Collaboration et al. (2016).

The addition of the passive tensor contribution does not improve the constraints on the amplitude but strongly affects the posterior distribution of the PMF spectral index. The results confirm that CMB data prefer lower amplitudes for positive spectral indices. The inclusion of the contribution of passive modes, thanks to their sensitivity to the slope of the PMF spectrum, strongly disfavours nearly scale-invariant magnetic spectra for amplitudes greater than very few nG as shown in the two dimensional posterior distribution in Fig. 6.11.

Figure 6.12 shows the degeneracy of the PMF amplitude with the four foreground parameters that are used in the Planck 2015 likelihood to describe the Poissonian term for undetected point sources. The figure shows how, differently from the easier models used for WMAP7 + SPT analysis, for the more complex model necessary to analyse the multifrequency data of Planck, the PMF contribution degenerates with the Poissonian term in the 2015 likelihood.

Together with only Planck data, an analysis with the combination of Planck 2015 and the BICEP2/Keck-Planck cross-correlation (BICEP2/Keck et al., 2015) is also presented. The results are compatible with those from only Planck data.

In Planck Collaboration et al. (2016) are derived the constraints on helical PMFs. Because the helical components of PMF decrease, the amplitude of the magnetic CMB fluctuations PMFs that are maximally helical are less constrained than non-helical fields, with $B_{1\,\mathrm{Mpc}} < 5.6\,\mathrm{nG}$.

Specific PMF models of interest are constrained to be: causally generated fields with a spectral index of $n_B = 2$ with $B_{1\,\mathrm{Mpc}}^{n_B=2} < 0.011\,\mathrm{nG}$ and fields with an almost scale-invariant power spectrum with $n_B = -2.9$ with $B_{1\,\mathrm{Mpc}}^{n_B=-2.9} < 2.1\,\mathrm{nG}$.

Alternative constraints are derived using the impact of PMF on the ionization history obtaining from Planck 2015 constraints which are below the nG level (see also Chluba et al., 2015).

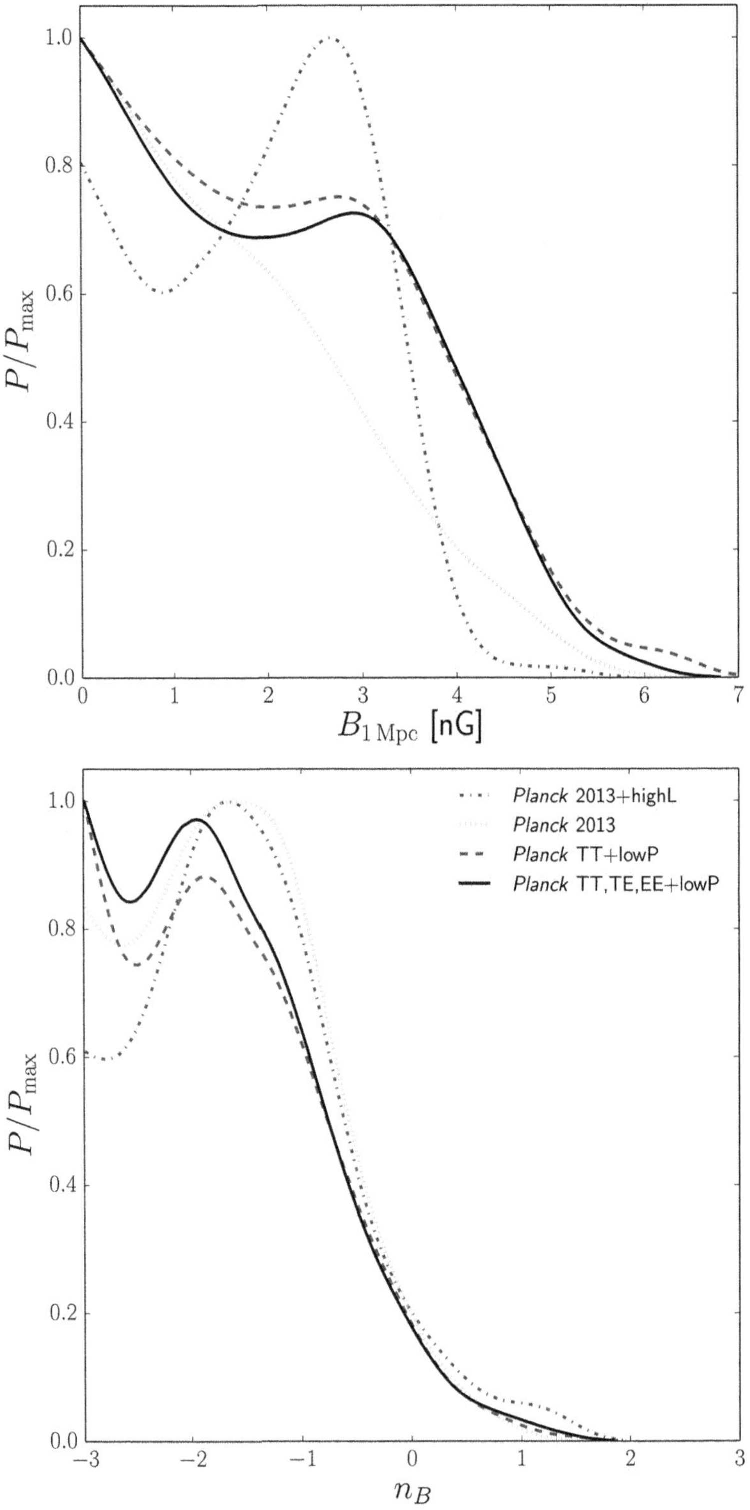

FIGURE 6.10. Comparison of the constraints on the smoothed PMF amplitude (*top*) and the spectral index (*bottom*) from the 2015 temperature and temperature plus polarization data with the 2013 results. Reproduced with permission from *Astronomy & Astrophysics*, © ESO; original source ESA and the Planck Collaboration.

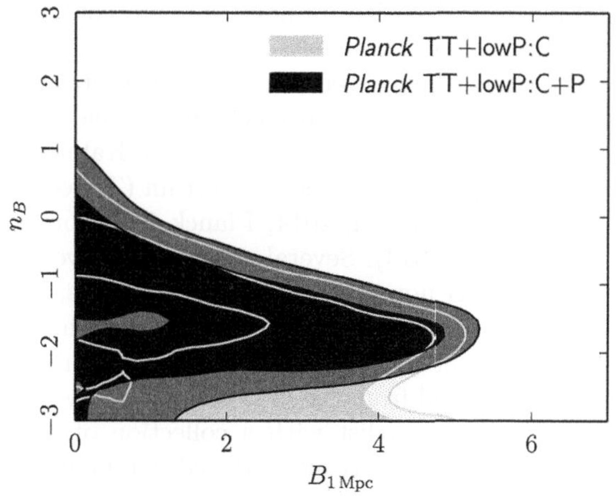

FIGURE 6.11. PMF amplitude versus the spectral index for the baseline Planck 2015 case. C+P denotes the case where both compensated and passive modes are considered, whereas C indicates the case with only compensated modes.The two contours represent the 68% and 95% confidence levels. Reproduced with permission from *Astronomy & Astrophysics*, © ESO; original source ESA and the Planck Collaboration.

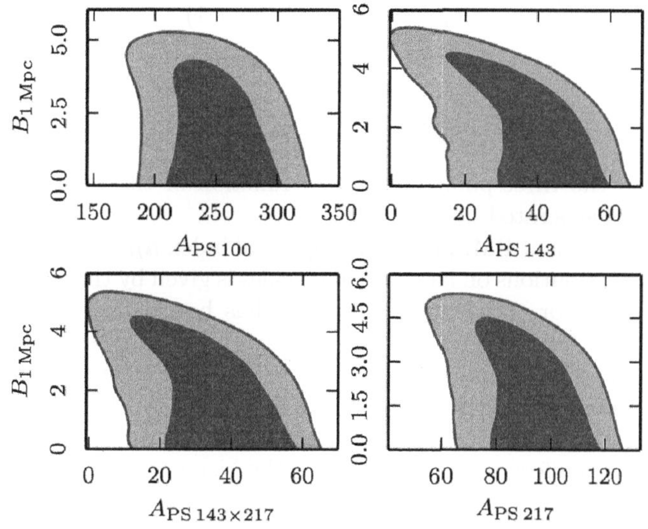

FIGURE 6.12. Two-dimensional posterior distributions of the PMF amplitude versus the parameter describing the Poissonian term of unresolved point sources for the three frequencies considered in the likelihood. The two contours represent the 68% and 95% confidence levels. Reproduced with permission from *Astronomy & Astrophysics*, © ESO; original source ESA and the Planck Collaboration.

6.8 The Imprints of Primordial Magnetic Fields on CMB Non-Gaussianities

Because the PMF contribution to cosmological perturbations is quadratic in the fields, this contribution is non-Gaussian and has a χ^2 distribution (Brown & Crittenden, 2005). PMF therefore provide a non-Gaussian contribution to CMB anisotropies (this peculiarity is common to topological defects as well; see Figueroa et al., 2010). The presence of a stochastic background of PMF with its non-Gaussian contribution induces a non-negligible imprint on higher statistical moments, as in the bispectrum and the trispectrum of CMB anisotropies. There have been several works in the literature predicting the magnetically induced scalar compensated mode bispectrum (Seshadri & Subramanian, 2009; Caprini et al., 2009; Cai et al., 2010; Shiraishi et al., 2010; Kahniashvili & Lavrelashvili, 2010), the magnetically induced passive-mode bispectrum (Trivedi et al., 2010; Shiraishi et al., 2011, 2012; Shiraishi & Sekiguchi, 2014; Planck Collaboration et al., 2016) and trispectrum (Trivedi et al., 2012, 2014). Several constraints have been therefore derived on the PMF amplitude by CMB non-Gaussianities constraints; see Seshadri & Subramanian, 2009; Caprini et al., 2009; Trivedi et al., 2010; Shiraishi et al., 2012; Trivedi et al., 2012; Planck Collaboration et al., 2016. The non-Gaussianity constraints on PMF are complementary to those derived from the CMB angular power spectra and represent a field in continuous development. A list with a collection of many of the most relevant constraints derived with the magnetically induced non-Gaussianity can be found in Table 6.2.

In the following we briefly describe some examples of the methods that may be applied to investigate magnetically induced non-Gaussianities. In particular, we will focus on two cases related with the magnetically induced bispectrum. The CMB anisotropy bispectrum is the three-point correlation function of the harmonic expansion coefficients $a_{l,m}$:

$$\begin{aligned}\langle a_{\ell_1 m_1} a_{\ell_2 m_2} a_{\ell_3 m_3} \rangle = & \frac{(4\pi)^3(-i)^{\ell_1+\ell_2+\ell_3}}{(2\ell_1+1)(2\ell_2+1)(2\ell_3+1)} \times \\ & \times \int \frac{d^3k\, d^3q\, d^3p}{(2\pi)^9} Y^*_{\ell_1 m_1}(\hat{k};\hat{e}) Y^*_{\ell_2 m_2}(\hat{q};\hat{e}) Y^*_{\ell_3 m_3}(\hat{p};\hat{e}) \\ & \times \langle \Theta^{(0)}_{\ell_1}(\eta_0,\mathbf{k}) \Theta^{(0)}_{\ell_2}(\eta_0,\mathbf{q}) \Theta^{(0)}_{\ell_3}(\eta_0,\mathbf{p}) \rangle\,; \end{aligned} \tag{6.52}$$

and it depends on the three-point correlation function of the temperature anisotropy, which in our case is generated by the PMF.

For nearly scale-invariant PMF, i.e. with a spectral index $n_B \to -3$, one of the possibly important PMF contributions on large angular scales is given by the magnetically induced passive tensor mode. For this reason this mode has been considered in several works (Trivedi et al., 2010; Shiraishi et al., 2011, 2012; Shiraishi & Sekiguchi, 2014; Planck Collaboration et al., 2016). The scalar and tensor passive magnetized modes evolve as the adiabatic solution with the power spectrum depending on the amplitude of the PMF and the ratio of the PMF generation time with the neutrino decoupling time. For these passive modes it is therefore possible to predict the bispectrum using the same methods applied for adiabatic initial conditions from inflation. As for inflation, in this case the dominant contribution is given by the squeezed-limit configuration, $\ell_1 \ll \ell_2 \approx \ell_3$, because of the local-type structure of the passive-mode non-Gaussianities. The estimate of the amplitude of the magnetically induced bispectrum can then be compared to the one derived from the data and used to constrain the PMF amplitude for nearly scale invariant spectral index; see Table 6.2 and Planck Collaboration et al. (2016).

The magnetically induced compensated modes require a completely different approach. The crucial difference with respect to the passive modes is given by the different evolution of compensated perturbations. While in the passive case the evolution is described with a primary inflationary-like approach, in this case we must consider all the magnetic contributions to the cosmological perturbations, which drastically change our working environment. In particular, it is not possible to determine an a priori dominant configuration for the magnetically induced non-Gaussianities. Therefore it is necessary to maintain the maximum generality when deriving the bispectrum. It has been shown by

Seshadri & Subramanian (2009) and Caprini et al. (2009) that the magnetically induced compensated-scalar bispectrum on large angular scales depends on the three-point correlation function of the magnetic energy density, which in turns depends on the six-point correlation function of the PMF (Caprini et al., 2009):

$$\begin{aligned} \langle \rho_{\rm B}(\mathbf{k})\rho_{\rm B}(\mathbf{q})\rho_{\rm B}(\mathbf{p})\rangle &= \frac{1}{(64\pi)^3}\int \frac{d^3\tilde{k}\, d^3\tilde{q}\, d^3\tilde{p}}{(2\pi)^9} \times \\ &\langle B_i(\tilde{\mathbf{k}})B_i(\mathbf{k}-\tilde{\mathbf{k}})B_j(\tilde{\mathbf{q}})B_j(\mathbf{q}-\tilde{\mathbf{q}})B_l(\tilde{\mathbf{p}})B_l(\mathbf{p}-\tilde{\mathbf{p}})\rangle\,. \end{aligned} \tag{6.53}$$

Caprini et al. (2009) derived a general expression for the three-point correlation function of the magnetic energy density that does not depend on the geometrical configuration. With this expression or using alternative approaches (Seshadri & Subramanian, 2009; Caprini et al., 2009; Trivedi et al., 2010, 2012; Shiraishi et al., 2012; Planck Collaboration et al., 2016) it is possible to derive an expression for the compensated bispectrum. This can be compared with the non-Gaussianity measurements to derive the constraints on PMF amplitude. In particular, in Caprini et al. (2009) and Planck Collaboration et al. (2016) an effective $f_{\rm NL}$ is derived as a function of the field amplitude. The $f_{\rm NL}$ can be used to constrain PMF amplitude with either WMAP9 or Planck measurements; the results are shown in Table 6.2.

6.9 Summary

In these lectures we have briefly summarized imprints on CMB anisotropies from PMF generated in the Early Universe. PMF frozen in the primordial plasma after nucleosynthesis would represent a new window in the physics of the Early Universe and therefore is a topic of great interest. The possible connection of PMF with the seeds for LSMF makes this topic even more interesting and suitable for a multidisciplinary approach.

CMB anisotropies data are an important source of information for PMF. Thanks to the PMF imprints on CMB anisotropies – and in general on structure formation – it is indeed possible to constrain the amplitude and the spectral index of the stochastic background of PMF. The progress in the observational experiments dedicated to CMB anisotropies at all scales in temperature and polarization generated a renewed interest in the theoretical predictions for cosmological observables, and made available a sensible improvement in the constraints on PMF from CMB. We have reviewed here how a stochastic background of PMF contributes to the Einstein–Boltzmann system of cosmological perturbations, and the different imprints of PMF on the CMB temperature and polarization anisotropies, on the power spectra and on the bispectra.

We have summarized different analysis and the derived constraints on the amplitude of a stochastic background of PMF with different CMB datasets in Table 6.1 and 6.2. These constraints show how PMF with an amplitude of the order of nG on the Mpc scale are coherently constrained by different observational probes. Much smaller amplitudes are allowed for positive spectral indices, such as $n = 2$, which for example is associated to causal generation mechanisms.

Acknowledgments

We wish to thank the organizers of the XXV Canary Islands Winter School of Astrophysics for their kind invitation, for their invaluable help in preparing the lectures and this summary. We are grateful to Mario Ballardini for a careful reading of this manuscript and many useful suggestions. We are grateful to Mario Ballardini, Chiara Caprini, Jens

Chluba, Francesco Paci, Antonio Riotto and Jose Alberto Rubino-Martin for many discussions and for collaboration on this subject.

REFERENCES

Addison G. E., Dunkley J., Spergel D. N., 2012, *MNRAS*, **427**, 1741

Amendola L., Finelli F., 2005, *PhRvL*, **94**, 221303

Ballardini M., Finelli F., Paoletti D., 2014, *arXiv*, arXiv:1412.1836

Bernet M. L., Miniati F., Lilly S. J., Kronberg P. P., Dessauges-Zavadsky M., 2008, *Natur*, **454**, 302

BICEP2/Keck, et al., 2015, arXiv, arXiv:1502.00612

Bonvin C., Caprini C., Durrer R., 2013, *PhRvD*, **88**, 083515

Brown M. L., et al., 2009, *ApJ*, **705**, 978

Brown I., Crittenden R., 2005, *PhRvD*, **72**, 063002

Bucher M., Moodley K., Turok N., 2001, *PhRvL*, **87**, 191301

Byrnes C. T., Hollenstein L., Jain R. K., Urban F. R., 2012, *JCAP*, **3**, 009

Cai R.-G., Hu B., Zhang H.-B., 2010, *JCAP*, **8**, 025

Calzetta E., Kandus A., 2002, *PhRvD*, **65**, 063004

Caprini C., Durrer R., Kahniashvili T., 2004, *PhRvD*, **69**, 063006

Caprini C., Finelli F., Paoletti D., Riotto A., 2009, *JCAP*, **6**, 021

Chiang H. C., et al., 2010, *ApJ*, **711**, 1123

Chluba J. and Thomas R. M, *MNRAS* **412**, 748

Chluba J., Paoletti D., Finelli F., Rubino-Martin J.-A., 2015, *MNRAS*, **451**, 2244.

Clarke T. E., Kronberg P. P., Böhringer H., 2001, *ApJ*, **547**, L111

Dolag K., Kachelriess M., Ostapchenko S., Tomàs R., 2011, *ApJ*, **727**, LL4

Durrer R., Caprini C., 2003, *JCAP*, **11**, 010

Dvorkin C., Wyman M., Hu W., 2011, *PhRvD*, **84**, 123519

Enßlin T. A., Vogt C., 2006, *A&A*, **453**, 447

Figueroa D. G., Caldwell R. R., Kamionkowski M., 2010, *PhRvD*, **81**, 123504

Finelli F., Gruppuso A., 2001, *PLB*, **502**, 216.

Finelli F., Gruppuso A., Venturi G., 1999, *CQG* **16**, 3923.

Finelli F., Paci F., Paoletti D., 2008, *PhRvD*, **78**, 023510

Fowler J. W., et al., 2010, *ApJ*, **722**, 1148

Gasperini M., Giovannini M., Veneziano G., 1995, *PhRvL*, **75**, 3796

Gelman A., Rubin D. B., 1992, *Statistical Science*, **7**, 457

Govoni F., Murgia M., Xu H., et al., 2013, *A&A*, **554**, AA102

Grasso D., Riotto A., 1998, *PhLB*, **418**, 258

Grasso D., Rubinstein H. R., 1995, *APh*, **3**, 95

Harrison E. R., 1970, *MNRAS*, **147**, 279

Hastings W. K., 1970, *Biometrika*, **57**, 97

Ichiki K., Takahashi K., Ohno H., Hanayama H., Sugiyama N., 2006, *Sci*, **311**, 827

Jarosik N., et al., 2011, *ApJS*, **192**, 14

Jedamzik K., Katalinić V., Olinto A. V., 1998, *PhRvD*, **57**, 3264

Jedamzik K., Katalinić V., Olinto A. V., 2000, *PhRvL*, 85, 700

Kahniashvili T., Lavrelashvili G., 2010, arXiv, arXiv:1010.4543

Kahniashvili T., Ratra B., 2005, *PhRvD*, **71**, 103006

Kahniashvili T., Ratra B., 2007, *PhRvD*, **75**, 023002

Kahniashvili T., Kosowsky A., Mack A., Durrer R., 2001, *AIPC*, **555**, 451

Kahniashvili T., Maravin Y., Kosowsky A., 2009, *PhRvD*, **80**, 023009

Kahniashvili T., et al., 2010, *PhRvD*, **82**, 083005

Kahniashvili T., Maravin Y., Lavrelashvili G., Kosowsky A., 2014, *PhRvD*, **90**, 083004

Keisler R., et al., 2011, *ApJ*, **743**, 28

Komatsu E., Seljak U., 2002, *MNRAS*, **336**, 1256

Kosowsky A., Kahniashvili T., Lavrelashvili G., Ratra B., 2005, *PhRvD*, **71**, 043006

Kunze K. E., Komatsu E., 2014, *JCAP*, **1**, 009

Kunze K. E., Komatsu E., 2015, arXiv, arXiv:1501.00142

Larson D., et al., 2011, *ApJS*, **192**, 16

Lewis A., 2004a, PhRvD, 70, 043011

Lewis A., 2004b, PhRvD, 70, 043518

Lewis A., Bridle S., 2002, *PhRvD*, **66**, 103511

Lewis A., Bridle S., 2011, *ascl.soft*, 1106.025

Lewis A., Challinor A., Lasenby A., 2000, *ApJ*, **538**, 473

Ma C.-P., Bertschinger E., 1995, *ApJ*, **455**, 7

Mack A., Kahniashvili T., Kosowsky A., 2002, *PhRvD*, **65**, 123004

Malyshkin L., Boldyrev S., 2007, *ApJ*, **671**, L185

Matarrese S., Mollerach S., Notari A., Riotto A., 2005, *PhRvD*, **71**, 043502

Mennella A., et al., 2011, *A&A*, *536*, AA3

Millea M., Doré O., Dudley J., et al., 2012, *ApJ*, **746**, 4

Neronov A., Vovk I., 2010, *Sci*, **328**, 73

Paoletti, D., 2007, Masters thesis, University of Bologna

Paoletti D., Finelli F., 2011, *PhRvD*, **83**, 123533

Paoletti D., Finelli F., 2013, *PhLB*, **726**, 45

Paoletti D., Finelli F., Paci F., 2009, *MNRAS*, **396**, 523

Paoletti D., Aghanim N., Douspis M., et al., 2012, *MNRAS*, **426**, 496

Planck Collaboration, et al., 2011a, *A&A*, **536**, AA13

Planck Collaboration, et al., 2011b, *A&A*, **536**, AA18

Planck Collaboration, et al., 2014, *A&A*, **571**, AA16

Planck Collaboration, Ade, P. A. R., Aghanim, N., et al., 2016, arXiv:1502.01594

Planck HFI Core Team, et al., 2011a, *A&A*, **536**, AA6

Planck HFI Core Team, et al., 2011b, *A&A*, **536**, AA4

Pogosian, L., Yadav, A., Ng, B., & Vachaspati, T. 2011, Astrophysics Source Code Library, 1106.018

Ratra B., 1992, *ApJL* **391**, L1.

Reichardt C. L., et al., 2009, *ApJ*, **694**, 1200

Reichardt C. L., et al., 2012, *ApJ*, **755**, 70

Schleicher D. R. G., Banerjee R., Klesser R. S., 2008, *PhRvD*, **78**, 083005

Seager S., Sasselov D. D. and Scott D., 2000, *ApJS*, **128**, 407

Serra P., Cooray A., Amblard A., Pagano L., Melchiorri A., 2008, *PhRvD*, **78**, 043004

Seshadri T. R., Subramanian K., 2009, *PhRvL*, **103**, 081303

Sethi S. K. and Subramanian K., 2005, *MNRAS*, **356**, 778

Shaw J. R., Lewis A., 2010, *PhRvD*, **81**, 043517

Shaw J. R., Lewis A., 2012, *PhRvD*, **86**, 043510

Shiraishi M., Sekiguchi T., 2014, *PhRvD*, **90**, 103002

Shiraishi M., Nitta D., Yokoyama S., Ichiki K., Takahashi K., 2010, *PhRvD*, **82**, 121302
Shiraishi M., Nitta D., Yokoyama S., Ichiki K., Takahashi K., 2011, *PhRvD*, **83**, 123003
Shiraishi M., Nitta D., Yokoyama S., Ichiki K., 2012, *JCAP*, **3**, 041
Subramanian K., Barrow J. D., 1998, *PhRvD*, **58**, 083502
Sunyaev R. A., Zeldovich Y. B., 1970, *Ap&SS*, **7**, 3
Sunyaev R. A., Zeldovich Y. B., 1972, *CoASP*, **4**, 173
Takahashi K., Ichiki K., Ohno H., Hanayama H., 2005, *PhRvL*, **95**, 121301
Tavecchio F., Ghisellini G., Foschini L., et al., 2010, *MNRAS*, **406**, L70
Tavecchio F., Ghisellini G., Bonnoli G., Foschini L., 2011, *MNRAS*, **414**, 3566
Taylor A. M., Vovk I., Neronov A., 2011, *A&A*, **529**, AA144
Tegmark M., et al., 2006, *PhRvD*, **74**, 12350
Trivedi P., Subramanian K., Seshadri T. R., 2010, *PhRvD*, **82**, 123006
Trivedi P., Seshadri T. R., Subramanian K., 2012, *PhRvL*, **108**, 231301
Trivedi P., Subramanian K., Seshadri T. R., 2014, *PhRvD*, **89**, 043523
Turner M. S., Widrow L. M., 1988, *PhRvD*, **37**, 2743
Vachaspati T., 1991, *PhLB*, **265**, 258
Vovk I., Taylor A. M., Semikoz D., Neronov A., 2012, *ApJ*, **747**, LL14
Widrow L. M., 2002, *RvMP*, **74**, 775
Wolfe A. M., Jorgenson R. A., Robishaw T., Heiles C., Prochaska J. X., 2008, *Natur*, **455**, 638
Xia J.-Q., Negrello M., Lapi A., et al., 2012, *MNRAS*, **422**, 1324
Xu H., et al., 2012, *ApJ*, **759**, 40